BRIDGES' DYNAMICS

By

GEORGE T. MICHALTSOS

PROFESSOR
Civil Engineering Department
National Technical University of Athens

IOANNIS G. RAFTOYIANNIS

ASSISTANT PROFESSOR
Civil Engineering Department
National Technical University of Athens

Cover photo: The Brooklyn Suspended Bridge in New York

CONTENTS

PREFACE

This book deals with the problems arising from the dynamic distress of a bridge. Up to now, there is not a book containing the main objects and factors related and affected the dynamic behavior of a bridge. The present book focuses on this goal.

After an extensive introduction, in which it is exposed not only the evolution of bridges but, also, an historical review of the research and the influence of each bridge destruction on this research (first chapter), the dynamic loadings are studied (second chapter). The dynamic wind loads, the aeroelasticity principles, the earthquake loadings, the air blast ones, the moving loads and their modeling are the objects of second chapter.

The third chapter deals with the dynamic analysis principles. Firstly, the needed mathematical concepts are exposed, as for example: elements of calculus of variations, the d' Alembert principle, Lagrange's equation, the Hamilton principle, the equations of Heilig, and the δ and H functions. Afterwards the general equations of dynamic equilibrium are concluded and the main methods for their solution are exposed, while the axial, bending and the coupled lateral-torsional free and forced vibrations are studied determining simultaneously the corresponding orthogonality conditions. It is also exposed the nature and the influence of the internal and external damping.

The fourth chapter deals with the problem of moving loads. After a short historical review the strict theory of the moving load-mass is exposed and the general equation of motion is found through the use of Hamilton principle, while the influence of a lot of factors are studied, as for example the damping, the speed, the load's eccentricity, the influence of a real (of two axles) vehicle, the influence of the mass of the load, and of the secondary mass-forces (Coriolis and centripetal). Finally, the real conditions of the support of a load are studied and also the influence of the deck's irregularities and deck's roughness.

The fifth chapter deals with the movement or rotation of bridge supports and the determination of the influence functions, while are studied the one, two, and three span beams. Finally the influence of the pylons' height and their different movements are studied.

In the sixth chapter is studied the bridge as a static system under dynamic loadings. They are studied: the one, two, and three spans beam, the arch bridges (circular and parabolic), the cable-stayed bridges (of radial and harp system) through a new analysis aspect, the suspended bridges, and the curved in plan bridges. For all the above static systems are given formulae for their eigenfrequencies, shape functions and orthogonality conditions.

The seventh chapter deals with the problem of aeroelasticity. Firstly the exact theory of Theodorsen is exposed and, after, the aerodynamic loads of bridges are studied and the main models are shown, while are also studied the buffeting forces. Finally, the factors for the 2D and 3D problems are given. Afterwards the distress of a bridge caused by galloping, torsional divergence, and flutter is studied. For the completeness of the chapter, elements of the theory of modeling are given and the way of bridge's study through the use of air-tunnels is briefly exposed.

In the eighth chapter, the mathematical theory of dynamic instability is developed for problems in plane (2D) or in space (3D). The basic equation of Mathieu-Hill is studied and the Bolotin's pioneer studies are exposed. Illustrative examples for the dynamic instability of the alone cable, or of a bridge pylon, or of a cable-stayed or a suspended bridge are given.

In the last chapter, the absorb systems are studied. The passive, active and semi-active systems are presented and their properties and operation are analyzed in detail. The alone absorber and the system bridge-absorber are mathematically probed, while the equations governing the behaviour of such a system are given and solved.

George T. Michaltsos

Ioannis G. Raftoyiannis

Civil Engineering Department

National Technical University of Athens

KEYWORDS

Chapter 1

Ancient bridges, stone bridges, historical bridges, keystone, arch-type bridges, dome-type bridges, wooden bridges, concrete bridges, steel bridges, modern bridges

Chapter 2

Dead loads, traffic loads, moving loads, vehicle modeling, snow loads, wind loads, aeroelastic loads, blast loads, earthquake loads, acceleration, deceleration, centrifugal forces, vehicle collision, support settlement, erection loads

Chapter 3

Dynamic principles, energy axioms, dynamic equilibrium, equations of motion, damping, Ritz method, Galerkin method, free vibration, axial motion, flexural motion, torsional motion, critical damping, orhtogonality conditions, forced motion

Chapter 4

Moving loads, moving mass, dynamic influence lines, velocity effects, damping effects, harmonic loads, vehicle modeling, vehicle mass, mass suspension, deck irregularities, surface irregularities

Chapter 5

Support motion, support settlements, longitudinal motion, vertical motion, transverse motion, influence functions, support sliding, overturning, piers

Chapter 6

Bridge modeling, bridge analysis, continuous systems, shape functions, frequency equations, bending vibrations, torsional vibrations, arched bridges, cable-stayed bridges, fan system, harp system, suspension bridges, curved-in-plane bridges

Chapter 7

Aeroelasticity, wind flow, aerodynamic forces, buffeting forces, galloping instability, torsional divergence, fluttering instability, flutter coefficients, scaled models

Chapter 8

Dynamic instability, Matheu-Hill equations, excitation parameter, Floquet solutions, critical eigenfrequencies, instability regions, instability boundaries

Chapter 9

Damping systems, passive control, active control, internal damping, external damping, elastomeric dampers, slipping dampers, viscous dampers, gravity pendulum, spring pendulum, complex pendulum, rolling pendulum

CHAPTER 1

Introduction

Abstract: This introductory chapter presents a brief historical review of bridge structures. From ancient times up to present days, bridge engineering is a continuously developing field of science although various construction materials have been used. From stone and wood in the past to concrete and structural steel at the present, various types of bridges have been designed and constructed. The most representative types of bridges are given schematically of in photographs.

HISTORICAL EVOLUTION OF BRIDGES

From Strabo 1.3.18*, one can conclude that the forerunners of bridges were dykewise passages like a sort of viaducts. Indeed, in the Greek language a bridge is called "**γέφυρα**" from the word "γαία"=earth, and the verbe "φυράω"= I make a dyke.

In Latin, a bridge is called **"ponte"** and **"pont"** from the Greek word "πόντος"=the boundless sea, or abyss, or passage way. In Anglo-Saxon, the word bridge comes from the old Norse word **brygga** = landing stage, gangway or movable pier that has the same root with the old Greek word (Homeric word) **βρύξ**= the bottom of the sea, the sea's abyss.

In the ancient times, the quantity of mined iron and its difficult elaboration did not allow its use in structural works and, of course, in bridge building. Therefore, wood and stone were the main materials in use until the Middle Ages.

Today, from the saved ruins we guess that the first bridges were constructed in Assyria, in Egypt and in Minoan Crete, although no historical reference regarding the above has been found anywhere.

The introduction and use in buildings of plate-covered drains and domes in Assyria and Minoan Crete is very old. In excavations in Babylon's Nineveh and Knossos, plate-covered sewers or sewers covered by domes of bricks made in 2,500 B.C. were found.

In the reconstruction of Photo **1** the historical bridge of Nabuchadnezzar the second, in Babylon (600 B.C.), is shown.

The oldest known existing bridges in the form of bridge-drain of small span (from 1 to 2.5m) are dated in the Minoan period in which we have the first significant morphological and static evolution.

Photo 1: Ancient bridge in Babylon

*"On the road to Syracuse there is a bridge that connects this town to the continental country, as the Roman poet Ibykos says, made from selected stone named **hand-picked**"

George T. Michaltsos and Ioannis G. Raftoyiannis

Photo 2: Ancient bridge over water flow

From the primitive plate-covered drain in Draconera of Argolis, Photo **2** (about 1,400 B.C.), until the most known "bridge of Kazarma", Photo **3**, on the side of the highway of Nafplion-Epidavros (about in 1,300 B.C.) one can see the progress in the span increase through the use of the corbel construction system, and finally the use of the functional keystone. According to the above corbel system, spans from 2.5 to 6 m were achieved, as one can see in the Eleftherna bridge of the later classic period, that is in use even today, with a span of 4.3 m (Photo **4**).

Photo 3: Ancient bridge in Epidavros

Photo 4: Ancient bridge in Eleftherna

The invention of keystone (or crown) led to the primitive evolution of the simple arch and later of the three-hinged one.

There are two reasons that ancient Greeks did not make arch bridges from stone with spans bigger than 10 to 15m. The first reason is the non-existence of big rivers in Greece and Asia Minor, while the second and most important reason was due to military purposes. Indeed, the Greeks did not wish to have "permanent" bridges but bridges with a wooden deck that could be removed fast or destroyed or burned when the enemy came.

This type of bridge with the above characteristics is the one called "Assos type", with a wooden instead of a stone deck and was dominant until the Roman conquest of Greece.

Some remains of this type of bridge (of the 4th century B.C.) are found in the town of Assos in Asia Minor with a total length of 52m and dense rhomboid hydrodynamic stone bases [3].

All saved bridges, mainly near or inside ancient cities, show the technical development and the perfection in the construction of stone bridges in the old years. In Photo **5**, the bridge in Rodino in Rhodes with a span of 8 m, erected in 227 B.C. still in use is shown.

Photo 5: Ancient bridge in Rhodes

In Photo **6** one can see the bridge on Kifissos, in Elefsis, with a total length of 30m, a width of 5.30m and four arched spans of 7m in diameter, for the middle arch, 4.30m for the end spans with the use of keystones. It was erected in 125 A.D. after the order of emperor Adrian for his initiation in the mysteries of Elefsis.

The Romans adopted the Greek arch and boldly enough built bridges with spans up to 30 m, while for bigger spans (35 to 40 m) they used wooden bridges.

The more frequent type of stone bridges was the one with a semi-circular dome of one center for the construction of bridges on roads. These bridges showed great strength and many of them have been saved until today.

Photo 6: Ancient bridge in Elefsis

It is noteworthy that the above bridges were so finely constructed and the dome-stones without voids between them so well fitted with each other that one cannot see the joints. Eight bridges were erected on the Tiber only. Two of them, the Fabricius and the Aelius bridges are still in use. In Photo **7** the Fabricius bridge is shown while in Photo **8** one of the best saved bridges, the Pont du Gard erected by Hyrippa, a friend of Augustus is shown.

Photo 7: Fabricius bridge

Photo 8: Pont du Gard bridge

During the Middle Ages the span limits changed only a little. In the 18th century the Frenchman Perronet constructed a stone bridge with a span of 54m, while the Swiss Grubenmann constructed a wooden bridge of 119m.

At the end of the 18th century starts, cautiously at first and then systematically, the use of iron in the construction of bridges and hence a new era in bridge building opens.

The first iron bridge in the world was made from cast iron by the factories "Goalbrookdale Iron Works" over the Severn river in Brosely - England (Fig. **1**) during 1776-79. This bridge has a span of 30.48m, arch height 12.10m, width of 7.30m and a total weight of 378 tons. This bridge is still in use.

Figure 1: The first iron bridge in Brosely - England

In 1824 the first railway iron bridge was built on the line Stockton-Darlington in England. The main beams of the bridge were made also from cast iron while they had a span of 3.80m each (Fig. **2**). From these old iron bridges, a railway bridge over the Tyne in Newcastle made in 1849 is worth noting (Fig. **3**), with 6 main spans of 38.10m each and two decks - one for trains and one for vehicles.

Figure 2: The first iron railway bridge in England

The substantial evolution of the iron industry, the multiple advantages of the iron materials and their alloys like cast iron, pulpous or liquefiable steel or the steels of high resistance compared to other structural materials, and the continuously increasing demands of transportation of goods and passengers were decisive factors for the generalization of the use of iron in bridge construction.

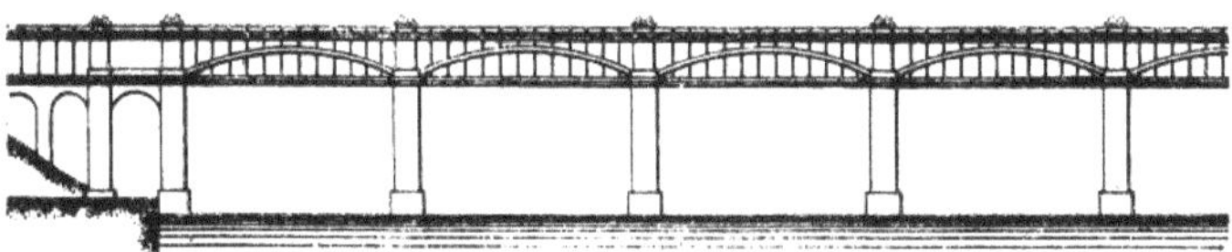

Figure 3: Railway bridge in Newcastle - England

The first awkward constructions were transformed into particularly aesthetic and impressive ones while the initially small spans increased substantially. The lattice beams offered special possibilities in bridge engineering.

In 1885 the Firth-of-Forth bridge of double lane was built, from pipe bars, with a free span of 521m of 104m height at supports and 13m in its middle (Fig. **4**).

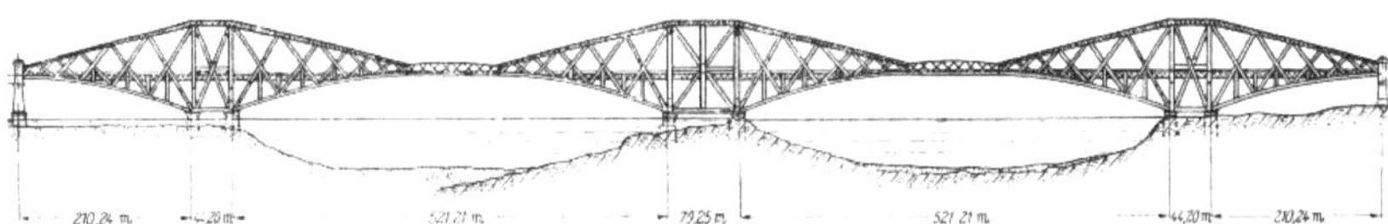

Figure 4: The Firth-of-Forth bridge

The bridge of Quebec on the St. Lawrence river in Canada (1917) is today one of bridges with the biggest span ever made from lattice beams (549m, Fig. **5**).

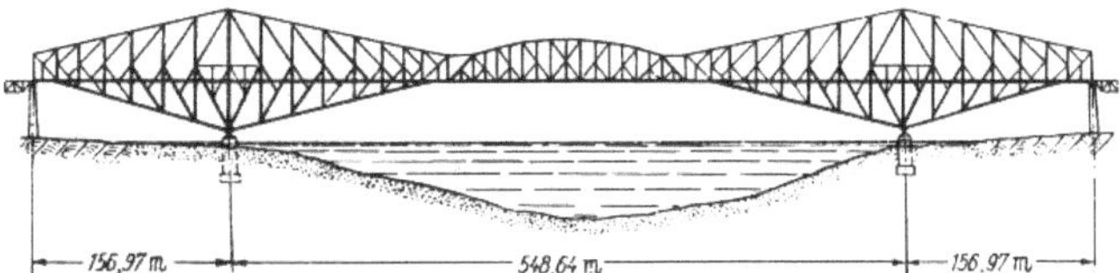

Figure 5: Quebec bridge

Parallel to the bridges from lattice beams, arches with significant possibilities and particular aesthetic results were also used. One should mention the bridge of Garabit (1884) that is one of the first arch bridges with a span of 165m, built by the famous French engineer G. Eiffel, and the Hell Gate bridge (1917) in N. York with a span of 298m (Fig. **6**).

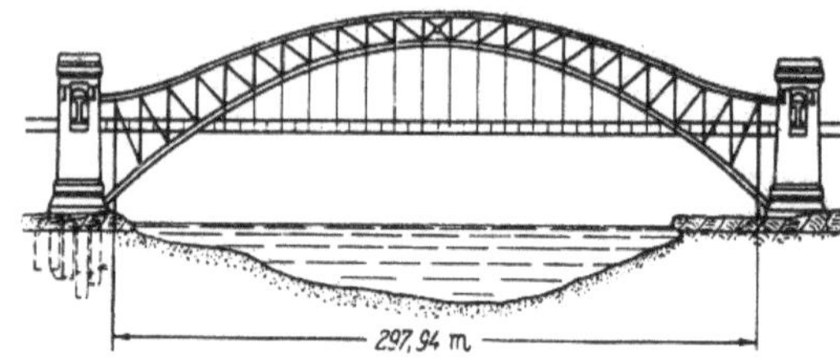

Figure 6: Hell-Gate bridge

Two of the biggest arch bridges worldwide are the Bayonne bridge on Kill van Kull river in N. York (1931) with a span of 510.50m (Fig. **7**), and the bridge at the port of Sydney in Australia (1932) with a span of 503m (Fig. **8**).

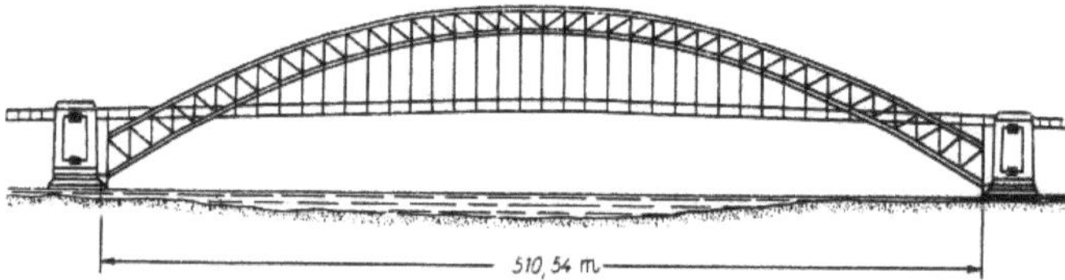

Figure 7: Kill van Kull bridge

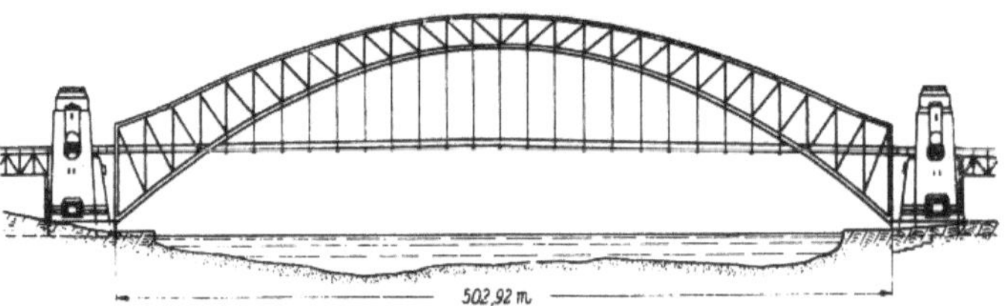

Figure 8: Sydney bridge

But the study and use of suspension bridges brought over the real revolution in bridge construction. This type of bridges existed in the past and it was used for small spans where, instead of parabolic cables, chains or small iron rods with length of 2 to 6 m were used. A characteristic example is the Menai bridge (1826) in England where chains from hammered iron were used, namely, jointed iron plates 3.50m long, specially elaborated at their ends, in order to be connected to each other with pins (Fig. **9**). The span of the bridge was 177m and its width 9m. The spans of suspension bridges were limited to 600m.

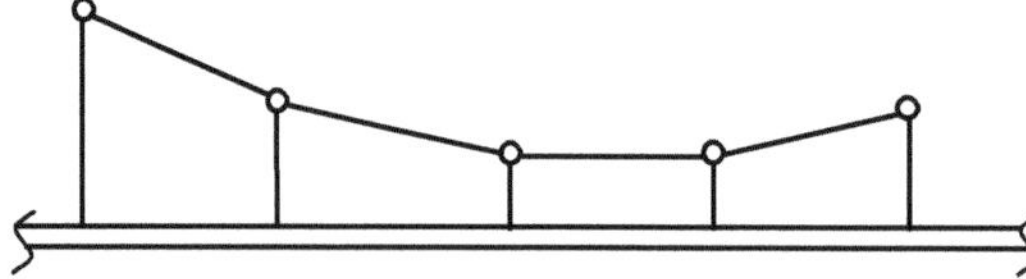

Figure 9: Menai bridge - suspension system

In 1931 the engineer O.H. Amman, after theoretical studies and in spite of fierce objections, dared the bridging of the Hudson river with a suspension bridge with a span bigger than 1000m. Later on, the George Washington bridge (N. York) was erected having a total length of 1450m, a middle span of 1067m, a width of 36.6m, and pylons height 198m (Fig. **10**). According to the initial design a second deck was placed to the bridge in 1959.

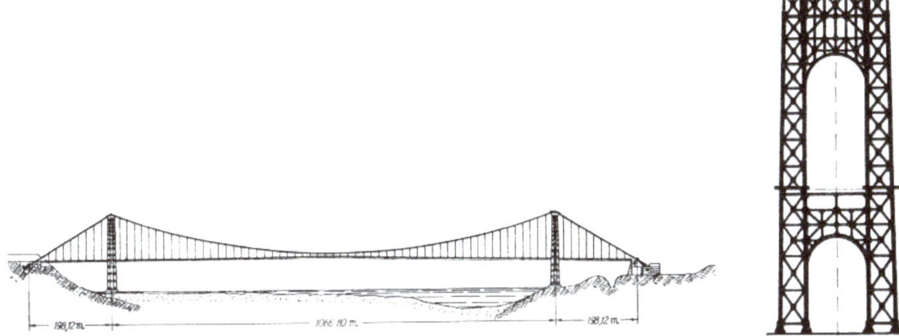

Figure 10: George Washington bridge

An equally famous suspension bridge, that was the longest one in the world for a long time, is the Golden Gate Bridge in S. Francisco (1937), with a main span of 1280m, a width 18.3m, and pylons height 227.5m (Fig. **11**).

Figure 11: Golden Gate bridge

Today, the longest suspension bridge is the Akashi-Kaikyo bridge in Japan (1998), with a middle span of 1990m (Photo **9**).

Photo 9: Akashi-Kaikyo bridge

Finally, another bridge type, which is often used today mainly as an alternative solution to suspension bridges for spans up to 700 – 800m, is the cable-stayed bridge system.

In Fig. **12**, the cable-stayed bridge of Bonn-Nord in Germany is shown.

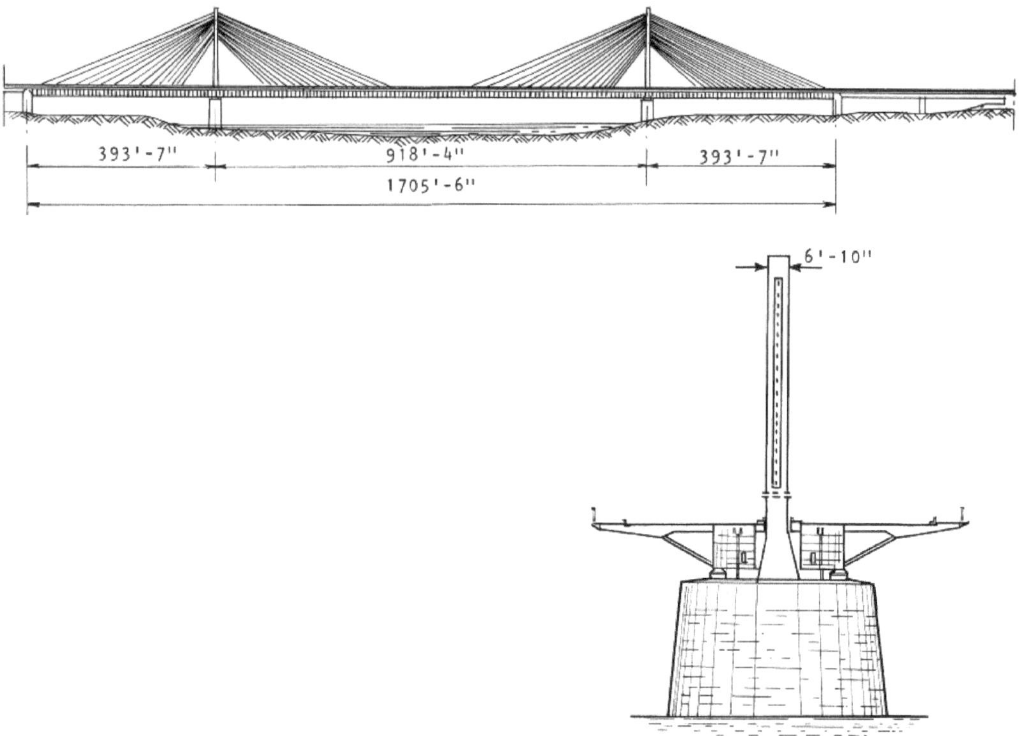

Figure 12: Bonn-Nord cable-stayed bridge

HISTORICAL REVIEW OF RESEARCH AND DESTRUCTIONS

Even if the terms "Historical review of research" and "Bridges' destructions" in the same title surprises us, we must accept that every great or serious destruction offered cause and initiative for serious theoretical and experimental research to the study of erection methods while they have added experience and knowledge as well.

Bridge destructions have helped us mainly to point out factors and phenomena that one is obliged to take into account in the bridge design, which designers did not sometimes imagine neither their existence nor their serious influence on the bridge's operation.

Until the Middle Ages, constructors followed the so called "rules of harmony-analogy" that were more architectural-aesthetic rules and less regulatory codes regarding the expected strength of the structure. Instinct, experience and sometimes boldness helped engineers in the construction of marvelous technical works. There were no data or statements regarding the safety margins or related to the loads that would cause the collapse of the structure. The constructors in the old periods ignored such factors and very often their ignorance "was rewarded" by catastrophes.

The famous architects Anthemios and Isidoros, although only in their third attempt had they had a successful design of the dome of Aghia Sophia, were rewarded with the discovery that the domes must not be extended beyond a parallel of about 55°. Indeed, the contemporary analysis of such shells proves that beyond the parallel of 51.82° tensile tensions that bring about the rupture of the spherical dome are developed (Fig. **13**).

The artist, engineer, mathematician and visionary Leonardo Da Vinci (1452-1519), one of the greatest figures of the Renaissance, proved that the strength of a beam of circular cross-section is analogous to the third power of its radius [5].

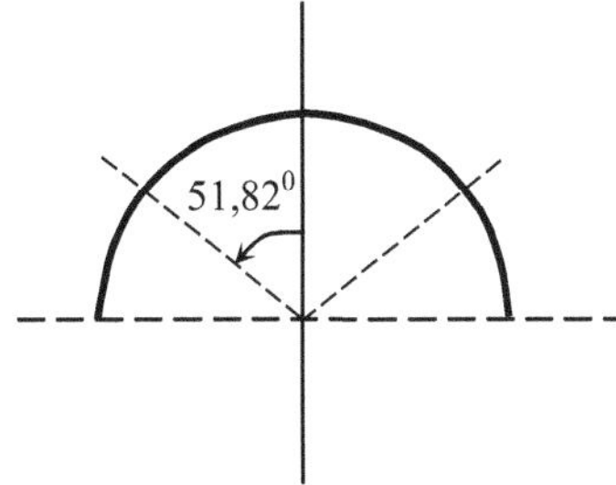

Figure 13: Spherical shell

But in 1638, Galileo in his work "Two New Sciences", first of all examined and analyzed systematically the strength of structures marking the end of the period of "rules of harmony-analogy" and raising the problem of the calculation of the strength of a cantilever.

This question might have been raised very earlier and not by Galileo in the margin of his attempt to develop practical design rules. The attempt of Galileo consisted in the determination of the strength of a laterally loaded beam as a function of its orthogonal cross-section's dimensions through a formula that would be able to calculate the strength of every other beam.

Galileo solved the problem in a strictly mathematical way and concluded that the rules of harmony-analogy cannot be applied exactly in many cases. If, for example, we double the dimensions of a cross-section, its strength will be extremely bigger than the double. Thus, the new branch of "Constructional Mechanics" was born and the idea of stresses started rising.

Two centuries later, Galileo's problem, *i.e.* the problem of the determination of the load under which a cantilever collapses, was modified into the problem of the determination of the tension of the beam.

Little by little these progresses with the use of logical mechanisms and the help of experiments on construction materials that were often used, allowed the drawing up of tables about the limit strength of those materials. The natural consequence was the attempt of joining these two values, namely the developed stresses with the allowed stresses. This was achieved through the safety factors that express the safety margin of a construction. The basic principles of Static and Dynamics were put by Newton in 1678 in his work "Philosophiae Naturalis Principia Mathematica", utilizing the meaning of centripetal force that Huygens expressed in 1673 in his work "Horologium Oscillatorium". So, when Newton expressed the rules of the composition of the forces and proved the rule of parallelogram, he was the first who separated the meaning of mass from that of weight, giving the first basic equation $F = m \cdot g$ that connects force to acceleration and mass.

The inner forces of a body must be in equilibrium with the externally applied ones. This is expressed through the equations of equilibrium. If these equations are sufficient and it is possible to solve them directly, then the first step is completed and one talks about a statically determined structure. But sometimes the equilibrium equations are not numerically adequate in order to be solved by themselves. Thus, the construction is statically undetermined. In this last case the equations have a lot of solution systems that are able to satisfy them, namely, there are a lot of possible equilibrium states, which means that there are a lot of ways through which the structure may undertake the applied loads. So, it was proved that the physical phenomena followed definite laws. One of them is the law of the minimal work that states that for all the possible equilibrium states, nature will select the one for which the consumption of the minimal energy is claimed.

Robert Hooke suspected the existence of this principle, when in 1675 in his work "The true Mathematical and Mechanical Form of all Manner of Arches for Building", gave the following condition for a correct design of an arch: "As a flexible line is suspended, likewise but turned backwards, the correct arch will stand".

Hooke was unable to investigate mathematically this succinct theorem that is shown in Fig. **14** (this sketch was made in 1718 by Poleni in his study for the dome of Saint Peter. The form of chain-line, that is loaded only by its own weight, is the same as that of an arch that has to undertake its loads but under pressure.

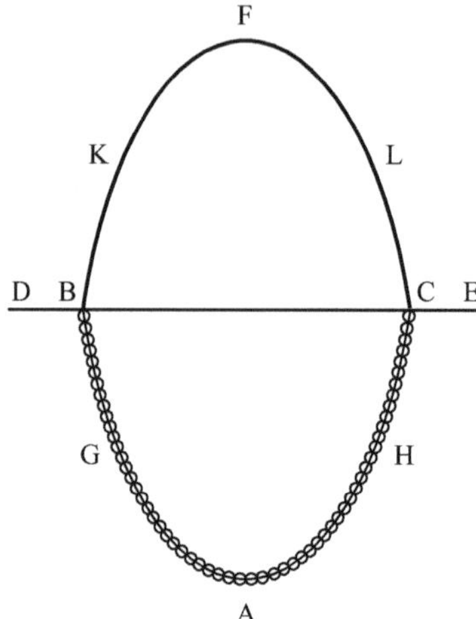

Figure 14: Hooke's arch theorem

As soon as Newton and Huygens founded, erected and placed the capping-stone of that excellent building, other mathematicians added to the décor [5]. Pierre Varignon (1654-1722) perfected the static one. Jean Bernouilli thought up the significant theorem of "Independent Velocities" and expressed the elasticity principle.

Amber, Poncelé and Stokes studied the Theory of Elasticity and pointed out the two basic states, namely the change of the shape without change of volume and the change of the volume without change of shape.

Daniel Bernouilli applied the theorem of kinetic energy. Wallis, Wren and Huygens expressed the laws of the impact of bodies and of the conservation of the quantities of movement.

D' Alembert, in 1743, in his work "Manual of Dynamics" introduced the acceleration as a derivative and gave, for dynamic phenomena, the first place to mass while he introduced the inertia forces and expressed the opinion that they appear secondarily as a reaction to other primary actions.

Euler, in 1760, in his work "Theoria motus corporum solidorum seu rigidorum" (Theory of movement of solid bodies), that was revised in 1790, determines the mass-centre or inertia-centre, and the moments of inertia and analyses the movement of a solid body into a movement of its inertia-centre and a rotation around an axis that passes from this centre. The classic differential equations of movement appeared for the first time.

"La Mechanique Analytique" by Lagrange in 1788 is the capping stone of the building of the great geometers of the 18th century. Lagrange bases the entire of his static on the principle of virtual movements. He introduces the multipliers, a simple and general method. Through them he determines the virtual work of reactions. Lagrange analyzes the four principles of dynamics:

1. The conservation of the kinetic energy (that is attributed to Huygens).
2. The conservation of Newton's gravity center.
3. The principle of fields.
4. The principle of the minimum of action quantity.

Analysis allowed Lagrange to express the general equations of the dynamics of systems with excellent simplicity and accuracy [5].

In 1826, Navier, studying the strength states, practically declared the obvious *i.e.* that the engineer is not practically interested in the construction's collapse stage (the reply that Galileo searched) given that everyone agreed that that stage should never happen. Navier brought back the claim of the Middle Ages that a construction must always stand.

But now this would have to be verified not only through the determination of some geometric data of the construction but also through the calculation of the stress of all the members of the costruction.

Amber and Poncelé formulated the principles of the elastic change of the shape of a body while Stokes pointed out, in 1845, the importance of their remarks.

The theory of elasticity had already been expressed by J. and D. Bernouilli. In the same problem were involved Poisson, Gauchy, Lamé and later Kelvin and Poincaré, while Coulomb, in 1780, studied the deformations caused by torsion.

The static had already strong foundations and it was possible to face statically determined and undetermined structures through a large number of analytical or graphic methods but only for static loads.

The dynamic character of loads, that especially in bridges is a substantial one, was neglected and in cases where there were strong dynamic phenomena they were faced through the so-called increasing factors that were named each time impact factors or dynamic factors *et al.*

On the other hand, except of the buckling of the simple column (Euler), no one suspected the more complicated problems of stability and the special conditions under which their appearance was possible.

The increasing of the bridges' spans resulted in having many more victims than in the older times in the case of an accident.

The first serious accident in the bridges' history, for which we have complete records, was the destruction of Brighton's Chain Pier bridge caused by a violent windstorm.

In 1854, the suspended bridge, Wheeling, in West Virginia was also destroyed by a fierce windstorm and then the engineers started being concerned. The famous G. Airy (Airy's strain functions) paid attention to that problem but through a wrong way, because he never suspected the aerodynamic situation that develops around the bridge.

In 1865 the cast-iron made railway bridge of Ashtabula in Ohio, USA was destroyed after a snowstorm with 100 victims. That destruction was attributed to the quality of the material that initiated, as a result, researches for the cast-iron improvement and also a shy turn to the structural steel.

In 1878 the Firth of Tay bridge in Dundee, Scotland, was destroyed while a train passed during a violent snowstorm, and 200 persons died. That destruction caused even more concern.

In 1891 the iron railway bridge in Birs in Switzerland was destroyed during the passage of a train resulting in 72 deaths. That was the first destruction were it was clear that it had not been caused by a natural phenomenon but as a result of bad design and calculations. This was the cause for the beginning of the study of buckling in the plastic area and Tetmajer gave us his empirical formulae while Engesser his theoretical ones [6].

From 1900 to 1917, during the erection of Quebec bridge in Canada, a lot of accidents had taken place, the more serious of which was caused by buckling and had 82 persons dead. In addition, it gave cause for the foundation of the theory of great deformations and also for the improvement of both the methods and the mechanical devices used for the assembly and erection of a bridge.

During that period research on the strength of a bridge, not only under static loads but also under dynamic-moving loads, also began. Willis, Stokes and Zimermann had treated the problem but under the condition that the mass of the bridge was negligible (1849-1896).

On the contrary, Krýlov (1905), and Timoshenko (1908) considered that the mass of the load was smaller than that of the bridge. Finally, Inglis (1934) setting the problem on its right basis, studied and gave solutions using harmonic analysis [7], while Ödmann and Koloûsek were the first ones that used the normal modes [8].

In 1940 the Tacoma bridge over Puget Sound, USA collapsed because of wind pressure (Photo **10**), while in 1944 the road bridge on the Mississippi in Chester, Illinois, USA, was drifted from its foundations and collapsed during a violent wind storm.

Photo 10: Collapse of the Tacoma bridge

These destructions, especially the Tacoma Bridge one, upset the dominating views for the design and analysis of a bridge subjected to wind actions. Extensive experimental researches started and it was proved that the problem was extremely complicated, depending on the combination of the dynamic characteristics of the bridge and the wind pressure ones (see chapter 7 on aeroelasticity).

After World War II, several State Organizations in USA, G. Britain, France, Germany, and some other European countries assigned the study and investigation of the dynamic problems of bridges through theoretical but mainly experimental studies to distinguished scientists, and recognized research institutes.

At the same time, individual researchers studied and are still studying a lot of phenomena and factors that are involved in these dynamic problems, factors that concern the bridge characteristics, the ground in which the bridge is founded, the deck's pavements but also factors connected with the moving vehicles' characteristics, like dimensions, number of axles, wheelbase, springs' and dampers' constants and, of course, with masses moving on the bridge.

Finally, a special sort of bridges comprises the pedestrian ones, because of their small weight and the possibility of producing large deformations (caused by tuning) in case of loading by huge crowd or military marches.

REFERENCES

[1] P. Bougia "*Ancient bridges in Greece and Coastal Asia Minor*" PhD Dissertation, 1996, Pensylvania University, Ed. Ann Arbor (in Greek).

[2] S. Vyzantios "*Lexicon of the Greek Language*", 1852, Printing of Andrea Koromila (in Greek).

[3] P. Bougia "*Mechanics and architecture connection. The case of pre-roman bridges in Greece and Asia Minor*", 1997, 1st Int. Conf. on Ancient Greek Technology (in Greek).

[4] J. Heyman "*The stone skeleton*", 1995, Cambridge University Press.

[5] S. Papadakis "*General History of Sciences*", 1958, Part B, D. Voyiatzis Publ. (in Greek).

[6] K.A. Reckling "*Plastizitäts theorie und ihre Anwendung auf Festigkeitsprobleme*", 1967, Springer-Verlag.

[7] C.E. Inglis "*A mathematical treatise on vibration in railway bridges*", 1934, Cambridge: The University Press.

[8] V. Koloûsek "*Dynamics of civil engineering structures*", 1956, Prague, NNTL.

CHAPTER 2

Distressing - Loading - Modelling of Vehicles

Abstract: This chapter deals with all load types imposed to bridge structures. Typical permanent and live loads such as self-weight, traffic loads, snow and wind loads and thermal loads are presented with reference to international codes for structural loadings. Special loads such as seismic loads, accidental loads, blast loads, support settlement, centrifugal forces etc. are also given. The designer must take into account all loads specified by the codes as well as special load cases due to structural type of the bridge.

INTRODUCTION

There is a clear separation and notable differences on the design and construction between steel structures and steel railway or viaduct bridges. These differences appear on the first stages of the bridge design, when various loadings are considered. We are of the opinion that the term "distresses" is better than the term "loadings" because during the design of a modern construction, factors such as, for example, fatigue are involved.

The differences are due to the nature of the loads which on the steel structural buildings do not have a special dynamic character (except, perhaps, in the case of high buildings), while in steel bridges the loads are mainly dynamical ones, with great and particular variety, regarding their nature, and, also, the way of their application.

For the design of a steel bridge we have to take into account all types of loadings, especially the dynamic ones. We note that except for the dead loads and those of snow, all the rest loads have a dynamic character.

Bridges suffer the action of a large variety of dynamic loadings or dynamic excitations that differ from each other in their origin and in their way of action. It is possible for the above excitations to be related to movements or rotations of the construction, while it is also possible for them to be related to forces or moments that act internally or externally on the construction.

In any case, a dynamic excitation is characterized by a force variable with time. The vibration of the foundation of a bridge-pylon during an earthquake is the most characteristic example of an imposed time-depended movement or rotation. The action of an air-blast produced by an explosion is a characteristic example of an externally imposed time-depended load. Finally, the development of inertia forces and rotations, when a structure vibrates with a changing speed of deformation, is a characteristic example of forces that are internally applied and time-depended.

The symbol $f(t)$, which denotes a time function will be called hereafter function of action or excitation function and will be used to symbolize the acting loadings or excitations.

In Fig. **1** a lot of such functions are shown. In Fig. **1(a)** one can see the vibration and the so-produced excitation by the action of a load with finite period that may be caused, for example, by a machine operation. In Fig. **1(b)** the action of an air-blast is shown, while in Fig. **1(c)** the action of an earthquake is shown.

Finally, in Figs **1(d)** and **1(e)**, the action of loads that are suddenly imposed (step-functions load) are shown, from which the first remains applied while the second one ceases at some time instant. If $f(t)=1$ the loads are called unit step-functions loads.

The valid codes always consider that the division of loads into categories according to their importance and their frequency of appearance is useful. But praxis has proven that all kinds of loads may produce significant destructions. There are many cases in which the most unusual loads, even if a special combination of the bridge's geometrical characteristics is required in order to become dangerous, have caused full destructions. The Tacoma Bridge is a well-known example of a bridge collapse due of the flutter phenomenon.

Classic books and national codes in use provide detailed instructions regarding bridge loadings and their values. In this chapter we will examine the main loadings and, particularly, the so-called dynamic loadings.

George T. Michaltsos and Ioannis G. Raftoyiannis

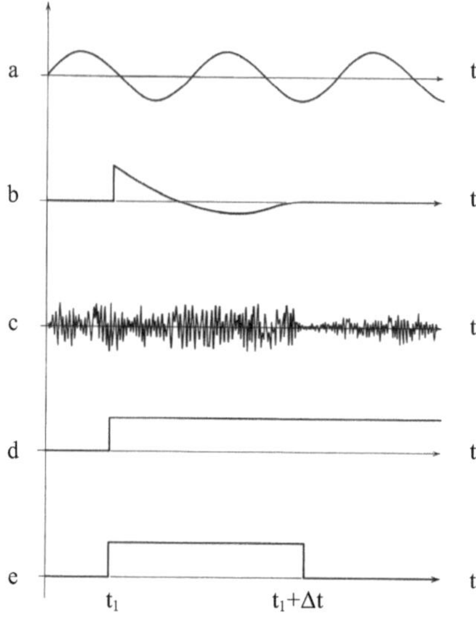

Figure 1: Characteristic cases of loading functions

DEAD LOADS

As dead loads we consider the ones acting continuously or for a long time.

The Self Weight [1]

As self weight of a bridge we consider the total weight of the beams and of their superstructures and also any kind of fixtures that are necessary for the use and the operation of the bridge or fixtures that serve networks for public use.

In all publications and relative text-books for design of steel bridges, we find a lot of empirical or semi-empirical formulas and rules for the estimation of the self weight of any type of bridge.

Snow Loads - Frost (Ice-Formation) [1]

The influence of the snow-load on a bridge is generally neglected. In any case, the weight of snow per square meter may be obtained from $g_S = 125 \cdot h$, where h is the thickness of the snow layer in meters on a horizontal and free surface and g_S in given in dN/m^2.

The case of frost, namely the crystallization of the snow, is practically negligible. But, in some special types of bridges and, mainly, in cable-stayed and suspension bridges, this loading affects the behavior of the hanging cables, but most of all, increases the surface attacked by wind. Manuals studying this kind of bridges provide also the proper design instructions.

LIVE LOADS

Live loads have a relatively limited time of action and due to their dynamic character they are divided into traffic loads, environment loads and destruction loads. With the term dynamic, we mean that the stress of the loads or their point of action or both, are changing with respect to time. As mentioned above, they are divided into:

a) The circulation loads or traffic loads, where every sort of circulating vehicles and the loads of pedestrian are included.

b) The loads that are developed by any environmental effects due to the bridge environment and the loads that are caused by wind pressure, small earthquakes (up to 6.5 to 7 on the Richter scale) and temperature changes.

c) The loads of destruction such as strong seismic excitations (over 7R), air-blasts that are produced by big explosions, settlement of the supports and distresses produced by impact of vehicles.

Moving Loads [2]

The moving loads depend on the sort of bridge they move on, *i.e.* if it is a road-bridge or a railway-bridge or a pedestrian one.

The modern codes give detailed instructions regarding the use, in every case, of heavy vehicles and their loadings per lane of circulation (for the road bridges, Fig. **2a**), or about ideal train models (railway bridges, Fig. **2b**), or about the variation of the uniformly distributed load in relation to the span of a pedestrian bridge (Fig. **3**), or about maintenance vehicles (Fig. **4**). In order for the design of a bridge to include the dynamic character of the above loads, the dynamical factor (or the older impact factor) has been employed, which is depended on the type of bridge and the characteristics of the bridge element examined. The above factor, in some codes, is included in the loads, while in other codes is given in a separate table. Its maximum value in both cases does not exceed the value 1.40 for road bridges or the value 1.50 for railway bridges. An exact analysis of the dynamic behavior of the modern bridges showed that the increase of the deformations is 1.50 to 2.50 times and in special cases 4 times higher than the ones obtained from a static analysis with the same loads.

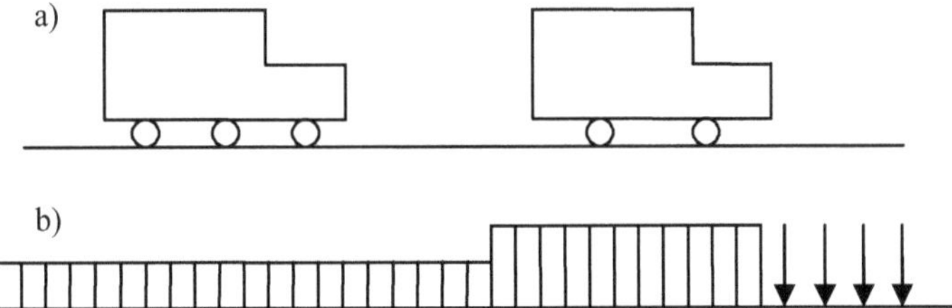

Figure 2: Modeling of moving loads

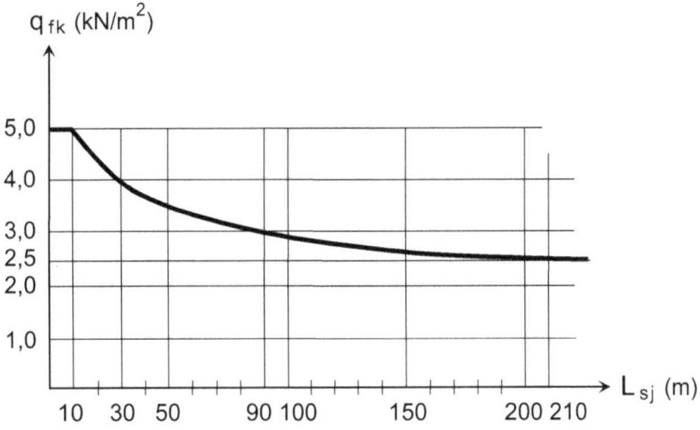

Figure 3: Reduction of live load in pedestrial bridges

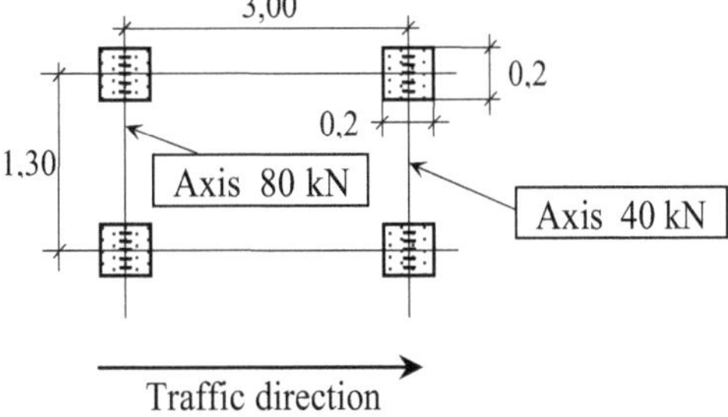

Figure 4: Maintenance vehicle model

For an exact analysis of the dynamic behavior of a bridge, the most possibly exact modeling of the bridge's static system and the model vehicle, has special significance. The correct choice of the vehicle's model has special significance for the accuracy of the results and the approach to the real bridge's dynamic behavior.

Regarding the expected accuracy of the results, the right choice of the vehicle's model is critical.

This choice depends on many different parameters. Among them the most important ones are:

- The use of the bridge, namely if it will be a road or a railway bridge.
- The material from which the bridge will be constructed (concrete or steel) and, therefore, the ratio of its weight to the ones of the traffic loads.
- The static system of the bridge.
- The density of traffic loads.
- The length (span) of the bridge (this is, perhaps, the main parameter).

In Fig. **5** one can see the main vehicle models that we use to study the dynamic behavior of a bridge.

Model **a** is the model of a single moving load P.

Model **b** is the model of a moving load $P = m \cdot g$, but whose mass is taken into account, *i.e.* the inertia forces developing on the moving load.

Model **c** is a more accurate approach than **b**, with the addition of a spring with constant k_V.

Model **d** includes, in addition to **c**, mass M_A of the axles and the modeling of the wheels through a spring with constant k_T.

Model **e** is an improvement of **c**, where a damper with constant c_V is added.

Model **f** is similar to model **e**, with the addition of a mass M_A for the axles of a spring with constant k_T and a damper with constant c_T for the wheels.

Model **g** is a simple biaxial vehicle model with springs and dampers, known also as Drosner model.

Model **h** is similar to the previous one with additional masses M_A for the wheel axles and springs k_T and dampers c_T in the wheels' position, known also as Veletsos model.

Model **i** is a biaxial lorry (or truck), with also a biaxial towed platform.

Model **j** is a railway vehicle for bridges with small and intermediate length.

Model **k** is a heavy railway vehicle, where its weight acts in a circular frequency ω_V.

Model **l** is a distributed moving load-mass for bridges of great length.

Model **m** is a distributed moving load-mass, having, in addition, distributed moments caused by the forces due to the rotatory inertia of the vehicles' masses.

Finally, model **n** is a distributed moving load, with an enforcement frequency Ω that is connected to the marching in phalanx (pedestrian bridges).

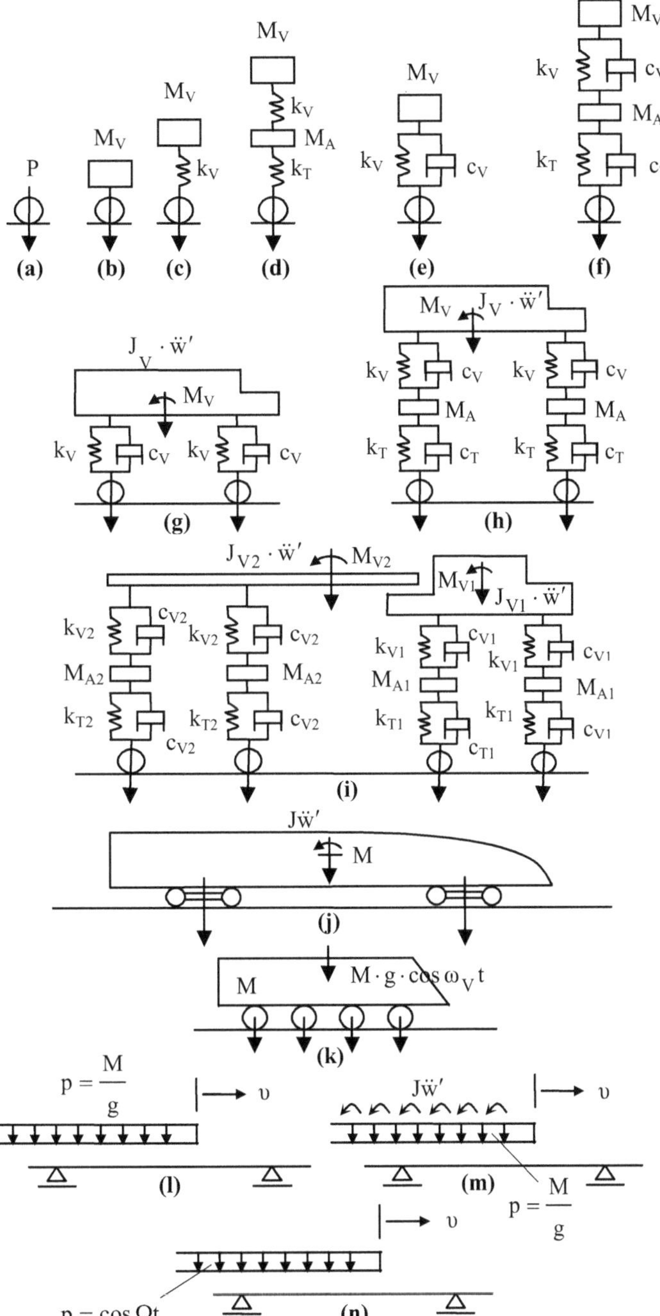

Figure 5: Modeling of most characteristic vehicles

For bridges with much greater length (compared to the length of a vehicle) the models of one-axle are preferred (models **a** to **f**). For intermediate length bridges, one must prefer models **g** to **k**. For bridges with big length and dense circulation, there are still the models **l** and **m**, which are used either alone or more usually, combined to one-axle models.

Finally, models **g, h, i, j** and **m** take into account the influence of the rotatory inertia J of the vehicles that, in addition, may be found from the approximate formula $J = M \cdot d^2 \cdot e$, where M is the vehicles' mass, d is the distance between its more distant axles (or the vehicle's length) and e is the so called shape coefficient of the vehicle, that usually has values from 0,08 to 0,10. Regarding the values used for the constants c and k, they differ from vehicle to vehicle. For a light vehicle, for example, they are: k=60.000N/m, c=1.000Nsec/m, while for a lorry they are: k=100.000N/m and c=5.000Nsec/m.

Wind Pressure

In order to understand and predict the behavior of a structure that is subjected to wind pressure, one must know the behavior and the wind "structure" (distribution of pressure). Wind behavior differs from place to place, depending on the general climate of the territory, the physical geography (terrain) of the location, the surface conditions of the place and the neighboring areas and many other unpredictable factors.

Wind, from a meteorological view, consists from movements of air masses within the atmosphere. Movements of great scale are governed, mainly, by differences in the temperature within the atmosphere, caused by the different solar temperatures of each location.

Because of the different quantity of energy per earth unit area that is coming from the sun and depends on the latitude of each location, a difference in the temperature and, therefore, a difference in the pressure arises, that in combination to the Coriolis forces and the centripetal ones (that are connected to the rotation of the earth), produce movements of the air masses.

The lower area of the atmosphere, where the majority of the civil engineering structures are, is known as the "planetary boundary layer" and in this layer the movement of the air is hindered by friction forces and obstacles that are on the terrain surface and, also, by the Reynolds tensions produced by interaction forces that are vertical to the main movement direction of the wind and are caused by turbulences.

A turbulence, which might exist due to mechanical or thermal origin, produces also unexpected fluctuations of the wind speed in a very wide spectrum of frequencies and amplitude commonly known as wind gusts.

The Wind as a Dynamic Load

a) Distribution by Height

As has been mentioned above, ground irregularities and friction forces decelerate the wind near the terrain surface. Thus, the lower layers of the air decelerate the higher layers resulting in gradual changes of the wind speed as we go higher until nullification of the shear forces. This nullification takes place in a height of about 2,000m (gradient height).

Beyond that height the wind speed is no more affected by the earth irregularities and depends only on the pressure of the atmosphere and the geographic position of the location under study. The quantity of the speed change in relation to height is called "wind shear".

The planetary boundary layer is considered to be composed by a number of layers each having different values of its parameters. From the above mentioned layers the most interesting ones, mainly for the design of structures, are the surface layer and the Ekman layer. The structure, the thickness and the names of the layers are shown in Fig. **6**.

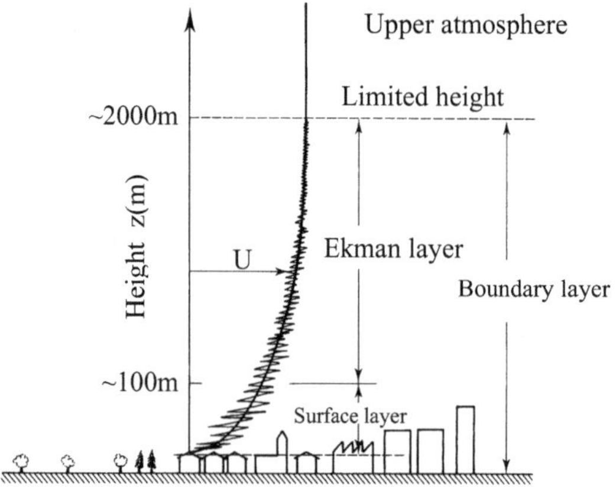

Figure 6: Atmosphere layers affecting the wind loads

The surface layer extends over 100m upwards. The change of the average speed of the wind within this layer in relation to height may be expressed by the logarithmic Prandtl formula-model:

$$\bar{V}(z) = \frac{u_0}{k} \cdot \ell n \frac{z}{z_0} \quad \textbf{(1)}$$

where $\bar{V}(z)$ is the average speed of the wind in height z, u_0 is the friction speed, k is the von Karman's constant equal to 0,4, and z_0 is the roughness length that is characteristic for each earth surface.

Typical values of z_0 for different types of surface are given in Table **1**.

Table 1: Typical values of the roughness length z_o and of the exponent a of formula (5) for different grounds sorts

Type of ground		**z_0 (m)**		α
mud, plane surface, ice	10^{-5}	to	$3*10^{-5}$	0,10
calm sea	$2*10^{-4}$	to	$3*10^{-4}$	
sand	$2*10^{-4}$	to	10^{-3}	
snow-covered surface	10^{-3}	to	$6*10^{-3}$	
sheared grass	10^{-3}	to	10^{-2}	0,13
low grass, steppe	10^{-2}	to	$4*10^{-2}$	
non cultivated area	$2*10^{-2}$	to	$3*10^{-2}$	0,19
high sprouting	$4*10^{-2}$	to	10^{-1}	
palm trees	10^{-1}	to	$3*10^{-1}$	
wood	10^{-1}	to	1	0,32
suburbs	1	to	2	
city	1	to	4	

The friction speed u_0 changes in relation to the earth roughness, the general wind speed and is connected to the developing shear on the surface layer. The following formula improves the Prandtl's initial formula eq(1) and extends its accuracy up to heights of 300m.

$$\bar{V}(z) = \frac{u_0}{k} \cdot \left(\ell n \frac{z}{z_0} + 5,75 \cdot \frac{z}{h} \right) \quad \textbf{(2)}$$

where $h = \frac{u_0}{6f}$ and $f = 2\Omega \sin\varphi$ is the Coriolis' parameter, that for an average latitude of ~40^0 takes the value $11,5*10^{-5}$ rad/s.

A more accurate expression of the average speed change by height of the wind is achieved, if we express it in relation to a height H, that is usually taken as H=10m and is the usual height of the anemometers. This expression eliminates the unknown friction speed u_0 and the only data required is the wind speed at height H:

$$\frac{\overline{V}(z)}{\overline{V}(H)}=\frac{\ell n\frac{z}{z_0}}{\ell n\frac{H}{z_0}} \tag{3}$$

More accurate is the following modified formula (where u_0 is also needed):

$$\frac{\overline{V}(z)}{\overline{V}(H)}=\frac{\ell n\frac{z}{z_0}+5,75\cdot\frac{z}{h}}{\ell n\frac{H}{z_0}+5,75\cdot\frac{H}{h}} \tag{4}$$

Over the surface layer that extends up to the limits of the planetary layer, the Ekman layer exists. In this layer, Coriolis' forces become bigger, while shear forces get smaller in relation to the increasing of the height. The shear forces become zero at height of about 2,000m, while the change of friction forces affects not only the wind speed but also its direction.

In a lot of cases, where simpler considerations of the distribution of the average speed of the wind in relation to the height are necessary, numerous researchers recommend the use of the following law-model that has a satisfactory accuracy:

$$\frac{\overline{V}(z)}{\overline{V}(H)}=\left(\frac{z}{H}\right)^{\alpha} \tag{5}$$

where α is obtained from Table **1**.

Finally, the change of α in relation to the surface roughness is given by the formula:

$$z_0=12,25\cdot e^{-\frac{1}{\alpha}} \tag{6}$$

b) The Influence of the Terrain Morphology

Most of the above presented are applicable only for a smooth surface with uniform roughness. When near the site of the work (the term near is depended on the scale of the work) there is an irregularity, as for example a hillock, the flow lines are disturbed and hence the wind speed in this disturbed layer increases (Fig. **7**). In this layer influenced by the irregularity, there is a special increase of the shear forces and thus it is called "inner layer". The flow field is shown in Fig. **7**. If we consider that the hill has height H and length L, the latter is determined as the distance of the top of the hill from a point where the height is H/2.

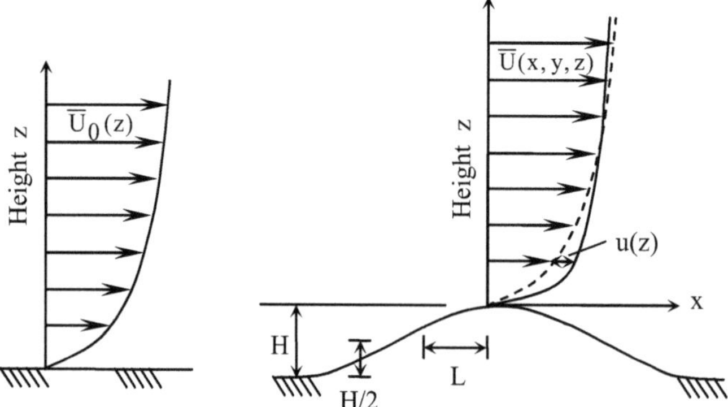

Figure 7: Wind load distribution in inner layer

We assume that the average wind speed at the top of the hill can be determined by the following relation:

$$\overline{V}(z)=\overline{V}_0(z)+u(z) \quad \textbf{(7)}$$

where $\overline{V}(z)$ is the average wind speed at height z over the top of the hill, $\overline{V}_0(z)$ is the wind speed before the arrival of the wind on the hill at height z and u(z) is the augmentation of the wind speed, which has been proved to be a linear function of the hill inclination and hence for its determination one can use a dimensionless augmentation Δu(z):

$$\Delta u(z)=\frac{\overline{V}(z)-\overline{V}_0(z)}{\frac{H}{L}\cdot\overline{V}_0(L)} \quad \textbf{(8)}$$

Easily, one can find:

$$u(z)=\frac{H}{L}\cdot\overline{V}_0(L)\cdot\Delta u(z) \quad \textbf{(9)}$$

Experimental studies have confirmed the above expressions. Finally, it has been proved that over the top of the hill the dimensionless augmentation of the speed is given by the following simple expression:

$$\Delta u(z)=\frac{1}{\left(1+\frac{z}{L}\right)^2} \quad \textbf{(10)}$$

c) The Wind Pressure [5]

The pressure of an air flow at some point of a solid that exists in the flow field (as for example the cylinder in Fig. **8**) can be expressed by the relation:

$$p=\frac{1}{2}\cdot\rho\cdot\upsilon^2 \quad \textbf{(11)}$$

where υ is the wind speed and ρ is the mass density of the air. For a normal pressure and a temperature $+15^{O}C$, it is $\rho=0{,}125kp.sec^2/m^4$ and hence:

$$p=\frac{\upsilon^2}{16} \quad \textbf{(12)}$$

where υ is given in m/sec^2 and p in kp/m^4.

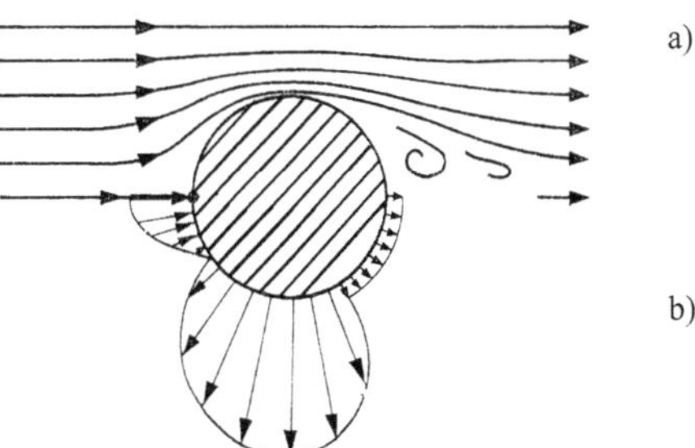

Figure 8: Flow lines and pressure distribution

The pressure force (at the front of the above solid) and, correspondingly the back suction force on the outside surface of the solid are depended on the shape of the cross-section. Fig. **8(b)** shows the distribution of the pressures on a cylinder. The resultant of this distribution gives the wind pressure:

$$w = p \cdot c \cdot F \quad \textbf{(13)}$$

where F is the attacked surface area, vertical to the wind direction and c is a dimensionless factor depending on the shape of the solid that is determined through experimental studies in air-tunnels. This value of c depends on the Reynold number:

$$R_e = \frac{\upsilon \cdot d}{\nu} \quad \textbf{(14)}$$

where d is the diameter, υ is the wind speed and ν is the kinematic constant of the air, that for normal conditions is: $\nu = 1{,}4 * 10^{-5} m/\sec^2$. Then eq(14) gives:

$$R_e = \frac{\upsilon \cdot d}{1{,}4 * 10^{-5}} = 2{,}86 * 10^{-5} \cdot d\sqrt{p} \quad \textbf{(15)}$$

The diagram of Fig. **9** shows the change of c in relation to R_e or to $d\sqrt{p}$.

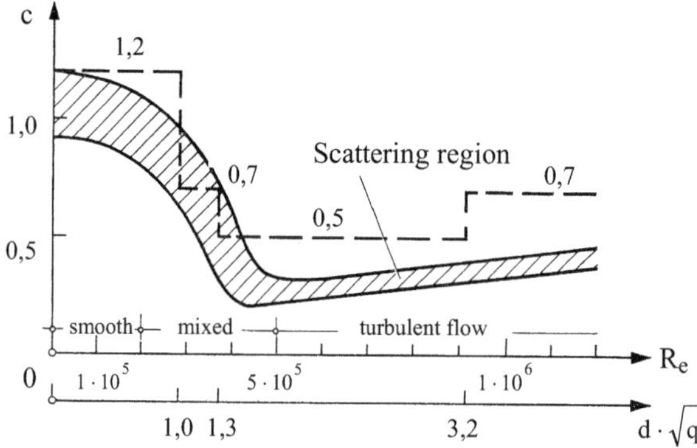

Figure 9: Variation of factor c versus R_e or $d\sqrt{p}$

Older codes were providing values for c according to the shape of the attacked surface. The foregoing shows that the max-p depends on the max-υ. These maximum speeds refer only to short time intervals. The checks, according to the codes, must take place taking into account these maximum pressures.

d) The Wind Pressure as a Dynamic Load

The instability of the air pressure and its change in relation to time, compel us to treat it as a dynamic load. The treatment of the wind as a dynamic load so far has been performed through the consideration of an augmentative dynamic factor φ, which increased the load from p to $p_\varphi = \varphi \cdot p$, and which depends on the eigenfrequency of the construction, its damping factor as well as the size and frequency of the repetition of appearance of the air gust. The first studies on factor φ were made by Rausch, who considered that the wind pressure is divided in two parts, a constant part p_0 (the basic loading) and another part (pushing loading) caused by air-gusts that are a function of time t. Thus, it is:

$$p(t) = p_0 + p_1 \cdot \sin\frac{2\pi}{T}t \quad \textbf{(16)}$$

where T is an arbitrary period. With initial conditions $\overline{x}(0) = \dot{\overline{x}}(0) = 0$ the equation of the forced motion of any static system under study, will take the following form:

$$\overline{x}(t) = x_0 + \frac{x_1}{1-\alpha^2} \cdot \left(\sin\omega t - \alpha \sin\omega_e t\right) \quad \textbf{(17)}$$

where x_0 is the deformation caused by the constant loading p_0, x_1 is the deformation caused by p_1 (considered also as constant), ω_e is the circular frequency of the structure and:

$$\alpha = \frac{\omega}{\omega_e} = \frac{T_e}{T} \quad , \quad \omega = \frac{2\pi}{T} \quad , \quad \omega_e = \frac{2\pi}{T_e}$$

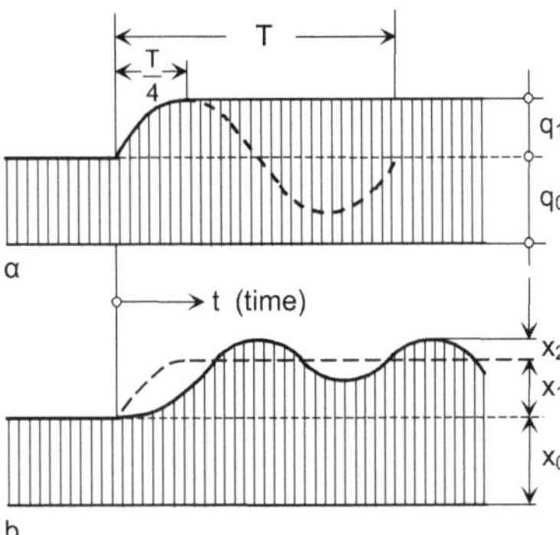

Figure 10: Dynamic response due to wind gusts

Employing the initial conditions, we find:

$$\beta = \frac{x_2}{x_1} = \frac{\alpha}{\left|1-\alpha^2\right|} \cdot \sqrt{1+\alpha^2 - 2\alpha \sin\frac{\pi}{2\alpha}}$$

and according to the determinations of φ we easily arrive at the following expression:

$$\varphi = \frac{p_0 + (1+\beta)\cdot p_1}{p_2 + p_1} = 1 + \beta \cdot \frac{p_1}{p}$$

or:

$$\varphi = 1 + \beta \cdot \left[1 - \left(\frac{\upsilon_0}{\upsilon}\right)^2\right] \tag{18}$$

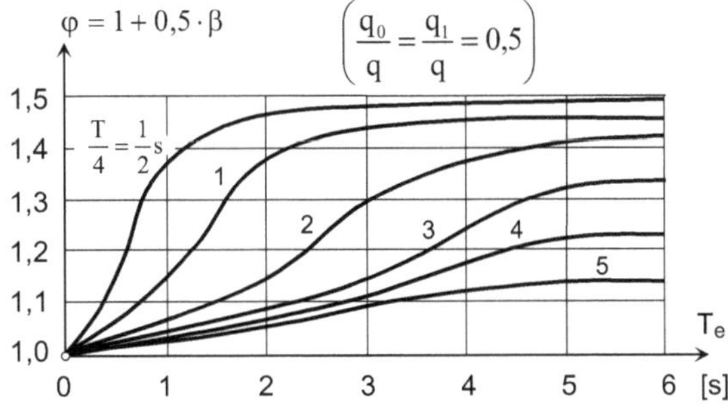

Figure 11: Calculation of φ vs T_e for $p_0 = p_1 = 0{,}5 \cdot p$

In Fig. **11**, the diagram for the determination of φ for $\frac{p_0}{p} = \frac{p_1}{p} = 0{,}5$ is given in relation to the eigenfrequency T_e of the structure. There are a lot of research studies in this field proposing similar expressions. Schlaich [6], for example, gives the diagram of Fig. **12**, where the damping factor C_e of the structure has been taken into account. The modern treatment of air gusts is to consider them as a function of time, but still the difficulty of the determination of the typical characteristics of such a loading exists, given that they are changing continuously and some times, with great divergences. An extreme view is to consider the most unfavourable values of the above characteristics, while a conservative one is to consider some unfavourable combination of their values. The final choice depends on the importance of the structure.

Wind pressure p can be described for short time intervals as a function of height z and time t by the relation:

$$p(z,t) = p_0\left(\frac{z}{10}\right)^{\alpha}\cdot\left(1+\frac{10}{10+z}\cdot\gamma\cdot\sin\omega t\right) \quad \textbf{(19)}$$

where p_0 is the wind pressure at a height of 10m from the surface of the terrain, as it is given by the codes prevalent each time, ω is the natural eigenfrequency of the wind, which usually extends from 0,1 to 5, α is given from Table **1** and γ is a coefficient that one can set from 0,10 to 1, where the higher values correspond to the instability and mostly the dynamic character of the wind. Values of $\gamma>1$ express fully unstable wind loadings with a changeable direction.

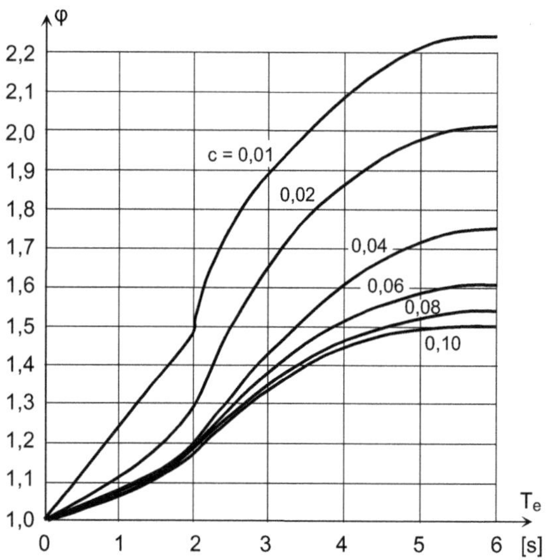

Figure 12: Calculation of φ vs T_e for various values of c

Aeroelasticity [7]

With the term aeroelasticity we mean the phenomena that are developed by the interaction of aerodynamic loadings and elastic deformations. This interaction is significant when the aerodynamic forces act on particularly flexible structures. The technological evolution, the recently produced new materials and the new construction methods, allowed the erection of specially high but flexible buildings and other flexible constructions, such as suspension and cable-stayed bridges, and in turn allowed the appearance of notable aeroelastic phenomena. For the study of the above phenomena and their influence on the structures, the relatively new field of applied mechanics or wind mechanics has been developed.

When aerodynamic loads are acting on a flexible structure, then, due to its deformations-displacements, the boundary conditions of the wind flow are also changing, thus affect the behavior of the structure. Because of this interaction of flow on the structure and vice versa, the dynamic loads are augmented analogous to the distribution of the masses, the rigidity (usually lateral-torsional) and the damping of the structure. This interaction creates the aeroelastic phenomena, namely, the diffusion of the turbulences, torsional divergence and flutter. The above phenomena cause the so-called aerodynamic instability, that affects the wind flow and the aeroelastic instability which, in turn affects the structure when the deformations of the bridge, because of the aerodynamic loads, tent to increase as they perform non-harmonical vibrations.

In the corresponding chapter later on, where stability problems will be studied, the developing actions and the way of their determination will be also analytically determined. In the present paragraph, the mechanism that causes these distresses will be exposed.

When the wind flow meets a solid body (Fig. **13**), then, of course, normal flow is disturbed. This disturbance causes turbulence around the body (in our case around the bridge section), which is the result of the influence of the structure on the flow and depends on the so called Strouhal number:

$$S = N \cdot \frac{b}{\upsilon} \tag{20}$$

where b is a dimension of the bridge along its transverse axis, υ is the flow speed, and N is the frequency of the turbulence's action that is given by the formula:

$$N = 2 \cdot \pi \cdot \delta \cdot \frac{\upsilon}{b} \tag{21}$$

where δ is the factor of the geometry of the structural shape.

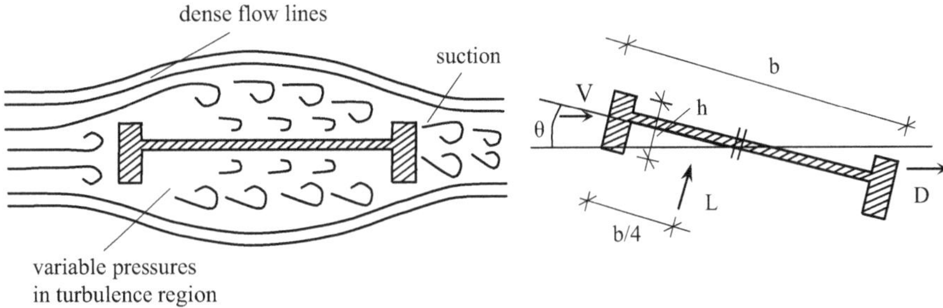

Figure 13: Wind turbulence causing dynamic excitation on a body

If the frequency of the fluctuations of the differences of the pressure (because of turbulences) approaches one eigenfrequency of the construction (in the transverse to the flow direction), then there will be an interaction of the dynamic behavior of the construction and of the dynamic characteristics of the flow, namely, there will be tuning and the construction will vibrate with amplitudes that will surpass the allowed ones. In such a field of frequencies the phenomenon of the "frequencies trapping" is observed, during which, the parameters of the structure keep under control the interaction of the diffusion of the frequencies and of the bridge's eigenfrequencies.

From a structural point of view, these floating differences of pressure mean the generation of forces, which act on the bridge. The sum of these forces is the "drag force", which is caused by the developed low pressure at the back of the bridge and, the "lift force" that acts vertically in the aerodynamic center (the point where the aerodynamic load is applied), which, usually, is located at the 1/4 of the bridges' width (Fig. **13**). This has as result the creation of torsional moments that in some cases are critical.

When the structure is flexible, the above developing forces have as a result the torsional divergence of the bridge by an angle θ. When the speed of the wind flow increases, the torsional moment increases as well, and that successively causes an increase to the effective angle of incidence θ, that in turn, increases the torsional moment. The wind speed for which the rotation moment surpasses the ability of the resistance of the bridge and creates an instability situation that in turn produces damage or even collapse of the bridge, is called "critical speed of divergence". Thus, a phenomenon of instability appears that is analogous (from statical point of view) to the column buckling phenomenon.

The critical flow speed or the critical value of the divergence speed is analogous to the critical load of Euler buckling. As the buckling load depends on the flexural rigidity of the column and the way of its action, similarly torsional divergence depends on the torsional rigidity of the structure and the way by which the aerodynamic moment is developed. There are cases where the above described phenomenon is developing only for a particular angle of incidence and wind flow speed, which we call pure torsional instability or pure torsional flutter. Usually though, the classical flutter that is the coupling of torsional and vertical vibrations appears (Fig. **14**).

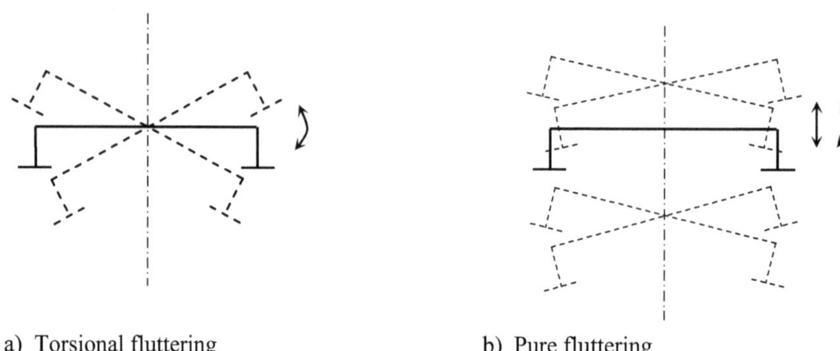

a) Torsional fluttering b) Pure fluttering

Figure 14: Fluttering due to wind loads

Special Cases

The cases in this paragraph concern mainly the influence of buildings neighbouring to the bridge that are affecting and changing or more generally disturbing the wind pressure around a moving train or a moving vehicle.

The size of this disturbance or of the produced wind pressure depends on the vehicle's speed, its aerodynamic shape, the shape of the neighbouring buildings and, mainly, its distance from the bridge.

a) Simple vertical surfaces (*i.e.* sound-proof panels or walls) parallel to the bridge:

The pressure q_{1K} is given by the diagram of Fig. **15**, in relation to the vehicle's speed to the height and the surface distance.

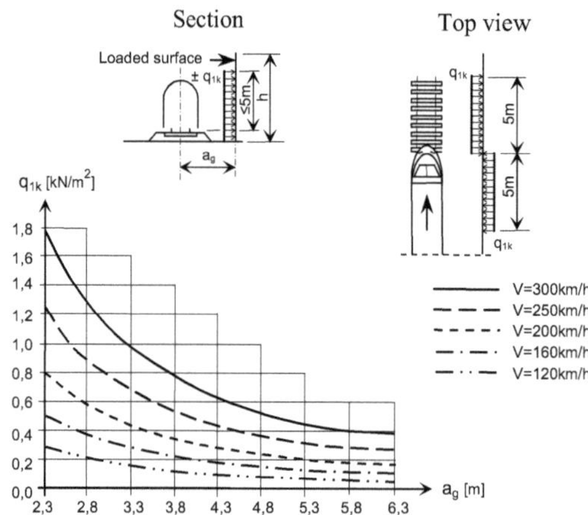

Figure 15: Wind pressure on vertical surfaces due to moving vehicles

b) Simple horizontal surfaces symmetrically laid to the bridge's axis [10]:

This case is the classical one of a bridge with two decks, where in the lower deck trains are passing. The pressure

q_{2K} is given by the diagram of Fig. **16**, in relation to the train's speed and to the distance between the two decks.

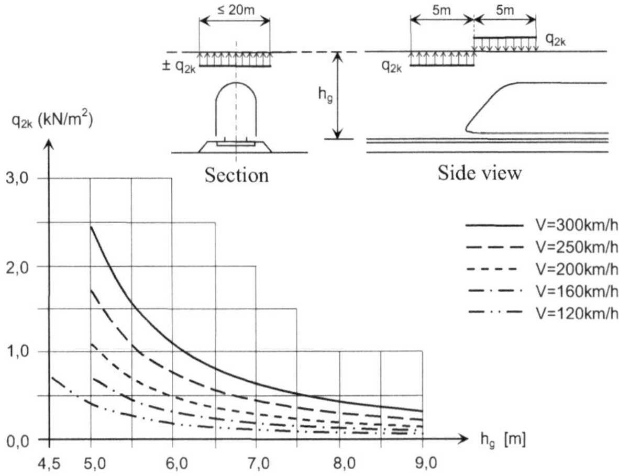

Figure 16: Wind pressure on top surfaces due to moving vehicles

c) Horizontal surfaces non-symmetrical to the bridge's axis [10]:

These surfaces may be sheds for example. The pressure q_{3K} is given in relation to the distance α_g of Fig. **17**. Analogous to the height h_g the actions are decreased by the factor K_3.

for $3{,}80\text{m}<h_g<7{,}50\text{m}$ $\quad K_3=\dfrac{7{,}50-h_g}{3{,}70}$

for $h_g \geq 7{,}50 \quad K_3=0$

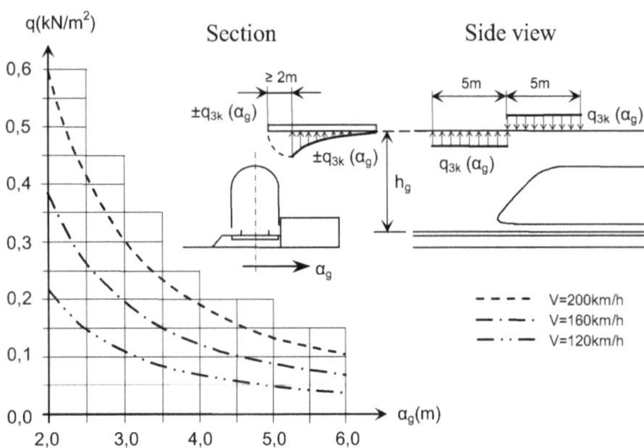

Figure 17: Wind pressure on top surfaces due to moving vehicles

Temperature Change

Even if this loading is not considered as dynamical one, it is classified as a live load.

It is obvious that statically determinate systems are not affected by the change of temperature through an appropriate layout of the moving supports or even if the developed frictions are decreased to their minimum.

Temperature change produces elongation or shortening of the bridge and of the elements from which the structure consists. Therefore, it is obvious that the statically undetermined systems are aggravated, some more and some less.

Some of these systems, especially, arches or suspension or cable-stayed bridges are affected even by uniform temperature changes.

The codes determine the erection temperature, which of course is different from place to place and, also, depends on the season of the erection. A temperature change +10 to +15°C may be considered generally accepted. The codes determine, in addition, the limits of the temperature changes. These limit values, obtained from recordings of temperatures in each country, we must add 15°C, given that steel beams, because of the solar rays, have a temperature bigger than that of the environment.

Finally, we have to study the case where some parts of the structure have different temperatures from the others, due to constructional reasons (*i.e.* when they are protected by an overlaid deck).

Destruction Loads

The Air-Blast

A dynamic disturbance, that requires special attention in both military and civil works is the airblast, which is caused by an explosion in the broader locality of the work.

The most characteristic kind of air blast is the one caused by nuclear or atomic weapons. Considering the power of a one kiloton weapon as a "reference power", we will arrive at conclusions for explosions with different power through scaled relations. We note that the power of one kiloton (1KT) is the one produced by an explosion of 1000 tons of TNT.

Generally, as explosion may be characterized the rapid release of a large amount of energy. The conventional explosions, such as from TNT, are produced from the re-arrangement of the atoms of the explosive material, while the nuclear explosions depend on the re-arrangement or redistribution of protons and neutrons within the atomic nuclei. The forces that bind protons and neutrons together are much greater than those that bind the atoms as a whole.

One can consider two basic kinds of nuclear interaction: fission and fusion.

In fission a free neutron enters the atom of a fissionable material (for example the materials with the heaviest nuclei such as some isotopes of uranium or of plutonium) and causes its fission.

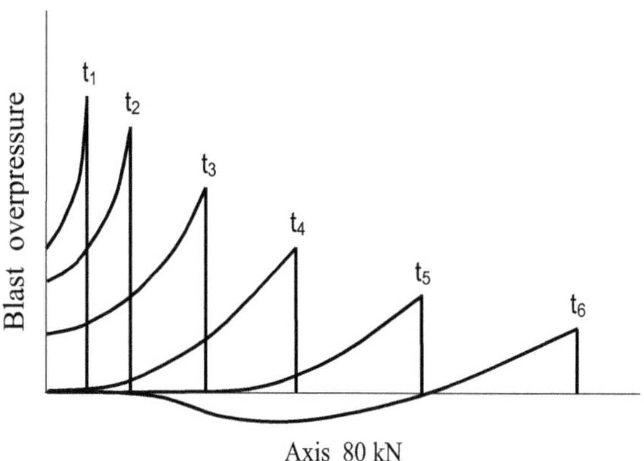

Figure 18: Pressure loads due to blasts

The complete fission of one pound (453gr) of uranium would cause the release of an amount of energy equivalent to 9,000 tons of TNT.

In the case of fusion of a pair of very light nuclei units (as for example the relative to the hydrogen isotope, deuterium or "heavy hydrogen") are united and form the nucleus of a heavier atom, such as helium. The released energy from the fusion of one pound of hydrogen would be equivalent to 26,000 tons of TNT. The realization of a fusion is accompanied by very high temperatures and therefore we speak of a thermo-nuclear process or method.

While both of these nuclear methods deal with the atom, the use of the expression "atomic weapons" for weapons based on the fission process and "thermo-nuclear weapons" for weapons based on the fusion process seems right.

For the purpose of the present unit, it is significant to know that the 50% of the released energy from such a weapon causes blast or shock. Such weapons may be stronger or weaker than those of one MT (1 megaton = 1.000.000 tons of TNT). Each one of the atomic bombs, that were dropped on Japan in 1945, was a weapon of about 20KT (kiloton = 1000 tons of TNT). The blast or the shock wave that accompanies an explosion is launched outward and moves from the point of the explosion, only a fraction of a second after the explosion occurs. The front of the spherical wave is called the front of the shock and it is like a wall of highly compressed air having an overpressure (the pressure is greater than one At, that is the normal pressure at sea level) much greater than that in the region behind it. This peak of overpressure decreases rapidly as the airblast moves away from the explosion point. After a short time, the pressure behind the wave front may drop below the atmospheric one (time t_6 in Fig. **18**). During this negative pressure (rarefaction or suction), a partial vacuum is created around the explosion point and the air is sucked in. Winds of high speed are directed away from the explosion point in the phase of high pressure (positive phase), while they are directed to the explosion point in the phase of negative pressure. It should be noted that the peak of the negative pressure is much lower than the peak of the positive pressure. In Fig. **19** we have a graphic demonstration of the meaning of the pressure-time curve. The curves and equations which will follow, as well as the comments and discussion on them, are based on the edition of "The Effects of Nuclear Weapons", published by the Atomic Energy Commission of U.S.A. [8].

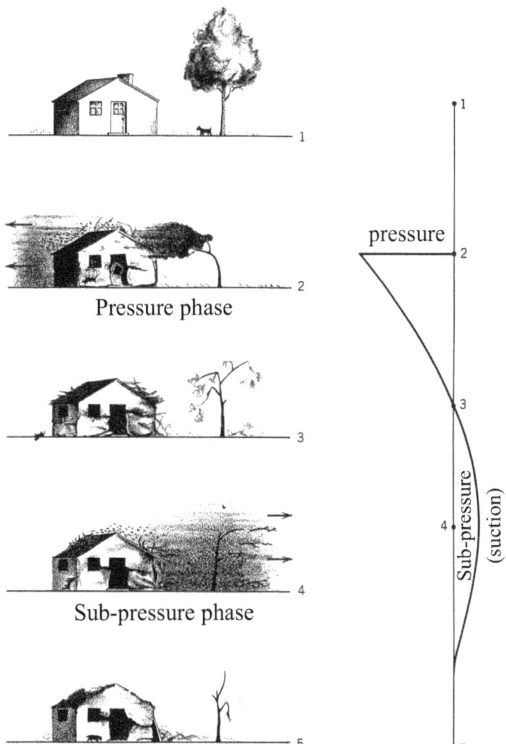

Figure 19: Pressure phases due to blast loads

When the shock wave starts from the explosion point, it moves regularly until it meets a solid of greater density than that of the normal atmosphere. Because of the impact, a reflected wave is sent back toward the explosion point. In the usual case of an explosion in the air, the initial wave and the reflected one might be moved as it is shown in Fig. **20**, where the continuous lines represent the initial one, while the dotted ones represent the reflected waves.

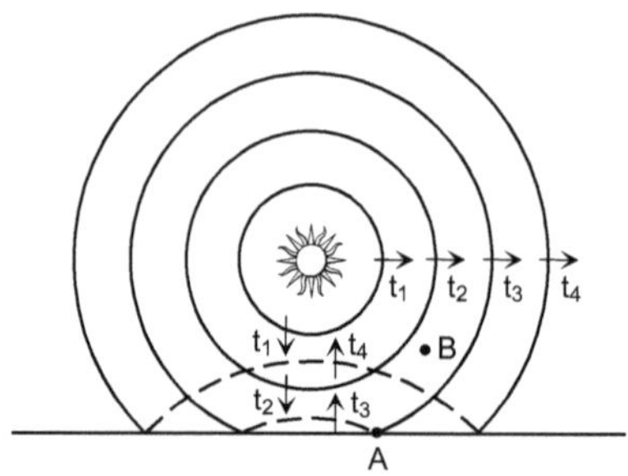

Figure 20: Reflection of pressure waves

The pressure at point A will be the sum of two pressures, one caused by the initial shock wave and another caused by the reflected wave. So, it is obvious that the pressure will be, perhaps, much greater than the one caused by the initial wave alone. If at some point below A, let it be point B, there is an object it will suffer two shocks: a shock caused by the initial wave and then a second shock caused by the reflected wave.

In Fig. **21** one can see the pressure curves for these two typical points (point A: Fig. **21(a)**, point B: Fig. **21(b)**).

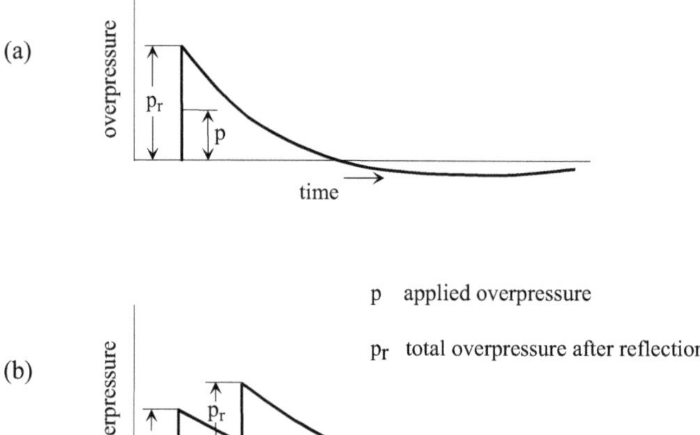

Figure 21: Pressure curves: a) in open space, b) due to reflection

The point on the ground surface that is exactly below of the explosion point, is called "ground zero" or GZ.

Since the velocity of the reflected wave is greater than that of the initial wave, it follows that, the reflected wave will catch up with the initial one at a point that has a distance from the GZ and a single vertical wave front, called "Mach Front" is created and moves horizontally along the surface of the ground. This interaction is shown in Fig. **22** for different distances from the G.Z.. In Fig. **22(c)** one can see the reflected wave, the initial one and the Mach Front, that meet at the junction point called "triple point".

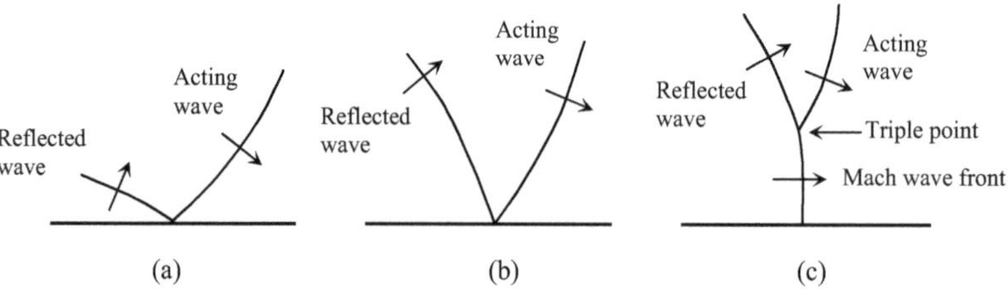

Figure 22: Acting and reflected waves and Mach Front

As the pressure wave passes over the ground surface, the triple point follows the path of Fig. **23**. Objects below this path will suffer one shock only (as shown in Fig. **22a**), while objects above this path, as for example airplanes or tops of tall buildings, will suffer two shocks (Fig. **22b**).

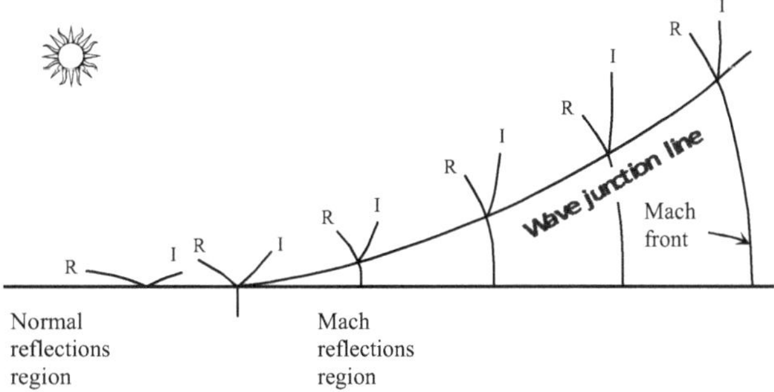

Figure 23: Normal and Mach waves front

Below the Mach path, the pressure and the wind speed are both directed horizontally, since the Mach front is practically vertical. The distance of the point at which the Mach front forms initially from the GZ point depends on the altitude and the power of the explosion. For a typical explosion of a 1MT weapon in the air, the Mach front begins at about 2km from the GZ point.

In the case of an explosion on the ground, the shock front moves along the ground surface, with an essentially vertical front near the ground (Fig. **24**).

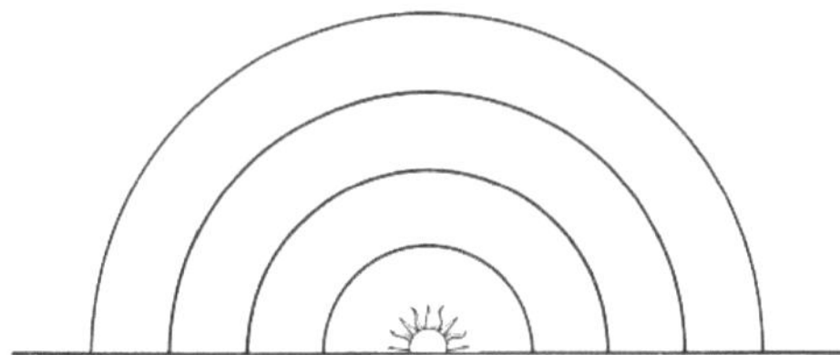

Figure 24: Wave front movement due to a Ground Zero blast

The succession of events caused by the shock of the air blast on a closed structure is shown in Fig. **25**, where the existence of an orthogonal in plan structure creates the "diffraction loading", namely, the side loadings and the drag loading, that are both caused by the developing turbulences around the structure.

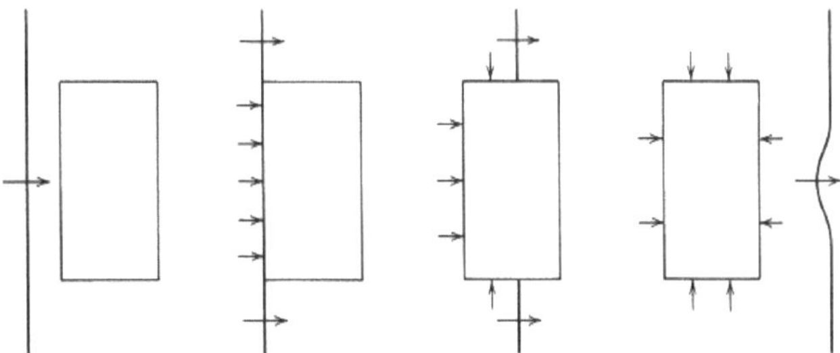

Figure 25: Over-pressure distribution due to a nuclear blast

The maximum overpressure developed from a weapon of 1KT varies in relation to the distance from GZ, according to the diagram of Fig. **26**.

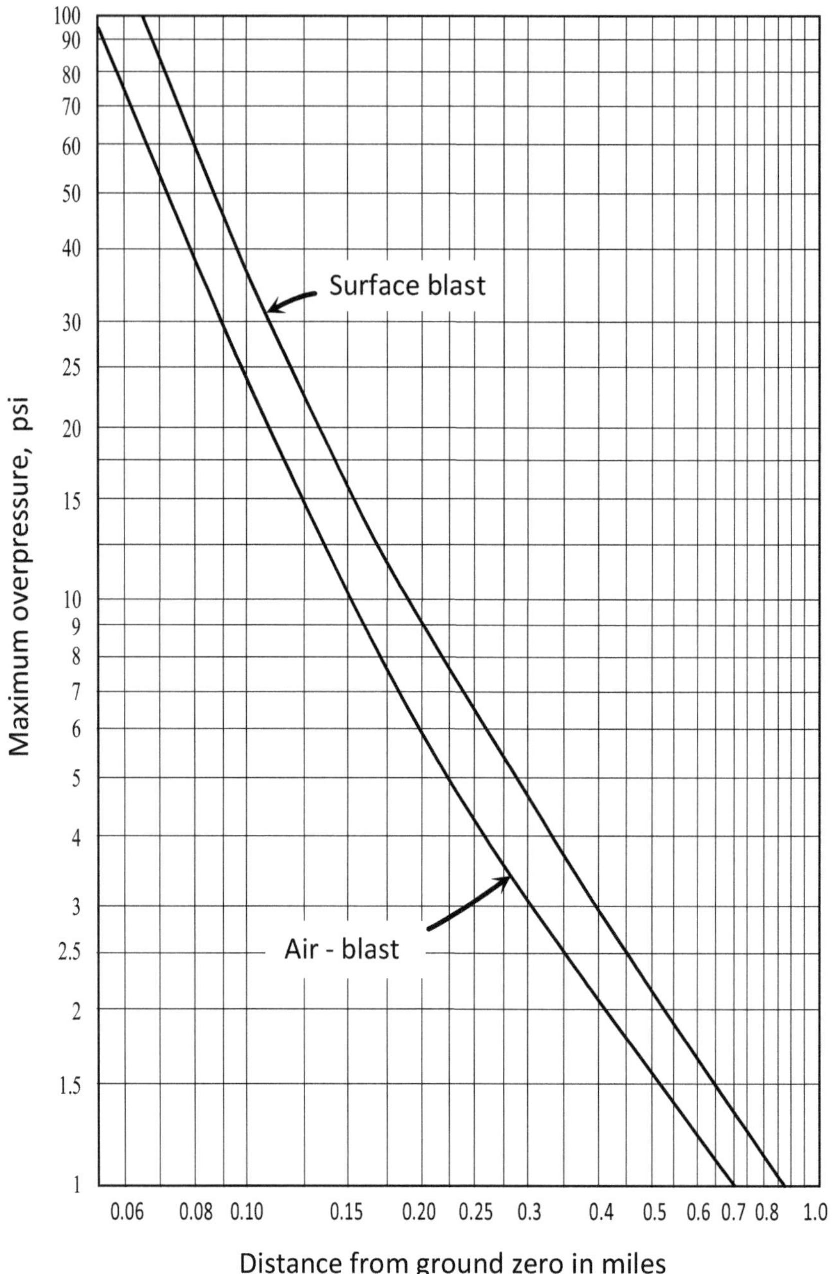

Figure 26: Pressure distribution due to air of surface blasts

For an explosion of different magnitude, we may use the above curve and the empirical expression:

$$\frac{d}{d_0} = \sqrt[3]{\frac{W}{W_0}} \tag{22}$$

where d is the distance from GZ, where the overpressure due to an explosion of power W exists, while d_0 is the distance from GZ where the same overpressure is developed and is caused by an explosion of power W_0. Let us consider, for example, that we are searching for the distance of the point at which an overpressure of 5psi (1psi *i.e.* pound per square inch=0,0703 kp/cm^2) is developed caused by an explosion of 1MT.

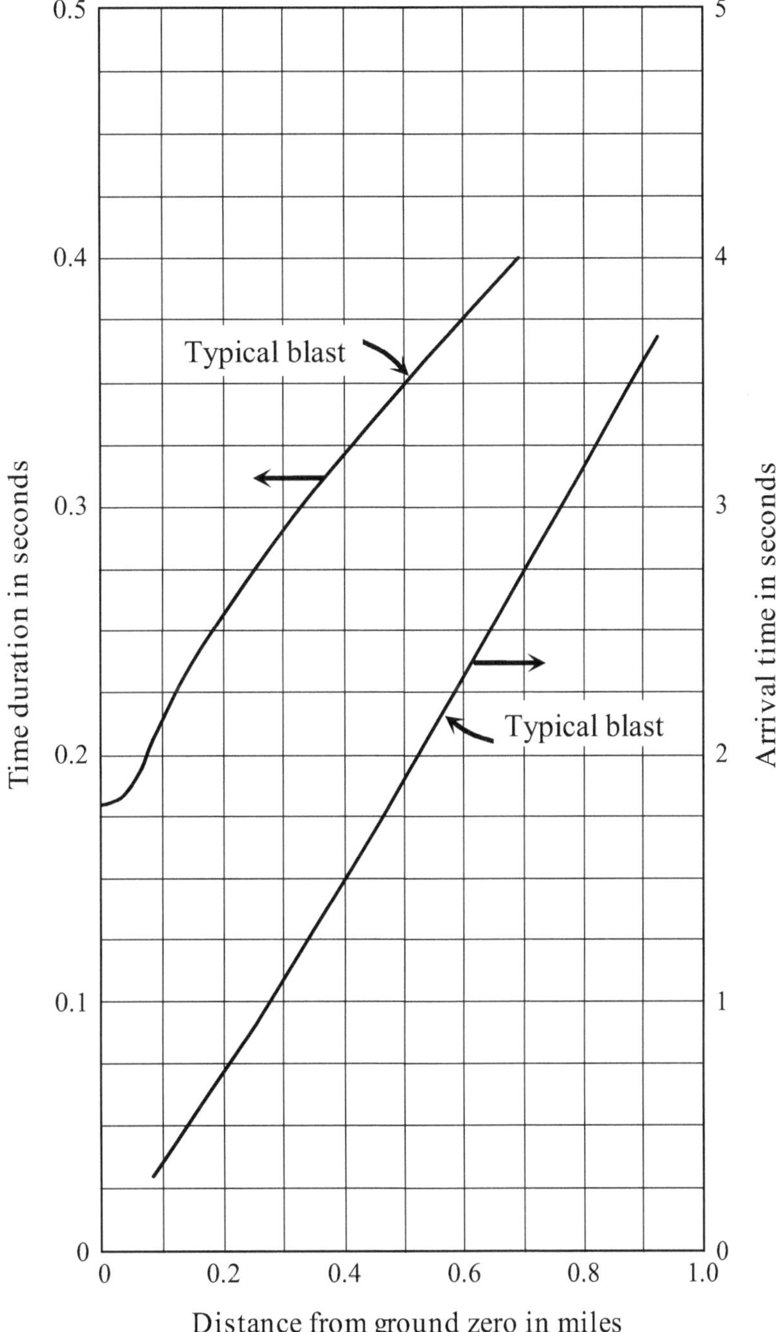

Figure 27: Time durations and arrival times due to typical blasts

From the diagram of Fig. **26** one can find that a pressure of 5psi for an explosion of 1KT, is developed at a point of a distance 0,28 miles from GZ. Thus, from eq(22) of Chapter 2 we find:

$$d = d_0 \sqrt[3]{\frac{W}{W_0}} = 0,28 * \sqrt[3]{\frac{1.000.000}{1.000}} = 2,8 \text{ miles}$$

The expression of the cube root may also be applied for the time needed for the arrival of the shock wave at a point:

$$\frac{t_+}{t_{0+}} = \sqrt[3]{\frac{W}{W_0}} \tag{23}$$

For an explosion of 1KT of power, the time duration t_{0+} of the positive phase in relation to the distance d_0 from GZ is given by the diagram of Fig. **27**. By the same diagram is also given the curve of the arrival time of the shock front.

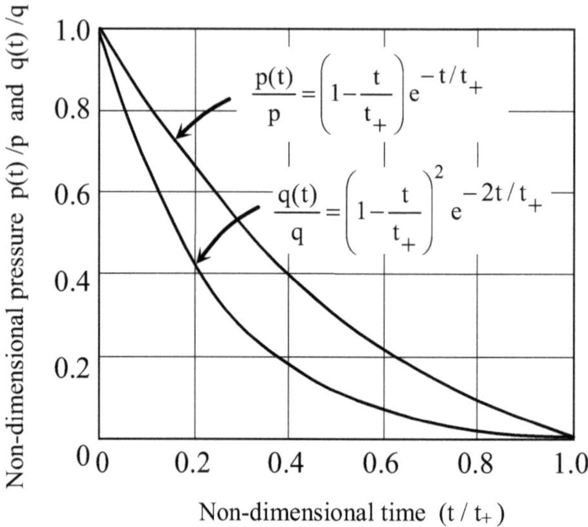

Figure 28: Blast time functions

Pressure p behind the wave front at any time t, may be expressed in relation to overpressure p_0 and to the time duration of the positive phase t_+ by the equation:

$$p(t) = p_0 \cdot \left(1 - \frac{t}{t_+}\right) \cdot e^{-\frac{t}{t_+}} \tag{24}$$

The maximum drag pressure q_0 is given approximately by the relation:

$$q_0 = 2 \cdot 5 \cdot \frac{p_0^2}{102.9 + p_0} \tag{25}$$

A more accurate expression for q_0 would include a coefficient C_d, which depends on the size and the shape of the designed structure. The variation for the drag pressure behind the wave front is approximately:

$$q(t) = q_0 \left(1 - \frac{t}{t_+}\right) \cdot e^{-\frac{2t}{t_+}} \tag{26}$$

Equations 24 and 26 of Chapter 2 are given in dimensionless form in Fig. **28**.

The pressure of the reflected wave (for an explosion in the air) is given approximately by the relation:

$$p_r = 2 \cdot p_0 \cdot \frac{102.9 + 4p_0}{102.9 + p_0} \tag{27}$$

We note that as the pressure p_0 decreases, pressure p_r approaches the value of $8p_0$, while for very small values of p_0, pressure p_r approaches the value of $2p_0$.

Earthquake [9]

Earthquake is called the violent movement of the ground that is produced by the disturbance of the balance of the rocks in the earth.

An earthquake is a pure dynamical phenomenon, but it is also unpredictable regarding its time of occurrence as well as its power. It is possible to foresee only its probability from statistical data (with a significant marginal error).

a) Types of Earthquakes

Earthquakes are divided into three categories regarding the depth of their focus: the "shallow earthquakes", namely those whose focus (see below) is up to a depth of 70km, the "intermediate earthquakes" having a focus depth up to 300km (or according to other researchers 450km) and, finally, the "deep earthquakes", having a focus depth of more than 300 or 450km. From the above three types, we are, mainly, interested in the shallow earthquakes.

b) Causes of Earthquake Generation

Earthquakes, according to the cause that gives rise to their appearance, are divided into the following three basic categories:

b1) Volcanic Earthquakes

They are the earthquakes that precede or accompany volcanic explosions and constitute about 7% of the shallow earthquakes (Fig. **29**).

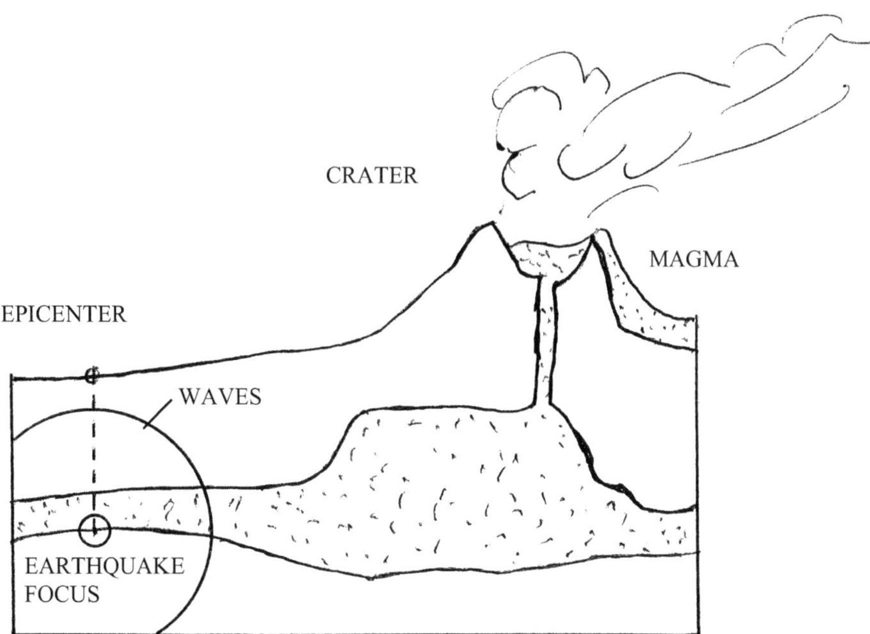

Figure 29: Seismic waves due to volcano action

b2) Earthquakes Caused by Cave Collapsing

These earthquakes are produced by the collapse of the domes of underground natural caves or, rarely, of underground rivers (see Fig. **30**). Only a percentage of about 3% of the shallow earthquakes falls into this category.

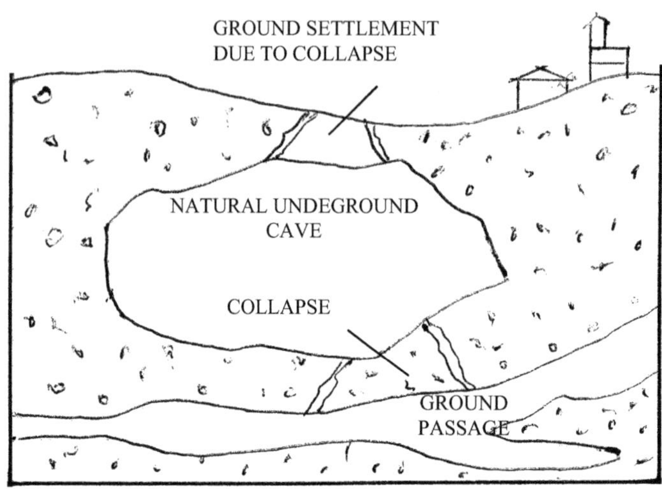

Figure 30: Seismic waves due to cave collapsing

b3) Tectonic Earthquakes

These earthquakes are caused by tectonic rearrangements of the lithospheric plates of the earth's crust and comprise about 90% of the earthquakes world-wide.

There are a lot of theories that attempt to explain the above phenomenon. The main and predominant theory is the one of the movement of tectonic plates. The views about tectonic plates are very old (F. Bacon 1620).

The solid crust of the earth is called "lithosphere" and has an average thickness of about 80km, when the average radius of the earth is 12.750km. The layer below the lithosphere is called asthenosphere.

The lithospheric plates are solid and have specific weight, lower than the one of asthenosphere that consists of viscous material (magma) of greater specific weight up to the depth of 300km, resulting in the lithospheric plates floating on the asthenospheric material. Within the asthenosphere carrying streams are created exactly like in the atmosphere and for the same reasons. The speeds of the movement are, of course, very low. The usual mechanism producing this sort of earthquake is the following:

The carrying streams cause a movement of the lithospheric plates that might be frontal (convergence or divergence of the plates) or tangential. Because of this movement, collision and friction of the plates are produced and the so emitted relative dynamic energy causes the deformation of the plates or their fracture resulting in the tremor caused by the above fracture-collapse and the production of a shocking-seismic wave (the earthquake).

Because of the earthquakes and the fracture of lithosphere seismic falls are created that, until the final arrangement of the ground, are permanent areas producing earthquakes.

In Fig. **31** one can see the main cause of earthquakes in Greece, especially in the broader area of South Aegean Sea, Crete and Libyan Sea.

c) Earthquake Data

c1) The focus or hypocenter of an earthquake (Fig. **29**) is the lithospheric point that is broken or rubbed because of its movement. It is obvious that the hypocenter is not a point but a whole area, the theoretical center of which is considered as the focus of the earthquake.

c2) The epicenter of the earthquake is the projection of the focus on the earth surface.

c3) The focus depth is the distance between the focus and the epicenter.

c4) The size of an earthquake is estimated by the so-called seismographs or accelerographs. The size of the earthquake's force is counted according to special scales.

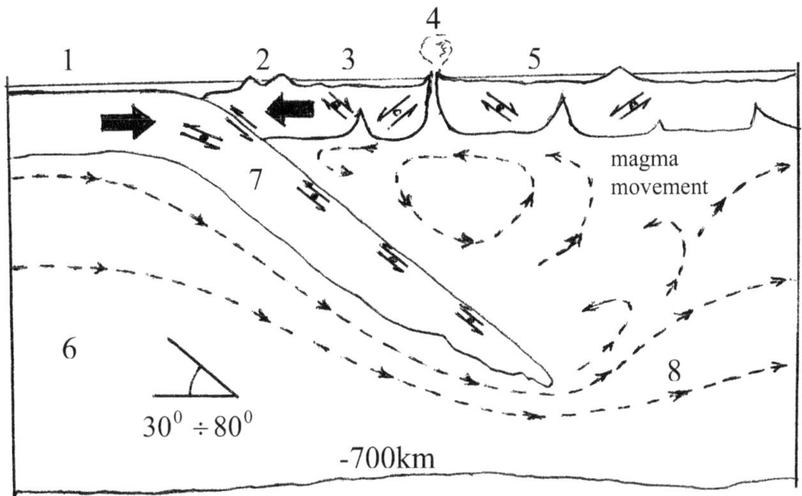

Figure 31: Seismic waves due to tectonic action

1. Mediterranean Sea
2. Crete
3. Cretan Sea
4. Santorini
5. Aegean Sea
6. Magma
7. Mediterranean plate
8. Prevailing movements

d) The Earthquake Magnitude

There are a lot of scales and variations for the measuring the magnitude of an earthquake. The two that are mainly prevalent are:

d1) The Mercalli-Sieberg Scale

It is a twelve-grade scale that measures an earthquake according to its results. Namely, it is basically a descriptive scale. For example, an earthquake of scale I is one "that is perceptible only by seismographs", while an earthquake of scale II is one from which "small bells ring, big things are upset, some tiles and chimneys fall, while buildings suffer some small damages", and finally an earthquake of scale V is the one "producing a destruction more than of 50%", e.t.c.

d2) The Richter Scale

It was conceived by Professor C.F. Richter in 1935 and has 10 degrees, it measures the acceleration of the ground and, indirectly, the energy released by an earthquake.

The size of an earthquake in Richter scale is expressed by the decimal logarithm of the maximum recorded width in microns ($1\mu=10^{-3}$mm) of a perfect seismograph, which is in a distance of about 100km from the earthquake's epicenter and is placed on a hard soil. The above conditions are not always fulfilled, and therefore it is needed the reduction of the receptions to the above conditions through suitable redaction formulae, which take into account the transmission and absorption of the seismic waves.

The greatest (measured) word-wide earthquake amounts at 8.9R. Finally we note that in order for an earthquake to cause some damages it must be greater than 5R.

e) Earthquake's Energy

The energy of an earthquake is given by the relation:

$$\log\left(\frac{E}{E_0}\right) = \alpha \cdot M_S \qquad \textbf{(28)}$$

where α=1.50, and $E_0 = 2,5 * 10^{11}$ is a basic energy in imaginary conditions given in erg, and M_S is the size of the studied earthquake in Richter degrees. In this case one can write:

$$E = E_0 * 10^{\alpha \cdot M_S} \quad \text{(erg)}$$

Considering two earthquakes differing 1R, the ratio of the released energy from each one of them will be:

$$\frac{E_{R+1}}{E_R} = \frac{10^{\alpha \cdot (M_S+1)}}{10^{\alpha \cdot M_S}} = 10^{1,5} = 31,62$$

Namely, an earthquake of 1R greater releases energy of 31.62 times greater. Considering now two earthquakes differing 0.1R we have:

$$\frac{E_{R+0,1}}{E_R} = \frac{10^{\alpha (M_S+0,1)}}{10^{\alpha \cdot M_S}} = 10^{0,15} = 1,41$$

namely, an earthquake of 0.1R greater releases 41% more energy.

f) Transmission-path of Seismic Waves

The way of transmission and the path of the seismic waves (i.e of waves that spread the seismic tremor and put in motion-vibration the particles of the soil) depends on two parameters.

1) the distance and 2) the way of reflection-refraction of the seismic waves through the soil layers of the earth crust, which meet during their course. It is obvious that the soils' lack of homogeneity affects the shape and the tremor that a seismograph receives.

Considering the imaginary case of a homogeneous soil from the epicenter to the recorder point, we see two kind of seismic waves (Fig. **32**):

1) The space waves, that are distinguished to the longitudinal or compression or first (P waves) that are the faster than all the waves and to the transverse ones or shear or seconds (S waves).
2) The surface waves that are divided to:
   a) Love waves (L), that are waves with only horizontal motion of the particles, vertical to the transmission direction, and have a speed of the same size with the one of S waves.
   b) The Rayleigh waves (R), that are waves of great period, where the soil particles move on elliptic orbits whose their plane is vertical and their speed is less than the ones of S waves.

Finally the transmission speed of P and S waves is given by the formulae:

$$\left.\begin{aligned} V_P &= \sqrt{\frac{E}{\rho} \cdot \frac{1-\nu}{(1-\nu-2\nu^2)}} \\ V_S &= \sqrt{\frac{G}{\rho}} \end{aligned}\right\} \qquad \textbf{(29)}$$

Where:

E: is the modulus of elasticity

G: is the shear modulus

ν: is the Poisson ratio

ρ: is the soil density

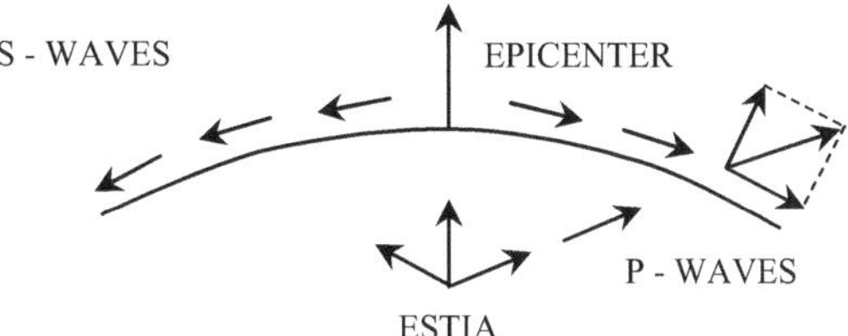

Figure 32: Seismic waves due to tectonic action

g) The Earthquake as a Disastrous Loading

The final seismic tremor, that a construction undertakes, is a complicate curve-function of time like the one shown in Fig. **33**. In the vertical axis the displacement or the speed or the acceleration of the vibrated soil particles is shown.

For each earthquake, from its seismograph, one can draw the "response spectrum", that is the enveloping curve of the behavior of oscillators of one degree to their eigenfrequency for a given damping.

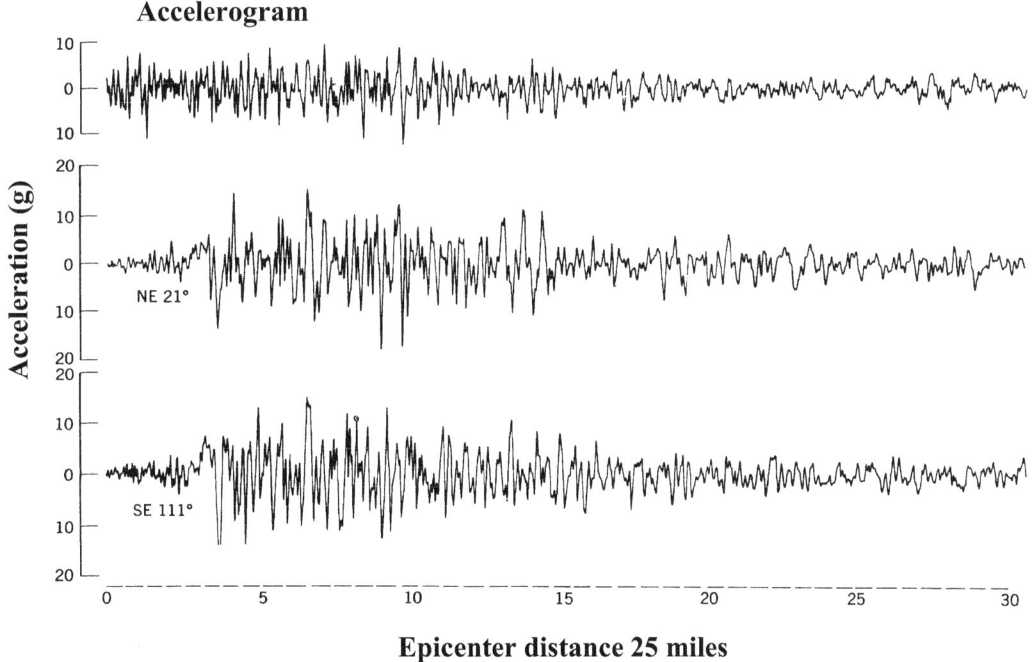

Figure 33: Typical accelerograms of an earthquake

Considering as "behavior" the maxima width of a vibration we get "the spectrum of relative displacements", considering the speeds "the speeds spectrum", or considering the accelerations "the accelerations spectrum".

In the praxis, we estimate firstly the "seismic hazard". As seismic hazard, we mean the possibility to take place in a given time period and in a particular area an earthquake equal or greater than an earthquake of definite size. Further, one can follow one of the following methods:

1) Modal superposition method. According this method, the free or the forced vibration of a multi-degree static system is consisted from eigenshapes or for the modal seismic behaviour. This method is valid for an elastic behaviour with small, relatively, damping.
2) Response spectrum approach. This method is based on the modal superposition and uses for seismic data the spectrum of seismic tremor. The maximum values of the seismic response are determined for each one modal, which are happened not necessarily simultaneously. The way of their combination is let on the designer decision.
3) The method of time-history analysis. The behavior of the structure is analyzed using either the modal superposition method or the directly integration of the differential equations of motion. In both cases it is necessary the accelerogram of the excitation while one can determine the response and the stress of the construction for every moment.

From the above it is obvious that the accelerogram of the excitation is a necessary data for the construction study. Let us consider now a part of the seismogram of Fig. **33**, as it is enlarged in Fig. **34**. Then, idealizing the seismic tremor, one can consider the maximum width d_i, the semi-period $T_i/2$ and the sinusoidal curve:

$$y(t) = d_i \cdot \sin \omega_i t \qquad \textbf{(30)}$$

where:

y: is the width of the vibration at time t

d: is the maximum width or semi-amplitude and

ω: is the circular frequency, equal to $\frac{2\pi}{T_i}$

The velocity and the acceleration of the vibrated soil particles will be:

$$\begin{aligned} \dot{y}(t) &= d_i \cdot \omega_i \cdot \cos \omega t = V_i \cdot \cos \omega t \\ \ddot{y}(t) &= -d_i \cdot \omega_i^2 \cdot \sin \omega t = -\omega_i^2 \cdot y(t) \end{aligned} \qquad \textbf{(31)}$$

A typical seismogram, like the one of Fig. **34**, it is obvious that is not possible to be expressed by only one equation of motion for the entire time of an earthquake and that the characteristics of such an equation (d_i and T_i) change from instant to instant. Then, we consider an imaginary seismic excitation, with amplitude the maximum amplitude given by the corresponding spectrum of relative displacements and period (usually) the dominating one.

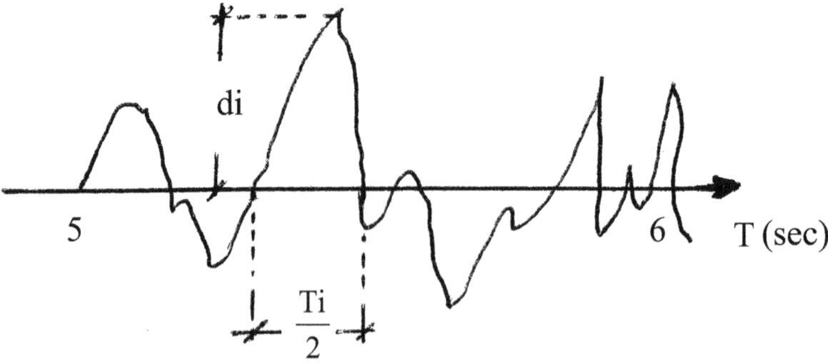

Figure 34: Typical diagram of ground motion

The seismic diagrams have different shapes, which is depended from the epicenter distance. Usually, the following equations are used:

1. For a distant earthquake, the following equation may be used:

$$f(t) = k \cdot t \cdot e^{-\beta \cdot t} \cdot \sin \Omega t \qquad \textbf{(32)}$$

where:

k: is a constant affecting the motion amplitude. It changes usually from 0,01 to 0,10.

β: is a constant affecting mainly the duration of the excitation and changes usually from 0,05 to 0,30.

Ω: is the natural frequency that affects mainly the acceleration and its values change from 1 to 20.

The equation of such a motion and its acceleration are plotted in Fig. **35**.

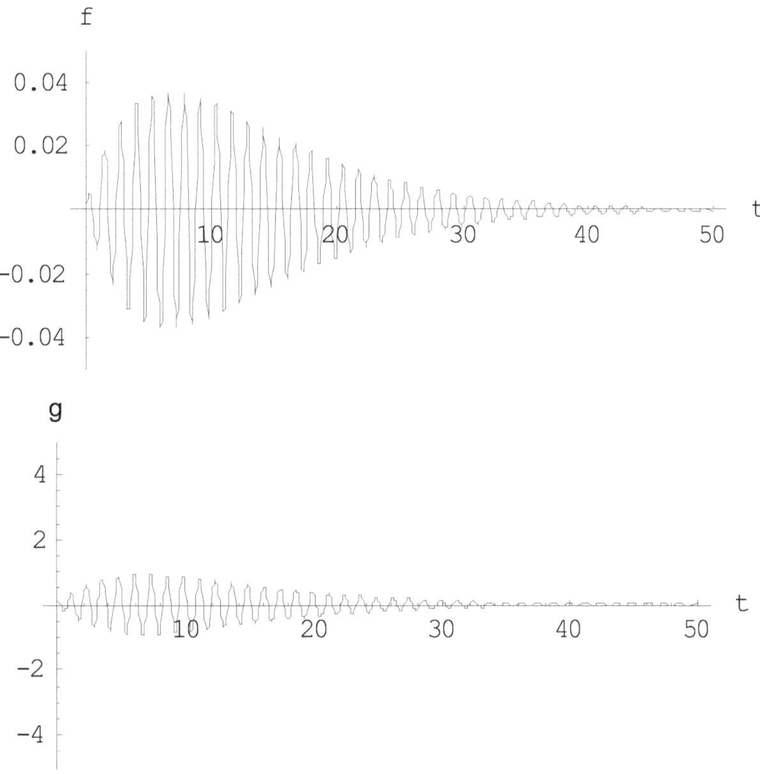

Figure 35: Distant earthquake ground motion

2. For a near earthquake, the following equation may be used:

$$f(t) = k \cdot e^{-\beta \cdot t} \cdot \sin \Omega\, t \qquad \textbf{(33)}$$

With constants as previously but with values:

k: from 0,01 to 0,10, β: from 0,15 to 0,50, and Ω: from 5 to 30.

The equation of such a motion and its acceleration are plotted in Fig. **36**.

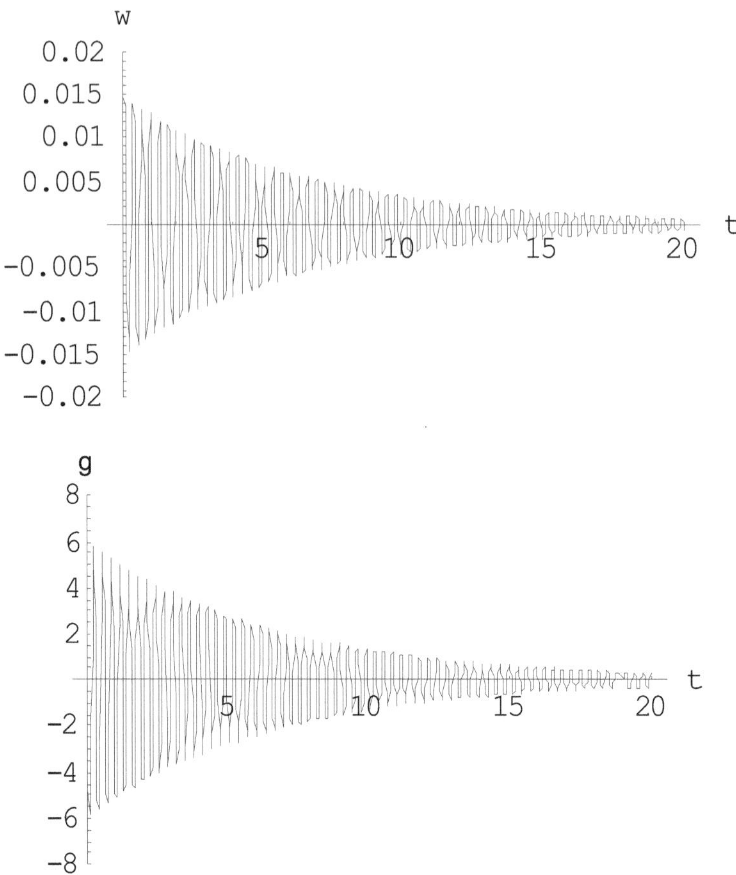

Figure 36: Near-source earthquake ground motion

Finally we point out that the examination of the strength of a bridge in seismic loads is a particularly complicated problem. There is a lot of factors that must be investigated, like: buildings on faults, the possible different behaviour of pylons especially for long span bridges where the seismic motion may be different between two distant foundations, the influence of the different height of pylons mainly in valley bridges and a numerous of secondary factors, which is possible to become critical for a combination of local and constructional conditions.

The Collision of a Vehicle [2]

The forces that are caused by a collision and act on the deck, the basis or other elements supporting the bridge apply according to Eurocode 1 at a height of 1.25 m above the deck level. These forces are equal to 1000KN, when act in parallel to the circulation direction, or 500KN, when act vertical to it.

For the dynamic analysis of such a loading, we advice an analysis according to the general theory of impact of a body to another of infinity mass, and bouncing forces with coefficient of absorption from 0.70 to 1.00.

Train's Derailment

The design philosophy is: in a derailment case, the disaster being limited to a minimum while in any case the bridge must be stable (without overturning or collapse).

Two cases are examined:

1st Case:

Derailment of one wagon, that remains on the bridge close to the rails.

The parts of the bridge on which the derailment happened are designed with the following equivalent loads q_{A1d}=50KN/m for a length of 6,40m (see UIC) in parallel to the rails and set on the most unfavorable place, within a width $1.5 \cdot S$ in both sides of the rail (Fig. **37**)

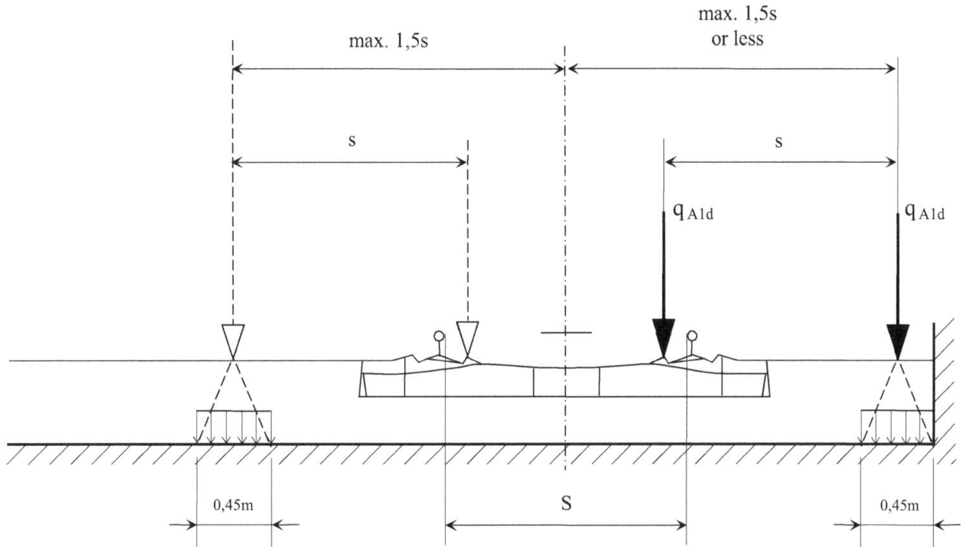

Figure 37: Accidental loading due to train derailment (case I)

2nd Case:

Derailment of more than one wagons that remain also on the bridge

The applied load is q_{A2d}=80KN/m for a length of 20m according to Fig. **38**.

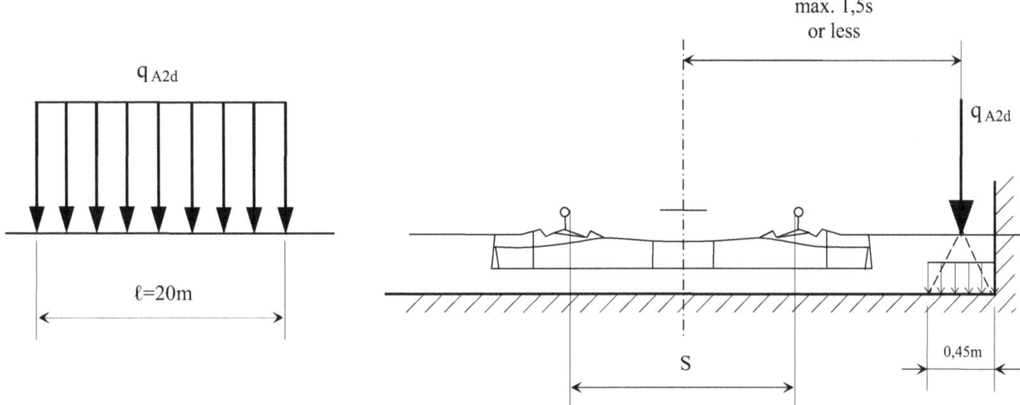

Figure 38: Accidental loading due to train derailment (case II)

SPECIAL LOADS

The special loads, according to the being in force codes, are taken as an rate of the vertical loads of circulation, accordingly of their shape, their geometry and the cross-section of the bridge, where incremental coefficients are included, taking into account the dynamic character of the moving loads.

This confrontation is for the benefit of safety. Recent theoretical and experimental researches proved that the action of such loadings are underestimated, because, mainly, we ignore either the character of the loads and the factors which affect their actions (*i.e.* suspensions), and also the behavior of a bridge under such special loads (f.e. circulation loads of very great speeds).

These special loads are studied in the corresponding chapters of this book.

Acceleration - Deceleration

The valid codes give these forces as a rate of the circulation loads and as a function of the length L of the studied part of the bridge. These forces are significant for a great ratio of the masses of circulated loads to the bridge's mass, namely may being critical for the railway bridges.

Furthermore, cotemporary studies proved that a vehicle accelerating on a bridge increases the, already, increased action caused by its velocity.

We note, finally, that for curved in plane bridges is needed a particular dynamic examination and we advise that the designer does not be content with the directions of a code, as long as they are severe.

Centrifugal Force

The centrifugal force that is developed in the case of a curved in plane bridge is particularly significant. Furthermore, in this case the influence is depended not only on the curvature but also on the ratio of the vehicles' masses to the bridge's one.

Two forces, whose is ignored even their existence, but recent researches proved that may being particularly significant, are the centrifugal and the Coriolis forces that are developed because of the vertical motion (and the sequential torsion) of a bridge.

Collision of Vehicles

This loading may be particular significant and must be studied during the design of the bridge.

As in previous section, we must taking into account the produced, because of the impact, forces and also the data for the absorbed energy that are given from experimental results in different tests (crash-tests).

Settlement of Supports

One must study not only the new equilibrium situation that comes from a statical point of view, but also the dynamic extra surge on the bridge at the instant and immediately after the settlement of a support.

A support settlement may be expressed by the relation:

$$\delta = \delta_0 \cdot f(t) \quad \textbf{(34)}$$

where $f(t)$ is a step load given by the diagram in Fig. **39**.

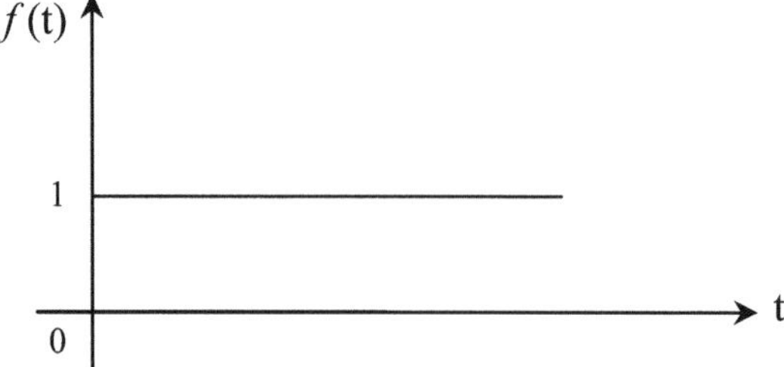

Figure 39: Support settlement time function

FATIGUE

The fatigue examination is particularly significant for the modern bridges, because of the dense circulation on the contemporary transport networks, and the increased circulation loads. The modern codes give a special emphasis in this examination, which is out of the goals of this book.

LOADINGS DURING ERECTION

The usual loadings during the erection of a bridge have always a statical character, and, particularly, because of the modern methods of erection, the bridge may be in each construction's step completely different from statical point of view. For example a simply supported bridge with big length constructed following the cantilever system from both its ends, behave as a cantilever.

The loads have, during the erection, a dynamic character only if are caused by an accident. For example the cracking of a cable in a cable stayed bridge during the erection acts as an impact force that produces the vibration of the constructed bridge. Such unusual cases must be taken into account with a particular severity.

REFERENCES

[1] Eurocode 1: Basis of design and actions on structures – Part 1: Basis of design, CEN, ENV 1991-1, 1994.

[2] Eurocode 1: Basis of design and actions on structures – Part 3: Traffic loads on bridges, CEN, ENV 1991-3, 1995.

[3] G.T. Michaltsos *"Parameters affecting the dynamic response of light (steel) Bridges"*, 2000, Facta Universitatis-Series: Mechanics, Automatic Control and Robotics, Vol. 2, No 10.

[4] L.L. Frerris *"Wind energy – Conversion Systems"*, 1990, Prentice Hill, New York.

[5] C. Petersen *"Abgespante Maste und Schornsteine – Statik und Dynamik"*, 1970, Bauingenieur – Praxis, Heft 76.

[6] J. Schlaich *"Beitrag zur Frage der Wirkung von Windstössen auf Bauwerke"*, 1966, Bauingenieur (41).

[7] B.K. Donaldson *"Analysis of Aircraft structures – An Introduction"*, 1993, Mc Graw-Hill, N. York.

[8] USAEC *"The effects of Nuclear Weapons"*, 1957, 1973+, 1986, Printing Office, Washington D.C.

[9] P. Karydis *"Notes on Seismic Technology"*, 1996, NTUA Publ., Athens (in Greek).

[10] Eurocode 8: Design provision for earthquake resistance of structures, Part 1.1 CEN, ENV 1998-1-1 (1994), Part 1.2 ENV 1998-1-2 (1994), Part 1.3 ENV 1998-1-3 (1995) and Part 2 ENV 1998-2 (1994).

[11] J. Ermopoulos *"Steel and Composite Bridges"*, 2000, Kleidarithmos Publ., Athens (in Greek).

CHAPTER 3

Principles of Dynamic Analysis

Abstract: In this chapter, the most important aspects related to dynamic analysis of structures are presented. Calculus of variations and energy principles such as d'Alembert principle, Lagrange equations of motion and Hamilton principle are given in brief form. The general form of equations of motion is presented along with the most common solution methods such as integral transformation or Ritz and Galerkin methods. The equations for free vibration in axial, bending and torsional mode are solved for various boundary conditions. The problem of forced vibrations is also presented for the above cases.

INTRODUCTIONAL CONCEPTS

The Calculus of Variations [1]

A functional is a function whose domain of definition is a set of functions. Thus, it is natural to search if the concept of the differentiation of a function can be extended to functionals.

Therefore, let us suppose that F(x,y,y′) is a functional defined by a set of functions {y(x)}, and let as try to determine an expression for the change in F corresponding to a fixed in advance change in y(x) for a fixed value of x. If y(x) is changed into the function $y(x)+\varepsilon\cdot n(x)$ with ε independent of x, we call $\varepsilon\cdot n(x)$ the **variation of y** and denote it by δy:

$$\delta y=\varepsilon\cdot n(x) \tag{1}$$

In addition, from the changed value of y we infer that the changed value of y′ is $y'(x)+\varepsilon\cdot n'(x)$ and therefore we have the relation:

$$\delta y'(x)=\varepsilon\cdot n'(x) \tag{2}$$

Correspondingly to these changes, we have the change of F:

$$\Delta F=F\left(x,y+\varepsilon\cdot n,y'+\varepsilon\cdot n'\right)-F\left(x,y,y'\right) \tag{3}$$

Expending the first right part term in a Maclaurin's series of ε, we have:

$$\Delta F=F\left(x,y,y'\right)+\left(\frac{\partial F}{\partial y}\cdot n+\frac{\partial F}{\partial y'}\cdot n'\right)\cdot\varepsilon+\left(\frac{\partial^2 F}{\partial y^2}\cdot n^2+2\cdot\frac{\partial^2 F}{\partial y\cdot\partial y'}\cdot n\cdot n'+\frac{\partial^2 F}{\partial y'^2}\cdot n'^2\right)\cdot\frac{\varepsilon^2}{2!}+.....-F\left(x,y,y'\right)$$

Neglecting powers of ε of higher order, we find:

$$\Delta F\cong\frac{\partial F}{\partial y}\cdot n\cdot\varepsilon+\frac{\partial F}{\partial y'}\cdot n'\cdot\varepsilon$$

or finally, defining this last expression as the **variation of function F**, and symbolizing it as δF, due to eqs (1) and (2) we have:

$$\delta F=\frac{\partial F}{\partial y}\cdot\delta y+\frac{\partial F}{\partial y'}\cdot\delta y' \tag{4}$$

By strict analogy with the differential of a function with three variables, we might have expected the definition:

$$\delta F=\frac{\partial F}{\partial x}\cdot\delta x+\frac{\partial F}{\partial y}\cdot\delta y+\frac{\partial F}{\partial y'}\cdot\delta y' \tag{5}$$

George T. Michaltsos and Ioannis G. Raftoyiannis

But, let us remember that the functional is the value of F(x,y,y′) at a particular value of x, which means that x does not vary in the calculation of δF and hence δx=0.

We remember that the differential δF_1 of a function F_1 is a first order approximation to the change in the function as x varies along one of the curves F_i of the family of curves F (curve F_1 in Fig. **1**).

Contrarily, the variation of a functional F is a first-order approximation to the change in the functional at a particular value of x as we go from curve to curve (Fig. **1**).

We note that variations can be calculated by the same rules that apply to differentials. Especially the following rules are valid:

$$\left.\begin{aligned} &\delta(F_1 \pm F_2) = \delta F_1 \pm \delta F_2 \\ &\delta(F_1 \cdot F_2) = F_1 \cdot \delta F_2 + F_2 \cdot \delta F_1 \\ &\delta\left(\frac{F_1}{F_2}\right) = \frac{F_2 \cdot \delta F_1 - F_1 \cdot \delta F_2}{F_2^2} \\ &\delta(F^n) = n \cdot F^{n-1} \cdot \delta F \end{aligned}\right\} \qquad \textbf{(6a,b,c,d)}$$

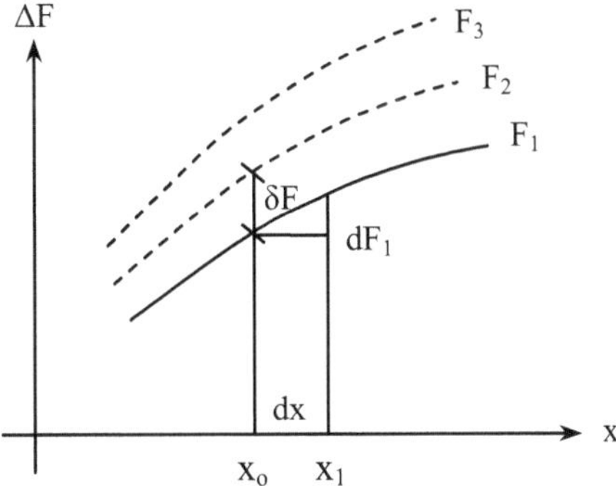

Figure 1: Assessment of function variation

The proof of the first of the above relations is easy. To prove the second one, we have:

$$\Delta(F_1 \cdot F_2) = F_1(x, y + \varepsilon \cdot n, y' + \varepsilon \cdot n') * F_2(x, y + \varepsilon \cdot n, y' + \varepsilon \cdot n') - F_1 * F_2$$

Thus, expanding again in series of powers of ε and because of eqs (1) and (2) we obtain:

$$\Delta(F_1 * F_2) = \left[F_1 + \left(\frac{\partial F_1}{\partial y} \cdot n + \frac{\partial F_1}{\partial y'} \cdot n'\right) \cdot \varepsilon + ...\right]\left[F_2 + \left(\frac{\partial F_2}{\partial y} \cdot n + \frac{\partial F_2}{\partial y'} \cdot n'\right) \cdot \varepsilon + ...\right] - F_1 \cdot F_2 =$$

$$= F_1 \cdot \left(\frac{\partial F_2}{\partial y} \cdot \delta y + \frac{\partial F_2}{\partial y'} \cdot \delta y'\right) + F_2 \cdot \left(\frac{\partial F_1}{\partial y} \cdot \delta y + \frac{\partial F_2}{\partial y'} \cdot \delta y'\right) = F_1 \cdot \delta F_2 + F_2 \cdot \delta F_1$$

Following a similar way, eqs (6c and d) are proved. From eq (1) we have:

$$\frac{d(\delta y)}{dx} = \varepsilon \cdot n' = \delta y' \qquad \textbf{(7)}$$

namely, taking the variation of a functional and differentiating with respect to the independent variable are commutative operations. We can consider functionals with more than one function. The variations of such functionals are determined by expressions similar to eq (4). For example, for the functional F(x,u,υ,u′,υ′) we have:

$$\delta F = \frac{\partial F}{\partial u}\cdot\delta u + \frac{\partial F}{\partial \upsilon}\cdot\delta\upsilon + \frac{\partial F}{\partial u'}\cdot\delta u' + \frac{\partial F}{\partial \upsilon'}\cdot\delta\upsilon' \quad \textbf{(8)}$$

We may, also, consider functionals that depend on functions with more than one variable. For instance, for the functional $F(x,y,u,u_x,u_y)$, whose value depends on the function u(x,y) for fixed x and y, we have:

$$\delta F = \frac{\partial F}{\partial u}\cdot\delta u + \frac{\partial F}{\partial u_x}\cdot\delta u_x + \frac{\partial F}{\partial u_y}\cdot\delta u_y \quad \textbf{(9)}$$

Finally, we look at a functional, which is expressed as a definite integral, say for instance the integral:

$$I(y) = \int_\alpha^\beta F(x,y,y')\cdot dx \quad \textbf{(10)}$$

We have at first that $\Delta I = I(y+\varepsilon\cdot n) - I(y)$ and because the limits of I(y) are independent on y, we get further:

$$\Delta I = \int_\alpha^\beta F(x,y+\varepsilon\cdot n,y'+\varepsilon\cdot n')\cdot dx - \int_\alpha^\beta F(x,y,y')\cdot dx = \int_\alpha^\beta \left[F(x,y+\varepsilon\cdot n,y'+\varepsilon\cdot n') - F(x,y,y')\right]\cdot dx = \int_\alpha^\beta \Delta F(x,y,y')\cdot dx$$

The variation of I is now defined as the expression resulting when ΔF in the last integral is replaced by the first-order approximation δF. Thus:

$$\delta I = \int_\alpha^\beta \delta F(x,y,y')\cdot dx \quad \textbf{(11)}$$

We will determine finally the necessary condition for the functional I to have an extreme value, namely, the condition under which δI=0 is valid.

Indeed using the results of the preceding discussion, we can write:

$$\delta I = \int_\alpha^\beta \delta F(x,y,y')\cdot dx = \int_\alpha^\beta \left(F_y\cdot\delta y + F_{y'}\cdot\delta y'\right)\cdot dx = \int_\alpha^\beta \left(F_y\cdot\delta y + F_{y'}\cdot\frac{d(\delta y)}{dx}\right)\cdot dx$$

and integrating the last term by parts we get:

$$\int_\alpha^\beta F_{y'}\cdot\frac{d(\delta y)}{dx}\cdot dx = \int_\alpha^\beta F_{y'}\cdot d(\delta y) = F_{y'}\cdot\delta y\Big|_\alpha^\beta - \int_\alpha^\beta \frac{dF_{y'}}{dx}\cdot\delta y\cdot dx$$

Because we have assumed that the variation $\delta y = \varepsilon\cdot n(x)$ vanishes at x=α and x=b, it follows that the integrated term is equal to zero. Thus:

$$\delta I = \int_\alpha^\beta \left[F_y - \frac{dF_{y'}}{dx}\right]\delta y dx$$

Therefore, the condition δI=0 (for variation $\delta y \neq 0$) implies the condition:

$$F_y = \frac{d(F_{y'})}{dx} \quad \textbf{(12)}$$

Energy Axioms

The D'Alembert Principle

J. Bernoulli and in the same period M. de l'Hôspital investigating the determination of the swinging center of the natural pendulum used the case of an equilibrating balance. Jean le Rond d' Alembert extended later this idea as a principle (d' Alembert principle) for the mechanic entire (Traité de Méchanique 1743). Lagrange, later on (1788), based on the d' Alembert principle its Analytic Mechanic [2].

D' Alembert considered the pendulum in Fig. **2** that had the masses m_1 and m_2 fixed on the solid rod OA in places being far r_1 and r_2 from the rotation point.

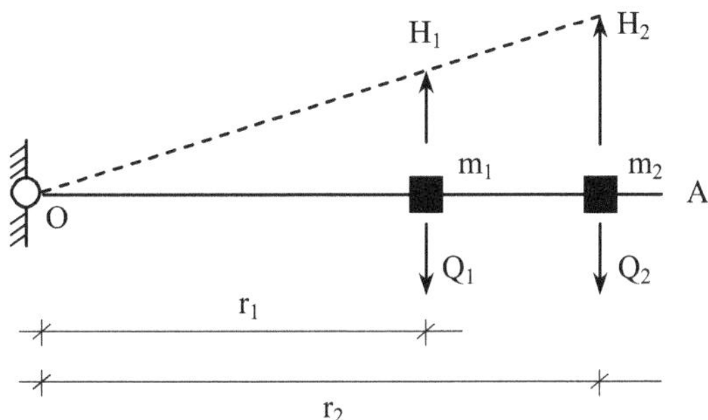

Figure 2: D'Alembert pendulum

Considering the rod OA in a horizontal position and letting it to fall free, because of its obligation to rotate around the point O, masses m_1 and m_2 will never get the common acceleration g (as they would get if they was free of obligations), but the following accelerations:

$$\gamma_1 = g \cdot r_1 \cdot \frac{m_1 \cdot r_1 + m_2 \cdot r_2}{m_1 \cdot r_1^2 + m_2 \cdot r_2^2} \quad , \quad \gamma_2 = g \cdot r_2 \cdot \frac{m_1 \cdot r_1 + m_2 \cdot r_2}{m_1 \cdot r_1^2 + m_2 \cdot r_2^2}$$

This is due to the imaginary forces H_1 and H_2, which have to act on the rotating rod in order for the system to equilibrate at every instant. D' Alembert called these forces **inertia forces** and found that they are equal to the each time mass m multiplied by the produced acceleration γ.

D' Alembert expressed its principle as follows:

The moving force F_κ and the inertia force $-m\gamma$ constitute each time a system of forces having a resultant equal to zero.

Thus, the following equation is valid:

$$\vec{F}_\kappa + \left(-m\vec{\gamma} \right) = 0 \tag{13}$$

We consider now a balanced system, consisted from n particles of mass m. If the force P_i (X_i, Ψ_i, Z_i) acts on each one of the above particles, the produced work for virtual displacements δx_i, δy_i, δz_i, must be zero.

Namely, it is valid the equation:

$$\sum_{i=1}^{n} \left[\left(X_i - m \cdot \ddot{x}_i \right) \cdot \delta x_i + \left(\Psi_i - m \cdot \ddot{y}_i \right) \cdot \delta y_i + \left(Z_i - m \cdot \ddot{z}_i \right) \cdot \delta z_i \right] = 0 \tag{14}$$

The above eq (14) expresses the general principle of virtual displacements, where the inertia forces of the system of n particles are taken into account.

The Lagrange Equations [3]

We suppose that we have determine κ independent to each-other parameters $q_1, q_2,\ldots, q_\kappa$, by which the position of each particle (from a system of n particles) is completely determined through equations having the form:

$$\left.\begin{array}{l} R_i = \bar{R}_i\left(t, q_1, q_2, \ldots, q_\kappa\right) \\ \text{with: } i = 1,\ldots,n \end{array}\right\} \tag{15}$$

The above independent to each-other parameters q_r are in a structure, for example, the possible deformations u, υ, w, θ.

Let, in addition, m_i be the masses of n particles. We suppose that **only one** from the parameters $q_1,\ldots,q_\kappa$ is changed. Because of the above change, the particles are displaced by:

$$\frac{\partial \bar{R}_1}{\partial q_r}\cdot\delta q_r \ , \quad \frac{\partial \bar{R}_2}{\partial q_r}\cdot\delta q_r \ ,\ldots\ldots, \quad \frac{\partial \bar{R}_n}{\partial q_r}\cdot\delta q_r$$

while, at the same time, the inertia forces $\bar{P}_1, \bar{P}_2,\ldots, \bar{P}_n$ appeared, are acting correspondingly on each particle. Thus, the work produced by forces $\bar{P}_i$ will be:

$$\left.\begin{array}{l} \delta \mathcal{E} = \sum_{i=1}^{n} \bar{P}_i \cdot \frac{\partial \bar{R}_i}{\partial q_r}\cdot \delta q_r = Q_r \cdot \delta q_r \\ \text{where we put:} \quad \sum_{i=1}^{n} \bar{P}_i \cdot \frac{\partial \bar{R}_i}{\partial q_r} = Q_r \end{array}\right\} \tag{16a,b}$$

If $\bar{\upsilon}_i$ is the vectorial speed of particle i, according to d' Alembert principle it will be: $\bar{P}_i = m_i \cdot \frac{d\bar{\upsilon}_i}{dt}$.

Thus: $Q_r = \sum_{i=1}^{n} \bar{P}_i \cdot \frac{\partial \bar{R}_i}{\partial q_r} = \sum_{i=1}^{n} m_i \cdot \frac{d\bar{\upsilon}_i}{dt}\cdot\frac{\partial \bar{R}_i}{\partial q_r}$ or finally:

$$Q_r = \frac{d}{dt}\left(\sum_{i=1}^{n} m_i \cdot \bar{\upsilon}_i \cdot \frac{\partial \bar{R}_i}{\partial q_r}\right) - \sum_{i=1}^{n} m_i \cdot \bar{\upsilon}_i \cdot \frac{d}{dt}\left(\frac{\partial \bar{R}_i}{\partial q_r}\right) \tag{17}$$

From eq (15) we have:

$$\begin{array}{l} \bar{\upsilon}_i = \frac{d\bar{R}_i}{dt} = \frac{\partial \bar{R}_i}{\partial t} + \frac{\partial \bar{R}_i}{\partial q_1}\cdot \dot{q}_1 + \frac{\partial \bar{R}_i}{\partial q_2}\cdot \dot{q}_2 + \ldots + \frac{\partial \bar{R}_i}{\partial q_\kappa}\cdot \dot{q}_\kappa \\ \text{for} \quad i = 1,\ldots\ldots,n \end{array}$$

and since $q_1,\ldots,q_\kappa$ are independent to each-other, we get:

$$\frac{\partial \bar{\upsilon}_i}{\partial \dot{q}_r} = \frac{\partial \bar{R}_i}{\partial q_r} \tag{18}$$

and thus:

$$\frac{\partial \bar{\upsilon}_i}{\partial q_r} = \frac{\partial}{\partial q_r}\left(\frac{d\bar{R}_i}{dt}\right) = \frac{d}{dt}\left(\frac{d\bar{R}_i}{\partial q_r}\right) \qquad \textbf{(19)}$$

Therefore, eq (17) because of eqs (18) and (19) becomes:

$$Q_r = \frac{d}{dt}\left(\sum_{i=1}^{n} m_i \cdot \bar{\upsilon}_i \cdot \frac{\partial \bar{\upsilon}_i}{\partial \dot{q}_r}\right) - \sum_{i=1}^{n} m_i \cdot \bar{\upsilon}_i \cdot \frac{\partial \bar{\upsilon}_i}{\partial q_r} \qquad \textbf{(20)}$$

Since the kinetic energy is given by the relation: $K = \frac{1}{2} \cdot \sum_{i=1}^{n} m_i \cdot \bar{\upsilon}_i^2$, eq (20) because of eq (16a) is written as follows:

$$\frac{d}{dt}\left(\frac{\partial K}{\partial \dot{q}_r}\right) - \frac{\partial K}{\partial q_r} = Q_r = \frac{\partial \mathcal{E}}{\partial q_r} \quad \text{with: } r = 1, ..., \kappa \qquad \textbf{(21)}$$

If the generalized forces Q_r are depended not only on the inertia forces, but a part of them is depended on the parameters q_κ **but not on $\dot{q}_\kappa$** (conservative forces), then the following negative work is produced:

$$dV = \sum Q_i \cdot dq_i \qquad \text{where: } Q_r = \frac{\partial V}{\partial q_r} \qquad \textbf{(22a,b)}$$

Substituting this part of Q_r from eq (22) into eq (21) we get:

$$\frac{d}{dt}\left(\frac{\partial K}{\partial \dot{q}_r}\right) - \frac{\partial K}{\partial q_r} + \frac{\partial V}{\partial q_r} = \bar{Q}_r = \frac{\partial \Omega}{\partial q_r} \quad \text{with: } r = 1, 2, ..., \kappa \qquad \textbf{(23)}$$

where Ω is the produced negative work by the part of the generalized forces, which have not potential, or in other words it is the work of the external forces.

Finally, it is proved that if Rayleigh damping forces exist (dissipation function) acting linearly with the form $c \cdot \dot{\vec{R}}_i \ (i = 1, ..., n)$, the corresponding energy term will be $F = \frac{1}{2} \cdot \sum_{i=1}^{n} c \cdot \dot{\vec{R}}_i^2$ while eq (23) becomes:

$$\frac{d}{dt}\left(\frac{\partial K}{\partial \dot{q}_r}\right) - \frac{\partial K}{\partial q_r} + \frac{\partial V}{\partial q_r} + \frac{\partial F}{\partial \dot{q}_r} = \frac{\partial \Omega}{\partial q_r} \quad , \ (r = 1, 2, ..., \kappa) \qquad \textbf{(24)}$$

Expression (24) is the complete form of the Lagrange equations, while some times, in literature, the expression (21) is met, which Lagrange had proved in his work "Mechanique analytique" in 1788.

The Hamilton Principle [4]

Considering that there is no loss of energy and thus F=0, eq (24) multiplied by δq_r and because of $K = \frac{1}{2} \cdot \sum m_i \cdot \bar{\upsilon}_i^2$ becomes:

$$\frac{d}{dt}\left(\sum_{i=1}^{n} m_i \cdot \bar{\upsilon}_i \cdot \frac{\partial \bar{\upsilon}_i}{\partial \dot{q}_r}\right) \cdot \delta q_r = \left(\frac{\partial K}{\partial q_r} - \frac{\partial V}{\partial q_r} + \frac{\partial \Omega}{\partial q_r}\right) \cdot \delta q_r$$

which because of eq (18) may be written as follows:

$$\left.\begin{array}{c}\frac{d}{dt}\sum_{i=1}^{n}\left(m_i\cdot\bar{\upsilon}_i\cdot\frac{\partial\bar{R}_i}{\partial q_r}\right)\cdot\delta q_r=\frac{\partial L}{\partial q_r}\cdot\delta q_r+\frac{\partial\Omega}{\partial q_r}\cdot\delta q_r\\ \text{where: } L=K-V\end{array}\right\}\qquad\textbf{(25a,b)}$$

The function L that represents the difference of kinetic and potential energies of the system is called the **Lagrangean function**, and the right hand side of eq (25a) represents the increment of this function corresponding to the assumed virtual displacement δq_r.

Let us consider now a time interval between two arbitrarily chosen limits t_1 and t_2 and assume that at every instant, some virtual displacement is given to the system, so that δq_r becomes a certain function of time, the magnitude of which always remains very small. We assume also that this function does not change abruptly but is a continuous function of time and also that its derivative with respect to time is always small. Let the curve mn in Fig. **3** represent values of the chosen coordinate q_r during actual motion of the system for the interval of time from t_1 to t_2. Then, as a result of the continuously changing virtual displacement q_r, we shall obtain another curve very close to the curve mn. One such possible variation of the curve mn is shown in Fig. **3** by the dotted line m_1n_1.

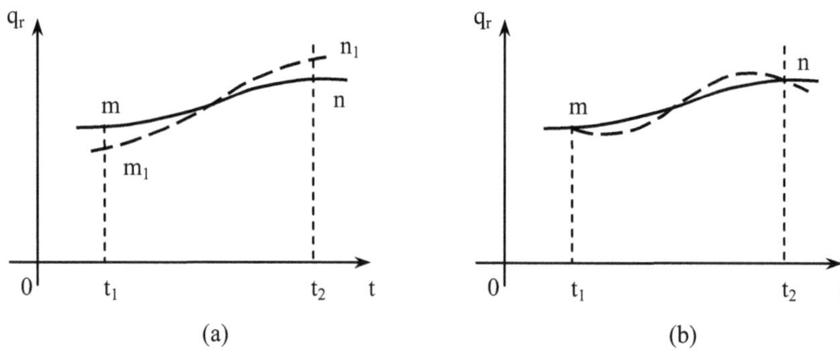

Figure 3: Basic concepts of calculus of variation

By superimposing the virtual displacements to the actual motion, not only the coordinates but also their derivatives will be slightly changed with respect to time.

If the selected coordinate q_r changes to $q_r+\delta q_r$, the corresponding velocity will change from $\dot{q}_r$ to $\dot{q}_r+\frac{d(\delta q_r)}{dt}$.

By assuming a function of time for δq_r, we a priori know the change of the coordinate q_r, and of the corresponding velocity $\dot{q}_r$ at any instant and we can calculate the corresponding value for the expression on the right-hand side of eq (25a).

Then, multiplying eq (25a) by dt and integrating from t_1 to t_2 we obtain:

$$\sum_{i=1}^{n}\left(m_i\cdot\bar{\upsilon}_i\cdot\frac{\partial R_i}{\partial q_r}\right)\cdot\delta q_r\Big|_{t_1}^{t_2}=\int_{t_1}^{t_2}(\delta L+\delta\Omega)\cdot dt\qquad\textbf{(26)}$$

We now introduce a certain limitation for the virtual displacement δq_r and assume that it vanishes for $t=t_1$ and $t=t_2$, as illustrated by the dotted line in Fig. **3b**. With this limitation, the left-hand side of eq (26) vanishes as well and we get:

$$\int_{t_1}^{t_2}(\delta L+\delta\Omega)\cdot dt=0$$

This equation holds for any trivial displacement defined by changes q_1, q_2,....,q_r, of the coordinates, provided that they vanish at time instants t_1 and t_2. We can also change the calculation sequence and perform first integration with respect to time and then evaluate the increment of the integral due to virtual displacements $\delta q_1,\ldots,\delta q_r$. In this manner, we obtain:

$$\delta\int_{t_1}^{t_2}(L+\Omega)\cdot dt = 0 \qquad \textbf{(27)}$$

Equation (27) represents Hamilton's principle, which states that the true motion of a system within a certain arbitrarily chosen time interval is characterized by the fact that the increment of the integral $\left(\int_{t_1}^{t_2}(L+\Omega)\cdot dt\right)$ vanishes for any continuously varying virtual displacement, provided this displacement vanishes at the limits $t=t_1$ and $t=t_2$, of the chosen interval.

The Equations of Heilig

The equilibrium equations for a beam with thin-walled cross-section, that is subjected to external loads (q_y, q_z) acting at the gravity center S, and also to a torsional moment m_x acting at the shear center M, are the following [5]:

$$\left.\begin{aligned}(EAu')' &= -q_x\\ (EJ_y w''_{s_0})'' &= q_z\\ (EJ_z \upsilon''_{s_0})'' &= q_y\\ (EC_M\theta'')'' - (GJ_d\theta')' &= m_x\end{aligned}\right\} \qquad \textbf{(28 a,b,c,d)}$$

where A(x) is the cross-sectional area, $EJ_y(x)$, $EJ_z(x)$ are the bending stiffnesses about axes y and z, respectively, $EC_M(x)$ is the warping rigidity with respect to the shear center M, $EJ_d(x)$ is the torsional stiffness for pure (St. Venant) torsion, w_{so}, υ_{so} are the displacements of the gravity center S along the axes z and y, and θ is the cross-section's rotation about the x axis.

During the deformation of the beam, the external loads q_y, q_z produce a torsional moment because of their movement from the initial position on the gravity center S to its new one S`. This moment may be added to the initially acting moments. From Fig. **4** we have:

$$\overline{AS} = \overline{SS'}\cos\alpha = r\cdot\theta\cdot\cos\alpha$$

Since the angle θ is very small, line $\overline{SS'}$ is practically vertical to line $\overline{MS}$ and therefore $\hat{\alpha}\cong\hat{\beta}$. Thus, we have:

$\overline{AS} = r\cdot\theta\cdot\cos\alpha = r\cdot\theta\cdot\cos\beta = z_M\cdot\theta$ and similarly:

$\overline{AS'} = \overline{SS'}\cdot\sin\alpha = r\cdot\theta\cdot\sin\alpha = r\cdot\theta\cdot\sin\beta = y_M\cdot\theta$, or finally:

$$\left.\begin{aligned}z_M + \Delta z_M &= z_M + y_M\cdot\theta\\ y_M + \Delta y_M &= y_M - z_M\cdot\theta\end{aligned}\right\} \qquad \textbf{(29a,b)}$$

Consequently, the actual moment will be the following:

$$m = m_x + q_y\left(z_M + y_M\cdot\theta\right) - q_z\left(y_M - z_M\cdot\theta\right)$$

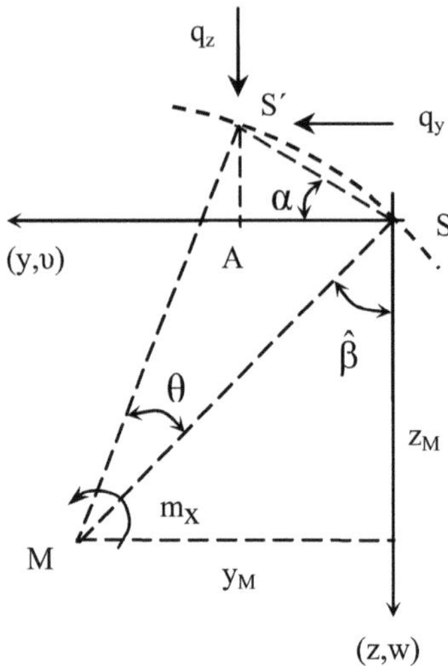

Figure 4: Point S displacement of a cross-section due to torsion

or because of eq (28a,b):

$$m = m_x + (z_M + y_M \cdot \theta) \cdot (EI_z \upsilon''_{S_0})'' - (y_M - z_M \cdot \theta) \cdot (EI_y w''_{S_0})'' \quad \textbf{(30)}$$

The determination of the coordinates of point S at its final position S\` is possible through a similar process (see Fig. **5**):

$$\left.\begin{aligned} \upsilon_S &= \upsilon_{S_0} + z_M \cdot \theta \\ w_S &= w_{S_0} - y_M \cdot \theta \end{aligned}\right\} \quad \textbf{(31a,b)}$$

Then, equation (30) can be written as:

$$m = m_x + (z_M + y_M \cdot \theta) \cdot \left[EJ_z \left(\upsilon''_S - z_M \cdot \theta'' \right) \right]'' - (y_M - z_M \cdot \theta) \cdot \left[EJ_y \left(w''_S + y_M \cdot \theta'' \right) \right]''$$

or after manipulations, and neglecting the higher order terms, we get:

$$m = m_x + \left[EJ_z \cdot z_M \cdot \upsilon''_S \right]'' - \left[EJ_z \cdot z_M^2 \cdot \theta'' \right]'' - \left[EJ_y \cdot y_M \cdot w''_S \right]'' - \left[EJ_y \cdot y_M^2 \cdot \theta'' \right]'' \quad \textbf{(32)}$$

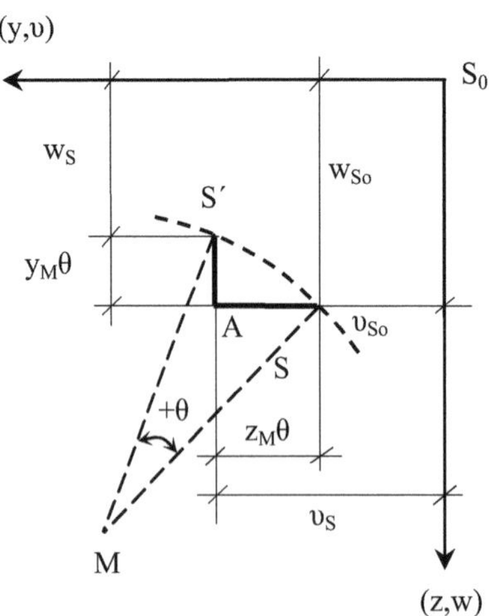

Figure 5: Notation for a cross-section rotation due to torsion

Therefore, equations (28), because of eq (31) and eq (32) are written:

$$\left.\begin{array}{l}[EI_y w_S'']'' + [EI_y y_M \theta'']'' = q_z \\ [EI_z \upsilon_S'']'' - [EI_z z_M \theta'']'' = q_y \\ [E(C_M + y_M^2 J_y + z_M^2 J_z)\cdot\theta'']'' + [EI_y y_M w_S'']'' - [EI_z z_M \upsilon_S'']'' - [GJ_d \theta']' = m_x\end{array}\right\} \quad \textbf{(33a,b,c)}$$

For a cross-section symmetrical to axis z, it will be y_M=0, and hence:

$$\left.\begin{array}{l}[EA\cdot u_S']' = q_x \\ [EJ_y \cdot w_S'']'' = q_z \\ [EJ_z \cdot \upsilon_S'']'' - [EI_z \cdot z_M \cdot \theta'']'' = q_y \\ [E(C_M + z_M^2 \cdot J_z)\cdot\theta'']'' - [EI_z \cdot z_M \cdot \upsilon_S'']'' - [GJ_d \cdot \theta']' = m_x\end{array}\right\} \quad \textbf{(34a,b,c,d)}$$

We note also that:

$$C_M + z_M^2 \cdot J_z = C_S \quad \textbf{(34e)}$$

where C_S is the warping constant of the cross-section with reference to the gravity center, which is often used to simplify further equations (34). Equations (34) or their more general form eqs (33) are known as the Heilig's equations [6].

The Symbolic Functions δ and H [1]

One of the commonly met functions that are particularly useful in mathematics and mechanics is the symbolic function δ(x-α) known as the Dirac's delta function, which is determined as follows:

$$\delta(x-\alpha) = \left.\begin{cases} 0 & \text{for } x \neq \alpha \\ \infty & \text{for } x = \alpha \end{cases}\right\} \quad \textbf{(35)}$$

with the significant property:

$$\int_{-\infty}^{\infty} \delta(x-\alpha)\cdot dx = \lim_{\varepsilon\to 0}\int_0^{\varepsilon} \delta(x-\alpha)\cdot dx = 1 \quad \textbf{(36)}$$

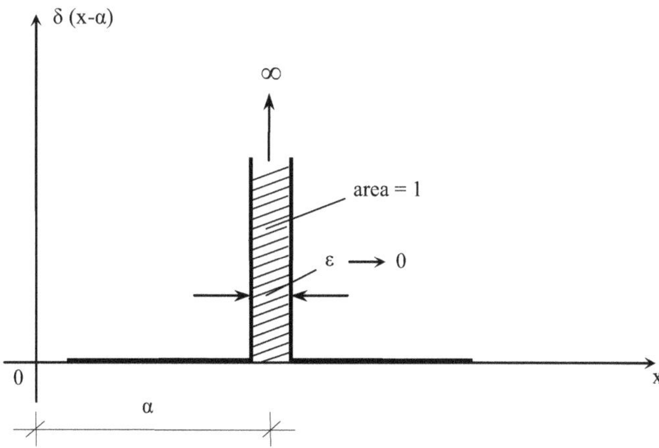

Figure 6: The Dirac's delta function δ

The Dirac's function may be expressed by the following Fourier integral:

$$\delta(x-\alpha)=\frac{1}{2\pi}\cdot\int_{-\infty}^{\infty} e^{iz\cdot(x-\alpha)}\cdot dz \qquad \textbf{(37)}$$

The following properties are valid:

$$\left.\begin{aligned}
&\kappa\cdot\delta(x-\alpha)=\delta\left(\frac{x-\alpha}{\kappa}\right)\\
&x\cdot\delta(x-\alpha)=0 \quad \text{for } x\neq\alpha\\
&f(x)\cdot\delta(x-\alpha)=f(\alpha)\cdot\delta(x-\alpha)\\
&\int_{-\infty}^{\infty} f(x)\cdot\delta(x-\alpha)\cdot dx=f(\alpha)\\
&\int_{-\infty}^{\infty} f(x)\cdot\delta'(x-\alpha)\cdot dx=-f'(\alpha)\\
&\int_{-\infty}^{\infty} f(x)\cdot\frac{d^n}{dx^n}\cdot\delta(x-\alpha)=(-1)^n\cdot\frac{d^n}{dx^n}\cdot f(\alpha)\\
&\delta'(x-\alpha)=-\frac{1}{x}\cdot\delta(x-\alpha)\\
&\frac{d^n}{dx^n}\cdot\delta(x-\alpha)=(-1)^n\cdot\frac{n!}{x^n}\cdot\delta(x-\alpha)\\
&\delta\left[\varphi(x)-\alpha\right]=\frac{1}{\varphi'(\alpha)}\cdot\delta(x-\alpha)\\
&\delta(\alpha x)=\frac{1}{\alpha}\cdot\delta(x)
\end{aligned}\right\} \qquad \textbf{(38)}$$

Another useful symbolic function is the Heaviside's unit function, which is determined as follows:

$$H(x-\alpha)=\begin{Bmatrix}0 & \text{for} & x<\alpha\\ 1 & \text{for} & x>\alpha\end{Bmatrix} \qquad \textbf{(39)}$$

The functions δ(x-α) and H(x-α) are related to each other as follows:

$$\left.\begin{aligned}
&\frac{dH(x-\alpha)}{dx}=\delta(x-\alpha) && \text{for } x=\alpha\\
\text{and}\quad &\frac{dH(x-\alpha)}{dx}=0 && \text{for } x>\alpha
\end{aligned}\right\} \qquad \textbf{(40)}$$

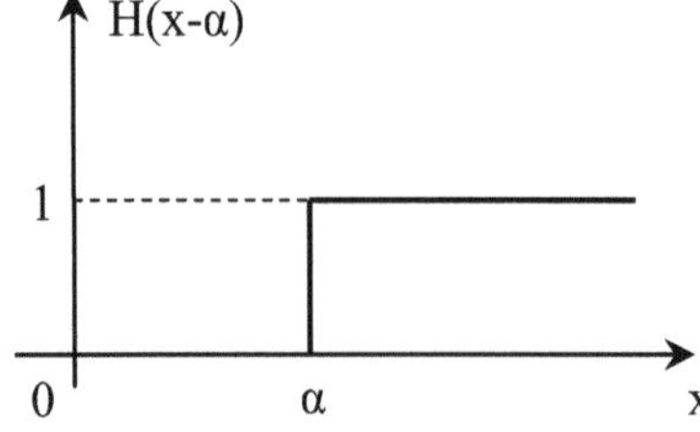

Figure 7: The Heaviside function H

Therefore, for possible differentiation we may overcome the non continuity-abnormality around the point $x=\alpha$:

$$H'(x-\alpha)=0 \quad \text{for} \quad x>\alpha \tag{41}$$

Loads and Symbolic Functions

The above symbolic functions make easier the application of any type of loads to equations (34), but most significantly make very easy the mathematical operation for the solution of problems.

The usually met loads are shown in Fig. **8(a** to **d**). The distributed load p(x) can be inserted without problem into the equilibrium equations which can be easily manipulated afterwards.

A concentrated load P applied at position x=α, can be inserted into the right-hand side of eq (34) in the form: $P\cdot\delta(x-\alpha)$.

The load p(x) in Fig. **8c**, which is a distributed load on the length $\alpha_1<x<\alpha_2$ can be inserted in the form:

$$p(x)\cdot\left[H(x-\alpha_1)-H(x-\alpha_2)\right]$$

To understand the way a concentrated moment M_y acting at position x=α is inserted into the equations, we use Fig. **8e**, where M_y is analyzed to an equivalent pair of forces $\frac{M_y}{\varepsilon}$ and $-\frac{M_y}{\varepsilon}$ with $\varepsilon\to 0$.

Thus, the equivalent loading of the two concentrated loads is inserted in the form:

$$\lim_{\varepsilon\to 0}\left[-\frac{M_y}{\varepsilon}\cdot\delta(x-\alpha)+\frac{M_y}{\varepsilon}\cdot\delta(x-\alpha-\varepsilon)\right]=-M_y\lim_{\varepsilon\to 0}\frac{\delta(x-\alpha)-\delta(x-\alpha-\varepsilon)}{\varepsilon}=$$
$$=-M_y\cdot\delta'(x-\alpha)$$

a)

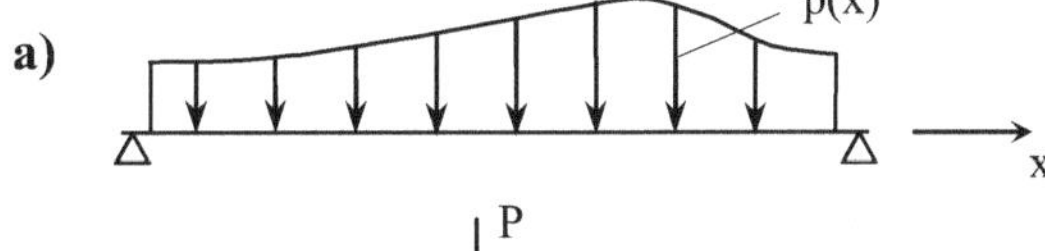

b)

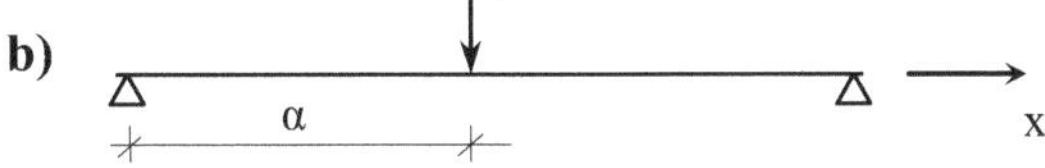

c)

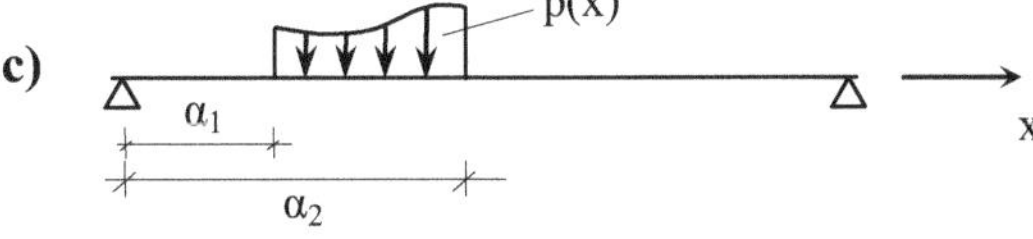

d)

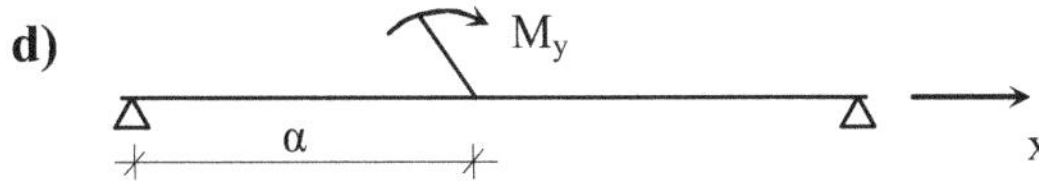

e)

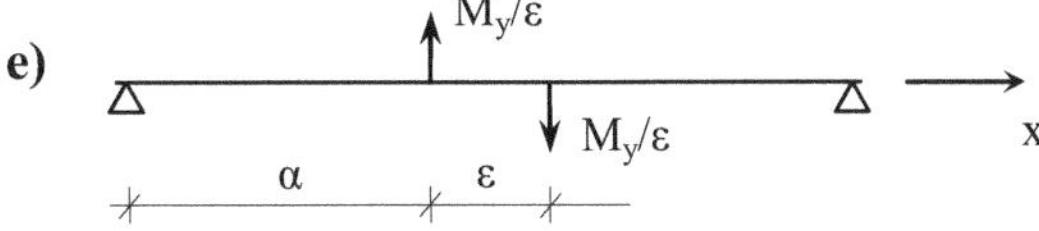

Figure 8: Various load cases for beams

The Problem of Damping

Internal Frictions (Passive Resistances)

From the observation of an oscillating solid body it is known that all periodic vibrations are characterized by a shift of phase between stresses and strains. As we know, damping manifests itself in the stress-strain curve by the hysteresis loop (Fig. **9**). The area L of the hysteresis loop represents the energy that is dissipated by the solid in order to confront the internal connection forces. This energy appears during one cycle as heat. The ratio of this energy to the potential energy V at maximum strain is an important constant of damping:

$$n = \frac{L}{V} \tag{42}$$

During the deformation of a beam, the above resistance against the motion of the beam appears, which is a function of the displacement and the velocity of motion. A mechanical model that has an analogous behavior is the Kelvin-Voigt model [7]. According to this model, the strain is related to the reduced deformation via the following relation:

$$\sigma = \varepsilon \cdot E + n \cdot \frac{d\varepsilon}{dt} \tag{43}$$

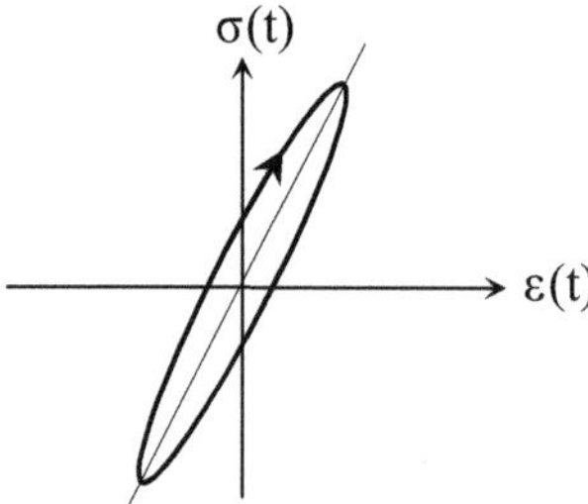

Figure 9: Stress-strain relation for energy dissipation

From Fig. **10** and for a positive value of ξ, we have negative angle φ. Thus, it is:

$$\tan\varphi = -\frac{\xi}{z} = -\frac{\partial w}{\partial x}$$

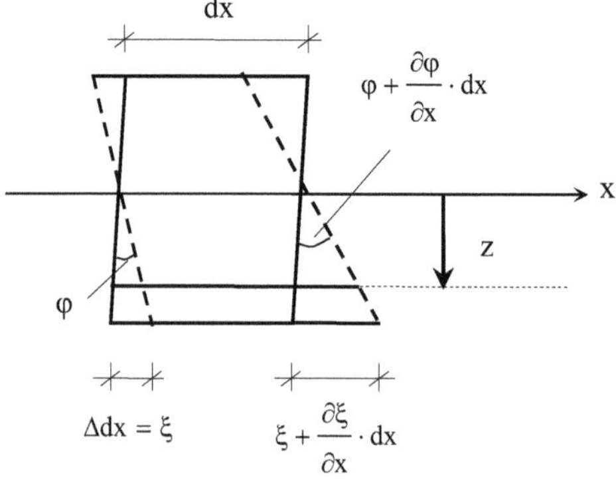

Figure 10: Deformation of an infinitesimal element dx

and therefore:

$$\xi = -\frac{\partial w}{\partial x} \cdot z \tag{44}$$

On the other hand, it is known that:

$$\left.\varepsilon = \frac{d(\Delta dx)}{dx} = \frac{d\xi}{dx} = -\frac{\partial^2 w}{\partial x^2}\cdot z \quad \text{and} \quad \dot{\varepsilon} = \frac{d\varepsilon}{dt} = -\frac{\partial^3 w}{\partial x^2 \cdot \partial t}\cdot z\right\} \qquad \textbf{(45a,b)}$$

Because of eqs (43) and (45), we get:

$$\sigma = -\left(E\cdot\frac{\partial^2 w}{\partial x^2} + n\cdot\frac{\partial^3 w}{\partial x^2\cdot\partial t}\right)\cdot z \qquad \textbf{(46)}$$

We know that $\sigma = \frac{M}{J}\cdot z$, and thus one can easily determine the new expression for the bending moment:

$$M = -EJ\cdot\frac{\partial^2 w}{\partial x^2} - n\cdot J\cdot\frac{\partial^3 w}{\partial x^2\cdot\partial t} \qquad \textbf{(47)}$$

External Frictions

The developed passive resistances of a system are distinguished in internal and external ones. The internal resistances due to the material nature (for the materials used in bridge-building), are satisfactorily expressed by the Kelvin-Voigt model of the above paragraph. The external resistances are due to the motion of the oscillating system and to the friction developed by the material in which the system moves (*i.e.* air or water). These resistances develop forces, which are considered as analogous to the speed of the moving system. These forces are equal to:

$$p_d = c\cdot\frac{\partial w}{\partial t} \qquad \textbf{(48)}$$

where: $c = c(x)$ is the coefficient of resistance which is possibly varying along the length beam. Usually though, we set c=constant.

THE GENERAL EQUATIONS OF DYNAMIC EQUILIBRIUM

Equations (34) for a dynamic loading, including the developed inertia and damping forces, are written as follows:

$$\left.\begin{array}{l}
\left[EA\cdot u_S'\right]' - \left[c_{ix}\cdot\dot{u}_S'\right]' - c_{ex}\cdot\dot{u}_S - m\cdot\ddot{u}_S = q_x(x,t) \\[2ex]
\left[EI_y\cdot w_S''\right]'' + \left[c_{iy}\cdot\dot{w}_S''\right]'' + c_{ey}\cdot\dot{w}_S - \left[\Theta_y\cdot\ddot{w}_S'\right]' + m\cdot\ddot{w}_S = q_z(x,t) \\[2ex]
\left[EI_z\cdot\upsilon_S''\right]'' - \left[EI_z\cdot z_M\cdot\theta''\right]'' + \left[c_{iz}\cdot\dot{\upsilon}_S''\right]'' + \\
\qquad + c_{ez}\cdot\dot{\upsilon}_S - \left[\Theta_z\cdot\ddot{\upsilon}_S'\right]' + m\cdot\ddot{\upsilon}_S = q_y(x,t) \\[2ex]
\left[EC_S\cdot\theta''\right]'' - \left[EI_z\cdot z_M\cdot\upsilon_S''\right]'' + \left[c_{i\theta}\cdot\dot{\theta}_S''\right]'' + \\
\qquad + c_{e\theta}\cdot\dot{\theta} - \left[GJ_d\cdot\theta'\right]' + \Theta_x\cdot\ddot{\theta} = m_x(x,t)
\end{array}\right\} \qquad \textbf{(49a,b,c,d)}$$

where: $\Theta_y = m\cdot\frac{I_y}{A}$, $\Theta_z = m\cdot\frac{I_z}{A}$, $\Theta_x = m\cdot\frac{I_y + I_z}{A}$ **(50)**

are the torsional-mass moment of inertia about axes y, z and x, respectively, while c_i and c_e are the coefficients of internal and external damping. We note that the internal damping is often much smaller than the external one and the inertia forces about y and z axes (but not about x axis) are neglected.

We observe that eqs (49 a,b) include only the unknowns u_S, w_S and therefore they may be solved independently, while eqs (49 c,d) compose a differential system with unknowns the deformations υ_S and θ.

Taking into account the above remarks, we will solve and study the following equations of dynamic equilibrium:

$$\left.\begin{aligned}
&\left[EA\cdot u_S'\right]' - c_x\cdot\dot{u}_S - m\cdot\ddot{u}_S = q_x(x,t)\\
&\left[EI_y\cdot w_S''\right]'' + c_y\cdot\dot{w}_S + m\cdot\ddot{w}_S = q_z(x,t)\\
&\left[EI_z\cdot\upsilon_S''\right]'' - \left[EI_z\cdot z_M\cdot\theta''\right]'' + c_z\cdot\dot{\upsilon}_S + m\cdot\ddot{\upsilon}_S = q_y(x,t)\\
&\left[EC_S\cdot\theta''\right]'' - \left[EI_z\cdot z_M\cdot\upsilon_S''\right]'' + c_\theta\cdot\dot{\theta} - \left[GJ_d\cdot\theta'\right]' + \Theta_x\cdot\ddot{\theta} = m_x(x,t)
\end{aligned}\right\} \qquad \textbf{(51a,b,c,d)}$$

where, index e in coefficients of external damping is omitted for simplicity.

The solution of a differential equation or differential system, corresponding to a real problem depends on the conditions through which the constants of integration are determined. Especially, in dynamics, the conditions are separated into two categories. The first ones, related to the geometry of the static system are the **boundary conditions.** The second ones, related to the dynamic prehistory of the system describe the situation at a given time instant and are called **time conditions**. The last ones, at t=0 are called **initial conditions.**

Exact Methods of Solution

There are a lot of solution methods for the above system. Frýba [8], follows the method of integral transformation applying a double transformation. The sinusoidal Fourier transformation for the variable x and the Corson-Laplace transformation for the variable t, are given as follows:

$$\left.\begin{aligned}
&W(i,t) = \int_0^{\ell} w_S(x,t)\cdot\sin\frac{i\pi x}{\ell}\cdot dx\\
&W(i,p) = p\cdot\int_0^{\infty} w_S(x,t)\cdot e^{-p\cdot t}\cdot dt
\end{aligned}\right\} \qquad \textbf{(52a,b)}$$

The above method has the serious advantage to eliminate the uncertainties appearing in cases where the excitation frequency coincides with one of the bridge's eigenfrequencies. However, this method considers that some data of the bridge like the eigenfrequencies and the shape functions are known. This is relatively easy for simple static systems, but becomes particularly difficult and sometimes impossible through the use of classic methods for more complicated static systems.

The Ritz Method [9]

Let us consider the function p(x), which satisfies the boundary conditions $G(L_1, L_2)$ at the ends of the interval L_1L_2. Furthermore, we consider the series $\varphi_n(x)$, (n=1,2,3,….), the terms of which are arbitrary or they are produced through some rule with the only restriction to satisfy the boundary conditions. We consider the summation:

$$\overline{p}(x) = \sum_n \alpha_n\cdot\varphi_n(x) \quad , \quad n = 1,2,3,..... \qquad \textbf{(53)}$$

where α_n are constant coefficients under determination.

We have to determine the above constants α_n so that $\overline{p}(x)$ coincides or approaches as possible the given function p(x) in the interval (L_1, L_2). The total square of the approach error is given by the relation:

$$\int_{L_1}^{L_2} (p-\overline{p})^2 \cdot dx = \int_{L_1}^{L_2} \left(p - \sum_n \alpha_n \cdot \varphi_n(x) \right)^2 dx \quad \textbf{(54)}$$

We will choose α_n so that the above error is minimum. The relative conditions are:

$$\frac{\partial}{\partial \alpha_n} \cdot \int_{L_1}^{L_2} (p-\overline{p})^2 \cdot dx = 0 \quad , \quad n = 1,2,3,.... \quad \text{or} \quad \int_{L_1}^{L_2} (p-\overline{p}) \cdot \frac{\partial \overline{p}}{\partial \alpha_n} = 0 \quad , \quad n = 1,2,3,.... \quad \textbf{(55)}$$

Observing that: $\frac{\partial \overline{p}}{\partial \alpha_n} = \varphi_n(x)$, eq (55) is written as:

$$\int_{L_1}^{L_2} p \cdot \varphi_n(x) \cdot dx - \int_{L_1}^{L_2} \left(\sum_m \alpha_m \cdot \varphi_m(x) \right) \cdot \varphi_n(x) \cdot dx = 0 \quad \textbf{(56)}$$

Choosing suitable functions $\varphi_n(x)$ for which the so called **orthogonality condition** is valid:

$$\int_{L_1}^{L_2} \varphi_n(x) \cdot \varphi_m(x) \cdot dx = \left. \begin{matrix} 0 & \text{for } n \neq m \\ \Gamma_n & \text{for } n = m \end{matrix} \right\} \quad \textbf{(57)}$$

we can determine the coefficients α_n by the relation:

$$\alpha_n = \frac{\int_{L_1}^{L_2} p(x) \cdot \varphi_n(x) \cdot dx}{\int_{L_1}^{L_2} \varphi_n^2(x) \cdot dx} \quad \textbf{(58)}$$

The Galerkin Method [9]

Let us consider the following differential equation:

$$\left. \begin{matrix} & F(w) = q \\ \text{with boundary conditions:} & G(L_1, L_2) = A \end{matrix} \right\} \quad \textbf{(59a,b)}$$

where F is a random differential function. For example, $F = \nabla^4$.

We consider additionally that:

$$w(x) = \sum_n \alpha_n \varphi_n(x) \quad (n = 1,2,3,....) \quad \textbf{(60)}$$

where $\varphi_n(x)$ satisfy the boundary conditions.

If $F(w) - \overline{F}\left(\sum_n \alpha_n \varphi_n \right) \neq 0$ is valid, then the total square error will be: $\int_{L_1}^{L_2} \left[F(w) - \overline{F}\left(\sum_n \alpha_n \varphi_n \right) \right]^2 dx$. In order for this expression to be minimum, the following conditions must be satisfied:

$$\frac{\partial}{\partial\alpha_m}\int_{L_1}^{L_2}\left[F(x)-\overline{F}\left(\sum_n \alpha_n\varphi_n\right)\right]^2 dx=0 \quad \text{or} \quad \int_{L_1}^{L_2} F\cdot\varphi_m(x)dx-\int_{L_1}^{L_2}\overline{F}\left(\sum_n \alpha_n\varphi_n\right)\cdot\varphi_m dx=0$$

and finally, because of eq (59a,) we get:

$$\left.\begin{array}{c}\int_{L_1}^{L_2}\overline{F}\left(\alpha_1\varphi_1+\alpha_2\varphi_2+.....+\alpha_n\varphi_n\right)\cdot\varphi_m dx=\int_{L_1}^{L_2} q\cdot\varphi_m dx \\ n=1,2,3,.....,n\end{array}\right\} \tag{61}$$

The above equations compose a linear homogeneous system of n-equations, from which the coefficients α_n are determined.

THE DYNAMIC CHARACTERISTICS OF BEAMS - THE FREE VIBRATION

The dynamic characteristics of a beam are the ones appearing during its free vibration. In order to solve the above system of eqs (51 a,b,c,d), we follow the method of separate variables. Therefore, we set:

$$\left.\begin{array}{r}u_S(x,t)=U(x)\cdot T_x(t)\\ w_S(x,t)=W(x)\cdot T_z(t)\\ \upsilon_S(x,t)=V(x)\cdot T_\theta(t)\\ \theta(x,t)=\Phi(x)\cdot T_\theta(t)\end{array}\right\} \tag{62a,b,c,d}$$

where U, W, V, and Φ are functions of x, while T are functions of time t. Note that the above displacements υ_S and θ have a common time function since the differential equations containing them are involved.

We consider, in addition, that the cross-section of the beam is constant along the axis x and therefore, eqs (51) for free vibration are written as:

$$\left.\begin{array}{l}EAu_S''-c_x\dot{u}_S-m\ddot{u}_S=0\\ EI_y w_S''''+c_y\dot{w}_S+m\ddot{w}_S=0\\ EI_z\upsilon_S''''-EI_z z_M\theta''''+c_z\dot{\upsilon}_S+m\ddot{\upsilon}_S=0\\ EC_S\theta''''-EI_z z_M\upsilon_S''''+c_\theta\dot{\theta}-GI_d\theta''+\Theta_x\ddot{\theta}=0\end{array}\right\} \tag{63a,b,c,d}$$

The Longitudinal Motion

Introducing eq (62a) into eq (63a) we get:

$$EAU''T_x-c_xU\dot{T}_x-mU\ddot{T}_x=0 \quad \text{or} \quad \frac{\ddot{T}_x+\frac{c_x}{m}\cdot\dot{T}_x}{T_x}=\frac{EAU''}{mU}=-\omega_x^2 \tag{64}$$

since the equality of two representations of the different variables implies that the above fractions are constant and independent from the variables.

Calling ρ the specific weight of the material, we get:

$$\left.\begin{aligned} &\ddot{T}_x + \frac{c_x}{m}\cdot\dot{T}_x + \omega_x^2 T_x = 0 \\ &U''(x) + \frac{\omega_x^2}{\kappa^2}\cdot U(x) = 0 \\ \text{where:}\quad &\kappa = \sqrt{\frac{EA}{m}} = \sqrt{\frac{E}{\rho}} \end{aligned}\right\} \qquad \textbf{(65 a,b,c)}$$

Constant κ is a characteristic constant called **speed of the propagation of the longitudinal elastic wave** and has dimensions of velocity.

The solution of (65a) is:

$$\left.\begin{aligned} &T_x(t) = e^{-\beta_x \cdot t}\left(A\cdot\sin\bar{\omega}_x t + B\cdot\cos\bar{\omega}_x t\right) \\ \text{where:}\quad &\beta_x = \frac{c_x}{2m}\ ,\quad \bar{\omega}_x = \sqrt{\omega_x^2 - \beta_x^2}\quad \text{with:}\ \left|\omega_x\right| > \left|\beta_x\right| \end{aligned}\right\} \qquad \textbf{(66 a,b,c)}$$

The solution of eq (65b) is:

$$U(x) = c_1\cdot\sin\frac{\omega_x}{\kappa}x + c_2\cdot\cos\frac{\omega_x}{\kappa}x \qquad \textbf{(67)}$$

Using the boundary conditions of the beam under study, one can determine the spectrum of eigenfrequencies ω_{xn} and the corresponding shape functions $U_n(x)$.

The Free Beam

For the free beam, for example, the boundary conditions are: $u'(0) = u'(\ell) = 0$, which introduced into eq (67) give:

$$c_1 = 0 \quad \text{and} \quad c_2\cdot\frac{\omega_x}{\kappa}\cdot\sin\frac{\omega_x \ell}{\kappa} = 0$$

In order that vibration exists, it must be $(c_2, \omega_x) \neq 0$ that gives $\sin\frac{\omega_x \ell}{\kappa} = 0$, or:

$$\left.\begin{aligned} &\omega_{xn} = \frac{n\pi\kappa}{\ell} = \frac{n\pi}{\ell}\cdot\sqrt{\frac{EA}{m}}\ ,\quad n = 1,2,3,\ldots \\ \text{and finally:}\quad &U_n(x) = c_{2n}\cdot\cos\frac{n\pi x}{\ell} \end{aligned}\right\} \qquad \textbf{(68a,b)}$$

Equation (68a) gives the eigenfrequencies spectrum ω_{xn} for the free vibrating beam in longitudinal motion, while eq (68b) gives the corresponding shape functions. Thus, it will be:

$$u_S(x,t) = \sum_n \cos\frac{n\pi x}{\ell}\cdot e^{-\beta_x\cdot t}\left(A_n\cdot\sin\bar{\omega}_{xn}t + B_n\cdot\cos\bar{\omega}_{xn}t\right) \qquad \textbf{(69)}$$

The Simply Supported Beam

For the simply supported beam, the boundary conditions are: $u(0) = u'(\ell) = 0$, which give:

$$c_2 = 0 \quad \text{and} \quad c_1\cdot\frac{\omega_x}{\kappa}\cdot\cos\frac{\omega_x \ell}{\kappa} = 0$$

In order that vibration exists, it must be $(c_1, \omega_x) \neq 0$ and hence $\cos\frac{\omega_x \ell}{\kappa} = 0$, or:

$$\left.\omega_{xn} = \frac{(2n-1)\pi\kappa}{2\ell} \quad , \quad n = 1,2,3,... \text{ and: } \quad U_n(x) = c_{1n} \cdot \sin\frac{(2n-1)\pi x}{2\ell}\right\} \quad \textbf{(70a,b)}$$

Finally, we get:

$$u_S(x,t) = \sum_n \sin\frac{(2n-1)\pi x}{2\ell} \cdot e^{-\beta_x \cdot t} \cdot (A_n \cdot \sin\overline{\omega}_{xn} t + B_n \cdot \cos\overline{\omega}_{xn} t)$$

The Pin-Supported Beam

For this beam, the boundary conditions are: $u(0,t) = u(\ell,t) = 0$, which give:

$$c_2 = 0 \quad \text{and} \quad c_1 \cdot \sin\frac{\omega_x \ell}{\kappa} = 0$$

In order that vibration exists, it must be $(c_1 \neq 0)$ and therefore: $\sin\frac{\omega_x \ell}{\kappa} = 0$ or

$$\left.\omega_x = \frac{n\pi\kappa}{\ell} \quad n > 1,2,3,... \quad \text{and:} \quad U_n(x) = c_{1n} \cdot \sin\frac{n\pi x}{\ell}\right\} \quad \textbf{(71a,b)}$$

Finally, we get:

$$u_S(x,t) = \sum_n \sin\frac{n\pi x}{\ell} \cdot e^{-\beta_x \cdot t} \cdot \left(A_n \cdot \sin\overline{\omega}_{xn} t + B_n \cdot \cos\overline{\omega}_{xn} t\right)$$

The Vertical Flexural Free Motion

Introducing eq (62b) into eq (63b), we get:

$$EI_y W''''T_z + c_y W\dot{T}_z + mW\ddot{T}_z = 0 \quad \text{or} \quad -\frac{\ddot{T}_z + \frac{c_y}{m} \cdot \dot{T}_z}{T_z} = \frac{EI_y W''''}{mW} = \omega_z^2 \quad \textbf{(72)}$$

The above system gives the following differential equations:

$$\left.\begin{array}{l} \ddot{T}_z + \frac{c_y}{m} \cdot \dot{T}_z + \omega_z^2 T_z = 0 \\ EI_y W'''' - m\omega_z^2 W = 0 \end{array}\right\} \quad \textbf{(73a,b)}$$

The solution of eq (73a) is:

$$\left.\begin{array}{l} T_z(t) = e_z^{-\beta \cdot t}\left(A \cdot \sin\overline{\omega}_z t + B \cdot \cos\overline{\omega}_z t\right) \\ \text{where:} \quad \beta_z = \frac{c_y}{2m} \quad , \quad \overline{\omega}_z = \sqrt{\omega_z^2 - \beta_z^2} \quad , \quad |\omega_z| > |\beta_z| \end{array}\right\} \quad \textbf{(74a,b,c)}$$

while the solution of eq (73b) is:

$$\left.\begin{array}{l} W(x) = c_1 \cdot \sin\lambda x + c_2 \cdot \cos\lambda x + c_3 \cdot \sinh\lambda x + c_4 \cdot \cosh\lambda x \\ \qquad \text{with :} \quad \lambda^4 = \dfrac{m\omega_z^2}{EI_y} \end{array}\right\} \tag{75a,b}$$

Using the boundary conditions of the static system under study, one can determine the eigenfrequencies spectrum ω_{zn} and also the shape functions W_n. For example, the boundary conditions for the simply supported beam are:

$$W(0) = W''(0) = W(\ell) = W''(\ell) = 0$$

which introduced into eq (75) give:

$$\left.\begin{array}{l} c_2 \qquad\qquad\qquad\qquad\qquad\qquad + c_4 = 0 \\ -c_2 \cdot \lambda^2 \qquad\qquad\qquad\qquad\quad + c_4 \cdot \lambda^2 = 0 \\ c_1 \cdot \sin\lambda\ell + c_2 \cdot \cos\lambda\ell + c_3 \cdot \sinh\lambda\ell + c_4 \cdot \cosh\lambda\ell = 0 \\ -c_1\lambda^2 \cdot \sin\lambda\ell - c_2\lambda^2 \cdot \cos\lambda\ell + c_3\lambda^2 \cdot \sinh\lambda\ell + \\ \qquad\qquad\qquad\qquad + c_4\lambda^2 \cdot \cosh\lambda\ell = 0 \end{array}\right\} \tag{76a,b,c,d}$$

In order for the above system to have a non-trivial solution, the determinant of the coefficients of the unknowns must be zero. Thus:

$$\begin{vmatrix} 0 & 1 & 0 & 1 \\ 0 & -1 & 0 & 1 \\ \sin\lambda\ell & \cos\lambda\ell & \sinh\lambda\ell & \cosh\lambda\ell \\ -\sin\lambda\ell & -\cos\lambda\ell & \sinh\lambda\ell & \cosh\lambda\ell \end{vmatrix} = 0 \tag{77}$$

Expanding the above determinant gives:

$$\sin\lambda\ell \cdot \sinh\lambda\ell = 0 \tag{77′}$$

If $\sinh\lambda\ell = 0$ then it will be λ=0, namely: ω_z=0 which means that motion does not exist. Therefore, it must be: $\sin\lambda\ell = 0$ or

$$\lambda\ell = n\pi \qquad (n = 1, 2, 3, ...)$$

and because of eq (75b):

$$\omega_{zn}^2 = \frac{n^4\pi^4 EI_y}{m\ell^4} \quad , \quad n = 1, 2, 3, ... \tag{78}$$

From eq (76a,b) it follows that c_2=c_4=0 and since $\sin\lambda\ell = 0$ and $\sinh\lambda\ell \neq 0$, eq (76c) gives: c_3=0. Therefore:

$$W_n(x) = c_{1n} \cdot \sin\frac{n\pi x}{\ell} \tag{79}$$

Equation (78) gives the eigenfrequencies spectrum of a simply supported beam in vertical flexural motion, while eq (79) gives the corresponding shape functions. Thus:

$$w_S(x,t) = \sum_n \sin\frac{n\pi x}{\ell} \cdot e^{-\beta_z \cdot t}\left(A_n \sin\overline{\omega}_{zn} t + B_n \cdot \cos\overline{\omega}_{zn} t\right) \tag{80}$$

The Lateral-Torsional Free Motion [10]

Introducing Equations (62c,d) into eqs (63c,d), we get:

$$\left.\begin{aligned} &EI_z V''''T_\theta - EI_z z_M \Phi''''T_\theta + c_z V\dot{T}_\theta + mV\ddot{T}_\theta = 0 \\ &EC_S\Phi''''T_\theta - EI_z z_M V''''T_\theta + c_{\theta x}\Phi\dot{T}_\theta - GI_d\Phi''T_\theta + \Theta_x\Phi\ddot{T}_\theta = 0 \end{aligned}\right\} \text{ or}$$

$$-\frac{\ddot{T}_\theta + \frac{c_z}{m}\dot{T}_\theta}{T_\theta} = \frac{EI_z V'''' - EI_z z_M \Phi''''}{mV} = \frac{EC_S\Phi'''' - EI_z z_M V'''' - GI_d\Phi''}{\Theta_x\Phi} = \omega_\theta^2 \tag{81}$$

Given that the following relation is valid:

$$c_\theta = c_z \cdot \frac{\Theta_x}{m} \tag{81a}$$

eq (81) gives the following system of differential equations:

$$\left.\begin{aligned} &\ddot{T}_\theta + \frac{c_z}{m}\cdot\dot{T}_\theta + \omega_\theta^2\cdot T_\theta = 0 \\ &EI_z V'''' - EI_z z_M \Phi'''' - m\omega_\theta^2 V = 0 \\ &EC_S\Phi'''' - EI_z z_M V'''' - GI_d\Phi'' - \Theta_x\omega_\theta^2\Phi = 0 \end{aligned}\right\} \tag{82a,b,c}$$

Equation (82a) has the solution:

$$\left.\begin{aligned} &T_\theta(t) = e^{-\beta_\theta\cdot t}\cdot\left(A\sin\bar{\omega}_\theta t + B\cos\bar{\omega}_\theta t\right) \\ \text{where:}\quad &\beta_\theta = -\frac{c_z}{2m} \;,\; \bar{\omega}_\theta = \sqrt{\omega_\theta^2 - \beta_\theta^2} \;,\; |\omega_\theta| > |\beta_\theta| \end{aligned}\right\} \tag{83a,b,c}$$

Multiplying eq (82b) by z_M and adding the outcome to eq (82c), we get:

$$EC_S\Phi'''' - GI_d\Phi'' - \Theta_x\omega_\theta^2\Phi - EI_z z_M^2\Phi'''' - m\omega_\theta^2 z_M V = 0$$

or finally:

$$V = \frac{1}{mz_M\omega_\theta^2}\cdot\left\{EC_M\Phi'''' - GI_d\Phi'' - \Theta_x\omega_\theta^2\Phi\right\} \tag{84}$$

Because of the above, equation (82c) is written:

$$\left.\begin{aligned} &\alpha\cdot\Phi^{(8)} + \beta\cdot\Phi^{(6)} + \gamma\cdot\Phi^{(4)} + \delta\cdot\Phi^{(2)} + \varepsilon\cdot\Phi = 0 \\ &\text{where:} \\ &\alpha = -E^2 I_z C_M \\ &\beta = EGI_z I_d \\ &\gamma = EC_S m\omega_\theta^2 + EI_z\Theta_x\omega_\theta^2 \\ &\delta = -GI_d m\omega_\theta^2 \\ &\varepsilon = -\Theta_x m\omega_\theta^4 \end{aligned}\right\} \tag{85a,b,c,d,e,f}$$

The characteristic of equation (85a) is:

$$\alpha \cdot \rho^8 + \beta \cdot \rho^6 + \gamma \cdot \rho^4 + \delta \cdot \rho^2 + \varepsilon = 0 \quad \textbf{(86)}$$

The above is a fourth order algebraic equation with respect to ρ^2, which can be solved analytically. Thus, the general integral of eq (85a) is given by the following relation:

$$\Phi(x) = \sum_i c_i \cdot e^{\rho_i \cdot x} + \sum_\kappa c_\kappa \cdot e^{\alpha_\kappa \cdot x} \cdot \left[\sin(\beta_\kappa x) + \cos(\beta_\kappa x)\right] \quad \textbf{(87)}$$

where $\pm\rho_i$ and $(\alpha_\kappa \pm j \cdot \beta_\kappa)$ are the real and the conjugate complex roots of

eq (86), respectively, while $j = \sqrt{-1}$ is the imaginary unit and $c_q\ (q = i + \kappa = 8)$ are the integration constants.

Introducing eq (87) into eq (84), we obtain the expression for V(x).

Using the boundary conditions into the expressions of $\Phi(x)$ and V(x), we obtain a linear homogeneous system with respect to the eight unknown constants $c_q\ (q = i + \kappa = 8)$. This eigenvalue problem gives the spectrum of eigenfrequencies $\omega_{\theta\eta}$, which results from the vanishing of the determinant of the factors of the unknowns in the above linear system.

Finally, from this system, one can find the seven constants with respect to the eighth constant (say for example the c_1). Finally, we derive the expressions of shape functions:

$$\left.\begin{aligned} \upsilon_S(x,t) &= \sum_n V_n \cdot e^{-\beta_\theta t}(A_n \sin\bar{\omega}_{\theta_n} t + B_n \cos\bar{\omega}_{\theta_n} t) \\ \theta(x,t) &= \sum_n \Phi_n \cdot e^{-\beta_\theta t}(A_n \sin\bar{\omega}_{\theta_n} t + B_n \cos\bar{\omega}_{\theta_n} t) \end{aligned}\right\} \quad \textbf{(88)}$$

The Special Case z_M=0

In the special case of a cross-section with doubly symmetric cross-section $(y_M = z_M = 0)$, eqs (63c,d) are uncoupled and take the form:

$$\left.\begin{aligned} EI_z\upsilon'''' + c_z\dot{\upsilon}_S + m\ddot{\upsilon}_S &= 0 \\ EC_S\theta'''' + c_\theta\dot{\theta} - GJ_d\theta'' + \Theta_x\ddot{\theta} &= 0 \end{aligned}\right\} \quad \textbf{(89a,b)}$$

It is obvious that the beam will be moving independently in bending along the y axis and rotation around the x axis. The first of eqs (89) has the form of eq (63b) and thus, it can be solved in a similar manner.

For example, for the case of a simply supported beam, it will be:

$$\left.\omega_{y_n}^2 = \frac{n^4\pi^4 EI_z}{m\ell^4} \quad , \quad n = 1,2,3,\ldots.. \quad \text{and} \quad V_n(x) = C_{1n} \cdot \sin\frac{n\pi x}{\ell}\right\} \quad \textbf{(90a,b)}$$

The first of eqs (90) gives the eigenfrequencies spectrum for a lateral free motion, while the second one gives the corresponding shape functions. The displacement υ_S is given by the relation:

$$\left.\upsilon_S(x,t) = \sum_n V_n(x) \cdot e^{-\beta_y \cdot t} \cdot \left(A_n \sin\bar{\omega}_{yn} t + B_n \cdot \cos\bar{\omega}_{yn} t\right) \quad \text{with} \quad \beta_y = \frac{c_z}{2m} \quad , \quad \bar{\omega}_{yn} = \sqrt{\omega_{yn}^2 - \beta_y^2}\right\} \quad \textbf{(91)}$$

In order to solve eq (89b), we employ the expression (62d) and we get:

$$EC_S\Phi''''T_\theta + c_\theta\Phi\dot{T}_\theta - GJ_d\Phi''T_\theta + \Theta_x\Phi\ddot{T}_\theta = 0 \quad \text{or} \quad -\frac{\ddot{T}_\theta + \frac{c_\theta}{\Theta_x}\cdot\dot{T}_\theta}{T_\theta} = \frac{EC_S\Phi'''' - GJ_d\Phi''}{\Theta_x\Phi} = \omega_\theta^2 \qquad (92)$$

which results the following differential equations:

$$\left.\begin{aligned} &\ddot{T}_\theta + \frac{c_\theta}{\Theta_x}\cdot\dot{T}_\theta + \omega_\theta^2 T_\theta = 0 \\ &EC_S\Phi'''' - GJ_d\Phi'' - \Theta_x\omega_\theta^2\Phi = 0 \end{aligned}\right\} \qquad (93a,b)$$

Equation (93a) has the solution:

$$T_\theta(x) = e^{-\beta_\theta\cdot t}\left(A\sin\bar{\omega}_\theta t + B\cdot\cos\bar{\omega}_\theta t\right) \quad \text{with} \quad \left.\beta_\theta = \frac{c_\theta}{2\Theta_x}\ ,\quad \bar{\omega}_\theta = \sqrt{\omega_\theta^2 - \beta_\theta^2}\ \right\} \qquad (94a,b,c)$$

Equation (93b) is written as:

$$\Phi'''' - \frac{GJ_d}{EC_S}\cdot\Phi'' - \frac{\Theta_x\omega_\theta^2}{EC_S}\cdot\Phi = 0 \qquad (95)$$

The above equation has the solution:

$$\left.\begin{aligned} &\Phi(x) = c_1\cdot\sin\lambda_1 x + c_2\cdot\cos\lambda_1 x + c_3\cdot\sinh\lambda_2 x + c_4\cdot\cosh\lambda_2 x \\ &\text{where:}\quad \lambda_1 = \sqrt{-\frac{GJ_d}{2EC_S} + \sqrt{\left(\frac{GJ_d}{2EC_S}\right)^2 + \frac{\Theta_x\omega_\theta^2}{EC_S}}} \\ &\qquad\qquad \lambda_2 = \sqrt{\frac{GJ_d}{2EC_S} + \sqrt{\left(\frac{GJ_d}{2EC_S}\right)^2 + \frac{\Theta_x\omega_\theta^2}{EC_S}}} \end{aligned}\right\} \qquad (96a,b,c)$$

Using the boundary conditions for the beam under study, we can find the eigenfrequencies spectrum $\omega_{\theta n}$ and the shape functions Φ_n. For the case of the simply supported beam, the boundary conditions are:

$$\Phi(0) = \Phi''(0) = \Phi(\ell) = \Phi''(\ell) = 0$$

which introduced into eq (96a) result the following system:

$$\left.\begin{aligned} &c_2 \qquad\qquad\qquad\qquad + c_4 = 0 \\ &-c_2\lambda_1^2 \qquad\qquad\qquad + c_4\lambda_2^2 = 0 \\ &c_1\sin\lambda_1\ell + c_2\cos\lambda_1\ell + c_3\sinh\lambda_2\ell \\ &\qquad\qquad + c_4\cosh\lambda_2\ell = 0 \\ &-c_1\lambda_1^2\sin\lambda_1\ell - c_2\lambda_1^2\cos\lambda_1\ell + c_3\lambda_2^2\sinh\lambda_2\ell + \\ &\qquad\qquad + c_4\lambda_2^2\cosh\lambda_2\ell = 0 \end{aligned}\right\} \qquad (97)$$

In order for the above system to have non-trivial solutions, the determinant of the coefficients must be zero, namely:

$$\begin{vmatrix} 0 & 1 & 0 & 1 \\ 0 & -\lambda_1^2 & 0 & \lambda_2^2 \\ \sin\lambda_1\ell & \cos\lambda_1\ell & \sinh\lambda_2\ell & \cosh\lambda_2\ell \\ -\lambda_1^2\sin\lambda_1\ell & -\lambda_1^2\cos\lambda_1\ell & \lambda_2^2\sinh\lambda_2\ell & \lambda_2^2\cosh\lambda_2\ell \end{vmatrix} = 0 \tag{98}$$

Expanding eq (98) we have:

$$(\lambda_1^2 + \lambda_2^2)^2 \cdot \sin\lambda_1\ell \cdot \sinh\lambda_2\ell = 0 \tag{99}$$

In order that motion exists, it must be $\lambda_1^2 + \lambda_2^2 \neq 0$ and $\sinh\lambda_2\ell \neq 0$.

Therefore, the following equation must be valid: $\sin\lambda_1\ell = 0$ or $\lambda_1 = \frac{n\pi}{\ell}$, $n = 1,2,3,....$ which due to eq (96b) gives:

$$\omega_{\theta n}^2 = \frac{n^4\pi^4 EC_S}{\Theta_x \ell^4} + \frac{n^2\pi^2 GJ_d}{\Theta_x \ell^2},\ n = 1,2,3,.... \tag{100}$$

From eq (97a,b) we have $c_2 = c_4 = 0$, and since from eq (97 c) we get $c_3 = -c_1 \cdot \frac{\sin\lambda_1\ell}{\sinh\lambda_2\ell}$, one finally obtains:

$$\Phi_n(x) = c_{1n} \cdot \left(\sin\lambda_{1n} x - \frac{\sin\lambda_{1n}\ell}{\sinh\lambda_{2n}\ell} \cdot \sinh\lambda_{2n} x \right) \tag{101}$$

Equation (100) gives the spectrum of eigenfrequencies for the simply supported beam in torsional free vibration, while eq (101) gives the corresponding shape functions. For θ(x,t), we get:

$$\theta(x,t) = \sum_n \Phi_n(x) \cdot e^{-\beta_\theta \cdot t}(A_n \sin\overline{\omega}_{\theta_n} t + B_n \cos\overline{\omega}_{\theta_n} t) \tag{102}$$

Critical Damping

From the previously examined free vibration problems we observe that in all the above cases of the previous paragraphs of the present chapter result (regarding the determination of the time function T_n) to differential equations of the same form (eqs 65a, 73a, 82a):

$$\ddot{T}_n + \frac{c}{m} \cdot \dot{T}_n + \omega_n^2 T_n = 0 \tag{103}$$

We accepted, in addition, that equations of the above form have the solution:

$T_n = e^{-\beta \cdot t} \cdot (A_n \cdot \sin\overline{\omega}_n t + B_n \cdot \cos\overline{\omega}_n t)$, which, however, is valid only for a definite area of values of the beam characteristics. We will try at first a study of the solution of eq (103).

The characteristic of the above equation (103) is:

$$\rho^2 + \frac{c}{m} \cdot \rho + \omega_n^2 = 0 \tag{a}$$

which has the following solution:

$$\rho = \frac{-\frac{c}{m} \pm \sqrt{\left(\frac{c}{m}\right)^2 - 4\omega_n^2}}{2} = -\frac{c}{2m} \pm \sqrt{\left(\frac{c}{2m}\right)^2 - \omega_n^2} \quad \textbf{(b)}$$

$$\text{We set}: \beta = \frac{c}{2m} \text{ and } \bar{\omega}_n = \sqrt{\omega_n^2 - \beta^2} \quad \textbf{(c)}$$

and observing that β has dimensions of frequency (sec^{-1}), we will call it **damping quantity.**

We will study, without loss of generality, the vertical free vibration of a beam. We distinguish the following cases:

a) If $\frac{c}{2m} < \omega_n$ then: $\frac{c}{2m} < \omega_n = \frac{n^2\pi^2}{\ell^2} \cdot \sqrt{\frac{EI}{m}}$ or

$$c < \frac{2n^2\pi^2}{\ell^2} \cdot \sqrt{EIm} \quad \textbf{(104)}$$

and the solution of (103) will be:

$$T_n(t) = e^{-\beta t}(A_n \cdot \sin\bar{\omega}_n t + B_n \cdot \cos\bar{\omega}_n t) \quad \textbf{(105)}$$

The above is a periodic function having the form in Fig. **11a**, and leads to a gradual decay of the motion. This type of damping is called **weak damping.**

b) If $\frac{c}{2m} = \omega_n$ then:

$$c = \frac{2n^2\pi^2}{\ell^2} \cdot \sqrt{EIm} \quad \textbf{(106)}$$

Equation (b) gives:

$$\rho_{1,2} = -\frac{c}{2m} = -\beta \quad \textbf{(d)}$$

and the solution of eq (103) will be:

$$T_n(t) = A_n \cdot e^{-\beta t} + B_n \cdot t \cdot e^{-\beta t} \quad \textbf{(107)}$$

The above is a non-periodic function having the form in Fig. **11b**. This type of damping is called **critical damping.**

c) If $\frac{c}{2m} > \omega_n$ then: $c > \frac{2n^2\pi^2}{\ell^2} \cdot \sqrt{EIm}$ **(108)**

and the solution of eq (103) is:

$$T_n(t) = e^{-\beta t}\left(A_n \cdot e^{\sqrt{\beta^2 - \omega^2} \cdot t} + B_n \cdot e^{\sqrt{\beta^2 - \omega^2} \cdot t}\right) \quad \textbf{(e)}$$

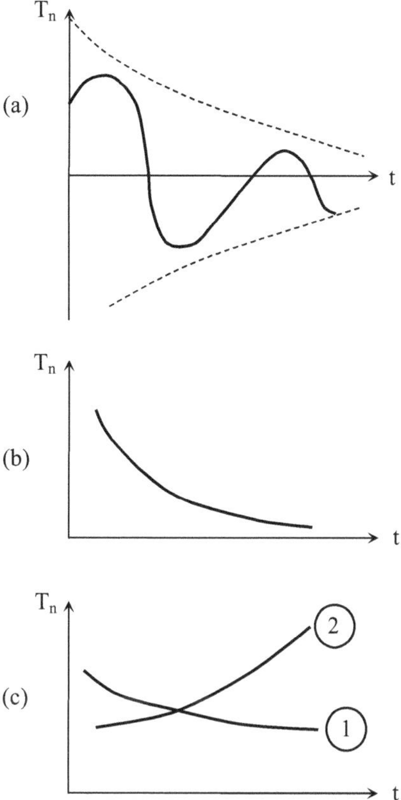

Figure 11: Subcritical, critical and supercritical damping

We know that the following relations are valid:

$$\left.\begin{aligned} e^{-\alpha} &= 1 - \frac{\alpha}{1!} + \frac{\alpha^2}{2!} - \frac{\alpha^3}{3!} + \ldots \\ \sinh\alpha &= \alpha + \frac{\alpha^3}{3!} + \frac{\alpha^5}{5!} + \ldots \\ \cosh\alpha &= 1 + \frac{\alpha^2}{2!} + \frac{\alpha^4}{4!} + \ldots \end{aligned}\right\} \quad \textbf{(f)}$$

Equation (e) is finally written as:

$$T_n(t) = e^{-\beta t}(\overline{A}_n \sinh\overline{\omega}_n t + \overline{B}_n \cosh\overline{\omega}_n t) \quad \textbf{(109)}$$

The above (109), is a **non periodic** function, which for some combinations of c, E, I, m (properties of the beam), but also of $\overline{A}_n, \overline{B}_n$ (determined by the time conditions) may give curves of the form (1) or also of the form (2) (Fig. **11c**).

Curves of the form (2) have a particular significance, since they correspond to vibration with continuously increasing amplitude and to the final collapse of the beam. This phenomenon is called **flutter** and it is usually inter-related to the phenomenon of aeroelasticity.

The Orthogonality Conditions

The above determined shape functions, because of their trigonometric form and the boundary conditions, satisfy the so called relations of orthogonality, which for the following cases are:

Longitudinal Free Vibration

Introducing the shape function U_n from eq (70) into eq (65b) we get:

$$U''_n + \frac{\omega^2_{xn}}{\kappa^2} \cdot U_n = 0$$

Multiplying the above by U_m and integrating the outcome from 0 to ℓ, we get:

$$\int_0^\ell U''_n U_m dx + \frac{\omega^2_{xn}}{\kappa^2} \cdot \int_0^\ell U_n U_m dx = 0$$

Performing integration by parts and taking into account the boundary conditions, we have:

$$\int_0^\ell U'_n U'_m dx + \frac{\omega^2_{xn}}{\kappa^2} \cdot \int_0^\ell U_n U_m dx = 0 \qquad \textbf{(110)}$$

Applying the same process for U_m, multiplying by U_n and integrating, we have:

$$\int_0^\ell U'_m U'_n dx + \frac{\omega^2_{xm}}{\kappa^2} \cdot \int_0^\ell U_m U_n dx = 0 \qquad \textbf{(111)}$$

Subtracting eqs (110) and (111) we get the following relation:

$$(\omega^2_{xn} - \omega^2_{xm}) \cdot \int_0^\ell U_n U_m dx = 0$$

which since $\omega_{xn} \neq \omega_{xm}$ gives:

$$\int_0^\ell U_m U_n dx = \begin{cases} 0 & \text{for } n \neq m \\ \Gamma_n & \text{for } n = m \end{cases} \qquad \textbf{(112)}$$

that is the corresponding orthogonality condition.

The Vertical Free Vibration

Introducing the shape function W_n into eq (73b) we have:

$$EI_y W''''_n - m\omega^2_{zn} W_n = 0$$

Multiplying the above by W_m and integrating from 0 to ℓ, we get:

$$EI_y \cdot \int_0^\ell W''''_n W_m dx - m\omega^2_{zn} \cdot \int_0^\ell W_n W_m dx = 0$$

Performing integration by parts, and taking into account the boundary conditions we have:

$$EI_y \cdot \int_0^\ell W''_n W''_m dx - m\omega^2_{zn} \cdot \int_0^\ell W_n W_m dx = 0 \qquad \textbf{(113)}$$

Applying the same procedure for W_m, we have:

$$EI_y \cdot \int_0^{\ell} W_m'' W_n'' dx - m\omega_{zm}^2 \cdot \int_0^{\ell} W_m W_n dx = 0 \tag{114}$$

Subtracting eqs (113) and (114), and given that $\omega_{zn} \neq \omega_{zm}$, we get the following relation:

$$m\int_0^{\ell} W_n W_m dx = \begin{cases} 0 & \text{for } n \neq m \\ \Gamma_n & \text{for } n = m \end{cases} \tag{115}$$

The Lateral-Torsional Free Vibration

Introducing the shape functions V_n and Φ_n into eq (82b), we get:

$$EI_z V_n'''' - EI_z z_M \Phi_n'''' - m\omega_{\theta n}^2 V_n = 0$$

Multiplying the above by V_m and integrating from 0 to ℓ, we have:

$$EI_z \cdot \int_0^{\ell} V_n'''' \cdot V_m dx - EI_z z_M \cdot \int_0^{\ell} \Phi_n'''' \cdot V_m dx - m\omega_{\theta n}^2 \cdot \int_0^{\ell} V_n V_m dx = 0$$

Integrating by parts and taking into account the boundary conditions, we finally obtain the following relation:

$$EI_z \cdot \int_0^{\ell} V_n'' V_m'' dx - EI_z z_M \cdot \int_0^{\ell} \Phi_n'' V_m'' dx - m\omega_{\theta n}^2 \cdot \int_0^{\ell} V_n V_m dx = 0 \tag{116}$$

Applying the same process but with V_m and Φ_m, we get the relation:

$$EI_z \cdot \int_0^{\ell} V_m'' V_n'' dx - EI_z z_M \cdot \int_0^{\ell} \Phi_m'' V_n'' dx - m\omega_{\theta m}^2 \cdot \int_0^{\ell} V_m V_n dx = 0 \tag{117}$$

Subtracting eqs (116) and (117) we have:

$$EI_z z_M \cdot \left(\int_0^{\ell} \Phi_n'' V_m'' \cdot dx - \int_0^{\ell} \Phi_m'' V_n'' \cdot dx \right) + m(\omega_{\theta n}^2 - \omega_{\theta m}^2) \cdot \int_0^{\ell} V_n V_m \cdot dx = 0 \tag{118}$$

Following the same procedure for eq (83c) we get:

$$EC_S \cdot \int_0^{\ell} \Phi_n'' \Phi_m'' dx - EI_z z_M \cdot \int_0^{\ell} V_n'' \Phi_m'' dx - GJ_d \cdot \int_0^{\ell} \Phi_n' \Phi_m' dx - \Theta_x \omega_{\theta n}^2 \cdot \int_0^{\ell} \Phi_n \Phi_m dx = 0$$

and

$$EC_S \cdot \int_0^{\ell} \Phi_m'' \Phi_n'' dx - EI_z z_M \cdot \int_0^{\ell} V_m'' \Phi_n'' dx - GJ_d \cdot \int_0^{\ell} \Phi_m' \Phi_n' dx - \Theta_x \omega_{\theta m}^2 \cdot \int_0^{\ell} \Phi_m \Phi_n dx = 0$$

or subtracting by parts:

$$EI_z z_m \left(\int_0^{\ell} V_n'' \Phi_m'' \cdot dx - \int_0^{\ell} V_m'' \Phi_n'' \cdot dx \right) + \Theta_x (\omega_{\theta n}^2 - \omega_{\theta m}^2) \cdot \int_0^{\ell} \Phi_n \Phi_m \cdot dx = 0 \tag{119}$$

Adding eqs (118) and (119), and given that $\omega_{\theta n} \neq \omega_{\theta m}$, the following orthogonality condition is obtained:

$$m\int_0^\ell V_n V_m dx + \Theta_x \int_0^\ell \Phi_n \Phi_m dx = \begin{cases} 0 & \text{for } n \neq m \\ \Gamma_n & \text{for } n = m \end{cases} \qquad \textbf{(120)}$$

The Special Case $z_M = 0$

In this case, one can easily find the following orthogonality conditions:

$$m\int_0^\ell V_n V_m dx = \begin{cases} 0 & \text{for } n \neq m \\ \Gamma_n & \text{for } n = m \end{cases} \qquad \textbf{(121)}$$

And

$$\Theta_x \int_0^\ell \Phi_n \Phi_m dx = \begin{cases} 0 & \text{for } n \neq m \\ \Gamma_n & \text{for } n = m \end{cases} \qquad \textbf{(122)}$$

THE FORCED MOTION

Equations (51) describe the problem of forced motion. Under the condition that the external loads are products of two functions, the one of x and the nother of t, and that the beam has a constant cross-section along the axis x, eqs (51) take the following form:

$$\left.\begin{array}{l} EAu_S'' - c_x \dot{u}_S - m\ddot{u}_S = q_x(x) \cdot f_x(t) \\ EI_y w_S'''' + c_y \dot{w}_S + m\ddot{w}_S = q_z(x) \cdot f_z(t) \\ EI_z \upsilon_S'''' - EI_z z_M \theta'''' + c_z \dot{\upsilon}_S + m\ddot{\upsilon}_S = q_y(x) \cdot f_\theta(t) \\ EC_S \theta'''' - EI_z z_M \upsilon_S'''' + c_\theta \dot{\theta} - GJ_d \theta'' + \Theta_x \ddot{\theta} = m_x(x) \cdot f_\theta(t) \end{array}\right\} \qquad \textbf{(123a,b,c,d)}$$

The general solution of the above equations is a summation of solutions of the corresponding homogeneous equations (that are the solutions of the free motion since the second members in the homogeneous equations are zero) plus some particular solutions of the full Equations (123). We search these particular solutions in the form:

$$\left.\begin{array}{l} u_S(x,t) = \sum_n U_n(x) \cdot P_{xn}(t) \\ w_S(x,t) = \sum_n W_n(x) \cdot P_{zn}(t) \\ \upsilon_S(x,t) = \sum_n V_n(x) \cdot P_{\theta n}(t) \\ \theta_S(x,t) = \sum_n \Phi_n(x) \cdot P_{\theta n}(t) \end{array}\right\} \qquad \textbf{(124a,b,c,d)}$$

where U_n, W_n, V_n, Φ_n are the corresponding known shape functions, while P_x, P_z, P_θ, are functions of time t to be determined.

Then, the general solution of eq (123) will be:

$$\left.\begin{aligned} u_S(x,t) &= \sum_n U_n(x)e^{-\beta_x \cdot t}\left(A_n \sin\bar{\omega}_{xn} t + B_n \cos\bar{\omega}_{xn} t\right) + \sum_n U_n(x)P_{xn}(t) \\ w_S(x,t) &= \sum_n W_n(x)e^{-\beta_z \cdot t}\left(A_n \sin\bar{\omega}_{zn} t + B_n \cdot \cos\bar{\omega}_{zn} t\right) + \sum_n W_n(x)P_{zn}(t) \\ \upsilon_S(x,t) &= \sum_n V_n(x)e^{-\beta_\theta \cdot t}\left(A_n \sin\bar{\omega}_{\theta n} t + B_n \cos\bar{\omega}_{\theta n} t\right) + \sum_n V_n(x)P_{\theta n}(t) \\ \theta(x,t) &= \sum_n \Phi_n(x)e^{-\beta_\theta \cdot t}\left(A_n \sin\bar{\omega}_{\theta n} t + B_n \cos\bar{\omega}_{\theta n} t\right) + \sum_n \Phi_n(x)P_{\theta n}(t) \end{aligned}\right\} \quad \textbf{(125a,b,c,d)}$$

The Longitudinal Forced Motion

Introducing the solution eq (124a) into eq (123a) we get:

$$EA \cdot \sum_n U_n'' P_{xn} - c_x \cdot \sum_n U_n \dot{P}_{xn} - m\sum_n U_n \ddot{P}_{xn} = q_x(x) \cdot f_x(t)$$

and given that U_n satisfies eq (65b), the above equation is written as:

$$-m\sum_n \omega_{xn}^2 U_n P_{xn} - c_x \sum_n U_n \dot{P}_{xn} - m\sum_n U_n \ddot{P}_{xn} = q_x(x) \cdot f_x(t)$$

Multiplying the above by U_n and integrating from 0 to ℓ, we get:

$$-m\omega_{xn}^2 \int_0^\ell U_n^2 dx \cdot P_{xn} - c_x \int_0^\ell U_n^2 dx \cdot \dot{P}_{xn} - m\int_0^\ell U_n^2 dx \cdot \ddot{P}_{xn} = f_x(t)\int_0^\ell q_x(x)U_n dx$$

$$\text{or} \quad \ddot{P}_{xn} + \frac{c_x}{m}\dot{P}_{xn} + \omega_{xn}^2 P_{xn} = -\frac{\int_0^\ell q_x(x)U_n dx}{m\int_0^\ell U_n^2 dx} \cdot f_x(t) \quad \textbf{(126)}$$

The solution of the above equation is given by the second Duhamel's integral as follows:

$$\left.\begin{aligned} P_{xn}(t) &= -\frac{\int_0^\ell q_x U_n dx}{m\bar{\omega}_{xn} \cdot \int_0^\ell U_n^2 dx} \cdot \int_0^t f_x(\tau) \cdot e^{-\beta_x \cdot (t-\tau)} \cdot \sin\bar{\omega}_{xn}(t-\tau) \cdot d\tau \\ \text{with:} \quad \beta_x &= \frac{c_x}{2m} \quad , \quad \bar{\omega}_{xn} = \sqrt{\omega_{xn}^2 - \beta_x^2} \end{aligned}\right\} \quad \textbf{(127)}$$

The Vertical Forced Motion

Introducing the solution eq (124b) into eq (123b) we get:

$$EI_y \sum_n W_n'''' P_{zn} + c_y \sum_n W_n \dot{P}_{zn} + m\sum_n W_n \ddot{P}_{zn} = q_z(x) \cdot f_z(t)$$

and given that W_n satisfies eq (73b), the above equation is written as:

$$m\sum_n \omega_{zn}^2 W_n P_{zn} + c_y \sum_n W_n \dot{P}_{zn} + m\sum_n W_n \ddot{P}_{zn} = q_z(x) \cdot f_z(t)$$

Multiplying the above by W_n and integrating from 0 to ℓ, we get:

$$m\omega_{zn}^2\int_0^\ell W_n^2dx\cdot P_{zn}+c_y\int_0^\ell W_n^2dx\cdot\dot{P}_{zn}+m\int_0^\ell W_n^2dx\cdot\ddot{P}_{zn}=f_z(t)\int_0^\ell q_zW_ndx$$

or $$\ddot{P}_{zn}+\frac{c_y}{m}\dot{P}_{zn}+\omega_{zn}^2P_{zn}=\frac{\int_0^\ell q_zW_ndx}{m\int_0^\ell W_n^2dx}\cdot f_z(t) \tag{128}$$

The solution of eq (128) is given by the second Duhamel's integral:

$$\left.\begin{array}{l} P_{zn}(t)=\dfrac{\int_0^\ell q_zW_ndx}{m\bar{\omega}_{zn}\int_0^\ell W_n^2dx}\cdot\int_0^t f_z(\tau)\cdot e^{-\beta_z\cdot(t-\tau)}\cdot\sin\bar{\omega}_{zn}(t-\tau)\cdot d\tau \quad \text{with}:\ \beta_z=\dfrac{c_y}{2m}\ ,\quad \bar{\omega}_{zn}=\sqrt{\omega_{zn}^2-\beta_z^2}\end{array}\right\} \tag{129}$$

The Lateral-Torsional Forced Motion

Introducing the solutions eqs (124c,d) into eqs (123c,d) we get:

$$EI_z\sum_n V_n''''P_{\theta n}-EI_zz_M\sum_n\Phi_n''''P_{\theta n}+c_z\sum_n V_n\dot{P}_{\theta n}+m\sum_n V_n\ddot{P}_{\theta n}=q_y(x)\cdot f_\theta(t)$$

$$EC_z\sum_n\Phi_n''''P_{\theta n}-EI_zz_m\sum_n V_n''''P_{\theta n}+c_\theta\sum_n\Phi_n\dot{P}_{\theta n}-GJ_d\sum_n\Phi_n''P_{\theta n}+\Theta_x\sum_n\Phi_n\ddot{P}_{\theta n}=m_x(x)\cdot f_\theta(t)$$

and given that V_n and Φ_n satisfy eqs (82 b and c), the above relations are written as:

$$\left.\begin{array}{l} m\sum_n\omega_{\theta n}^2V_nP_{\theta n}+c_z\sum_n V_n\dot{P}_{\theta n}+m\sum_n V_n\ddot{P}_\theta=q_y(x)\cdot f_\theta(t)\\ \Theta_x\sum_n\omega_{\theta n}^2\Phi_nP_{\theta n}+c_\theta\sum_n\Phi_n\dot{P}_{\theta n}+\Theta_x\sum_n\Phi_n\ddot{P}_{\theta n}=m_x(x)\cdot f_\theta(t)\end{array}\right\} \tag{130a,b}$$

Multiplying eq (130a) by V_n and integrating from 0 to ℓ, and eq (130b) by Φ_n and integrating similarly from 0 to ℓ, and using eq (81a) we get:

$$\sum_n\left(\int_0^\ell mV_n^2dx\right)\cdot\left(\omega_{\theta n}^2P_{\theta n}+\frac{c_z}{m}\dot{P}_{\theta n}+\ddot{P}_{\theta n}\right)=f_\theta(t)\cdot\int_0^\ell q_y(x)V_ndx$$

$$\sum_n\left(\int_0^\ell\Theta_x\Phi_n^2dx\right)\cdot\left(\omega_{\theta n}^2P_{\theta n}+\frac{c_z}{m}\dot{P}_{\theta n}+\ddot{P}_{\theta n}\right)=f_\theta(t)\cdot\int_0^\ell q_x(x)\Phi_ndx$$

Adding the above expressions and using the orthogonality condition eq (128) we conclude to the following equation:

$$\ddot{P}_{\theta n}+\frac{c_z}{m}\dot{P}_{\theta n}+\omega_{\theta n}^2P_{\theta n}=\frac{\int_0^\ell q_y(x)V_n(x)dx+\int_0^\ell m_x(x)\Phi_n(x)dx}{m\int_0^\ell V_n^2dx+\Theta_x\int_0^\ell\Phi_n^2dx}\cdot f_\theta(t) \tag{131}$$

The solution of eq (131) is given by the second Duhamel's integral:

$$\left.\begin{aligned} &P_{\theta n}(t)=\frac{\int_0^{\ell} q_y(x)V_n(x)dx+\int_0^{\ell} m_x(x)\Phi_n(x)dx}{\bar{\omega}_{\theta n}\cdot\left(m\int_0^{\ell} V_n^2dx+\Theta_x\int_0^{\ell}\Phi_n^2dx\right)}\cdot\int_0^t f_\theta(\tau)\cdot e^{-\beta\cdot(t-\tau)}\cdot\sin\bar{\omega}_{\theta n}(t-\tau)\cdot d\tau \\ &\text{with:}\quad \beta=\frac{c_z}{2m}\quad,\quad \bar{\omega}_{\theta n}=\sqrt{\omega_{\theta n}^2-\beta^2}\end{aligned}\right\}\qquad \textbf{(132)}$$

The Special Case z_M=0

In this special case, eqs (123 c and d) are uncoupled and take the form:

$$\left.\begin{aligned} EI_z\upsilon'''' + c_z\dot{\upsilon}_S + m\ddot{\upsilon}_S &= q_y(x)\cdot f_y(\theta) \\ EC_z\theta'''' + c_\theta\dot{\theta} - GJ_d\theta'' + \Theta_x\ddot{\theta} &= m_x(x)\cdot f_\theta(\theta)\end{aligned}\right\}\qquad \textbf{(133a,b)}$$

We are searching for solutions of the form:

$$\upsilon_S(x,t)=\sum_n V_nP_{yn}(t)\quad \text{for (133a)}$$

$$\text{and}\quad \theta(x,t)=\sum_n \Phi_nP_{\theta n}(t)\quad \text{for (133b)}$$

Following a similar process like the one in the previous paragraph and employing the orthogonality conditions (121) and (122) we find:

$$\left.\begin{aligned} &P_{yn}(t)=\frac{\int_0^{\ell} q_nV_ndx}{m\bar{\omega}_{yn}\int_0^{\ell} V_n^2dx}\cdot\int_0^t f_y(\tau)\cdot e^{-\beta_y\cdot(t-\tau)}\cdot\sin\bar{\omega}_{yn}(t-\tau)\cdot d\tau \\ &\text{where:}\quad \beta_y=\frac{c_z}{2m}\quad,\quad \bar{\omega}_{yn}=\sqrt{\omega_{yn}^2-\beta_y^2}\end{aligned}\right\}\qquad \textbf{(134)}$$

and

$$\left.\begin{aligned} &P_{\theta n}(t)=\frac{\int_0^{\ell} m_n\Phi_ndx}{\Theta_x\bar{\omega}_\theta\int_0^{\ell}\Phi_n^2dx}\cdot\int_0^t f_\theta(\tau)\cdot e^{-\beta_\theta\cdot(t-\tau)}\cdot\sin\bar{\omega}_{\theta n}(t-\tau)\cdot d\tau \\ &\text{where:}\quad \beta_\theta=\frac{c_\theta}{2\Theta_x}\quad,\quad \bar{\omega}_{\theta n}=\sqrt{\omega_{\theta n}^2-\beta_\theta^2}\end{aligned}\right\}\qquad \textbf{(135)}$$

REFERENCES

[1] C. Wylie *"Advanced Engineering Mathematics"*, 1975, Mc Graw- Hill, London.

[2] K. Georgikopoulos *"Solid Mechanics"*, 1968, T.C.G. Publ., Athens (in Greek).

[3] G.L. Rogers *"Dynamics of framed Structures"*, 1959, J. Wiley & Sons Inc., London.

[4] S.P. Timoshenko *"Advanced Dynamics"*, 1948, Mc Graw-Hill, N. York.

[5] G.T. Michaltsos *"Thin-walled Metal Structures – Calculation Procedures"*, 2008, Symeon Publ., Athens (in Greek).

[6] G. Heilig *"Torsions-und Biegeeingeschwigungen von dűnwandigen Trägern mit beliebiger offener Profilform bei Anwesenheilt von Vorlasten"*, 1950, Diss. T.H. Darmstadt.

[7] A.N. Kounadis *"Dynamics of Continuous Elastic Systems"*, 1983, NTUA Publ., Athens (in Greek).

[8] L. Frýba *"Vibration of Solids and Structures under moving Loads"*, 1990, Noordhoff Inter. Publishing, Croningen.

[9] L.S. Jacobsen, R.S. Ayre *"Engineering Vibrations"*, 1958, Mc Graw-Hill, N. York.

[10] D.S. Sophianopoulos, G.T. Michaltsos *"Combined torsional-lateral vibrations of beams under vehicular loading"*, 1999, Facta Universitatis, Vol.2, No 9.

CHAPTER 4

The Problem of Moving Loads

Abstract: This chapter presents the special cases of loads with or without mass moving over a bridge structure. More specifically, the cases of concentrated loads and uniform loads are treated taken into account the effect of damping. Dynamic influence lines are presented for each case taking into account the effect of load mass as well. The influence of deck irregularities is also presented in brief form.

GENERAL

Moving loads present an important influence in the dynamic behavior of flexible and inflexible structures and structural parts and may cause intense oscillations to such types of structures, specifically for big speeds. The characteristic feature of these loads is that they are altered so much in time as well as in space.

The oscillation of a three-dimensional body can in general be described by a functional relation of a position vector r(x,y,z,t) and an external load P(x,y,z,t) in the form: $L\left[r(x,y,z,t)\right]=P(x,y,z,t)$, where symbol L denotes a linear or nonlinear differential equation. With the initial and the boundary conditions, the aforementioned equation (or a group of such equations) characterizes the behavior of the body.

The problem of moving loads appeared and was first studied in the beginning of the 19th century, when the first railway bridges were erected and in general it constitutes one from the first problems of structural dynamics. That time, the following two considerations prevailed. According to the first consideration, the influence of a moving load was similar to that of impact, while the according to the second the rapid passage of a vehicle through a bridge would not constitute sufficient time for the deformation of the structure.

Jules Vern, a fervent partisan of scientific thought, in his novel "Tour of world in 80 days" describes the safe passage of a train through a destroyed bridge, which could not be justified until recently without the development of the theory for plastic resistance of materials.

Theoretical Research

The problem of moving loads was first studied theoretically and experimentally by R. Willis [1], who considered the mass of the beam being small in comparison to the mass of the individual load. With the same assumptions, G.G. Stokes [2] & H. Zimmermann [3] also studied the same problem.

A.N. Krýlov [4] and then S.P. Timoshenko [5], studied the same problem, but considering on the contrary small the mass of the load in comparison to the mass of the beam. Then, numerous researches have dealed with this problem following various approaches. A.N. Lowan and N.G. Bondar [6] employed the Green functions, while S.P. Timoshenko solved also the problem of a harmonic load moving with constant speed. The problem was studied for realistic load to beam mass ratios firstly by H. Saller [7] and then by H.H. Jeffcott [8], H. Steuding, A. Schallenkamp, V.M.Muchnikov and M.Y. Ryazonoua. Special attention should also be paid to C.E. Inglis [9] study, who employed harmonic analysis for the solution of all practically important cases of railway bridge subjected to loads resulting from the passage of steam moving trains. With the help of normal modes analysis, S.T. Ödmann and V. Kolûsek [10] solved complicated dynamical problems for statically indeterminate systems.

A. Hillerborg [12], by employing the Fourier method and other differential methods, provided a first solution to the problem of one load moving with constant speed on a system supported by elastic springs. Further progress to this direction became possible with the use of computers. J.M. Biggs, H.S. Suer & J.M. Louw [13] solved the problem using the method by Inglis, while T.P. Tung, L.E. Goodman, T.Y. Chen & N.M. Newmark using the Hillerborg method, and the obtained solutions were applied to bridge dynamics.

George T. Michaltsos and Ioannis G. Raftoyiannis

Experimental Research

The dynamic behavior of structures under the effect of moving loads, has been investigated over the last 160 years by both theoretical and experimental methods. There exist a vast number of studies, which renders impossible to report all of them.

As a starting line in the study of the dynamic loading of bridges phenomenon, it is considered the study of Willis. In the frame of a committee appointed by the queen of England in 1847, Willis conducted an entire series of experiments, in order to assess the influence of speed in the phenomenon of oscillation of structures [14].

Another pioneer, Deslandres [15], has showed the influence of frequency and resonance. Deslandres, in 1892, was led to the conclusion that even small oscillation amplitudes due to loads can cause big deformations when their frequency is equal to the fundamental frequency of the bridge. He also concluded that the biggest reaction of the bridge could not be calculated, since damping was dependent on the oscillation amplitude. This fact can be easily verified. In 1889, Sonleyre was among the first researchers who studied the phenomenon and concluded that the dynamic factor is increased as the length of bridge increases. The first relation for the above findings, was given in 1893, by Melan [16].

Great Britain was the first country in which big scale systematic experiments were conducted in the decade of 20s. The study of dynamic behavior of railway bridges progressed particularly due to the numerous inquiring programs, that financed railway organizations. The second international congress on bridges in 1928 in Vienna, provided the opportunity to assemble and evaluate the knowledge up to then.

In the decade of 50s, Van Eeman [17] carried out in the USA a series of experiments with general content on truss bridges. In these experiments a large number of vehicles had been used, while methods for the precise determination of the speed and the vehicle position on the bridge were developed for the first time. From the proceedings 124 of the HRB Bulletins in 1956, it appears that in the decade of 50s a big progress took place in the USA.

Edgerton & Beecroft [18] in a pioneering work were the first who used inductive trail recorders (divergence) and elongations, a method that is in use even nowadays. They have concluded that the paving abnormalities are considerably affecting the oscillations of bridges.

AASHO [19,20] Road Test was an extensive experimental program with no similar in the past and probably not in the future. The responsibility for conducting the experiments, that cost 27 millions US $, belong to the Highway Research Board (HRB). In August 1956 started the constitution of the experimental unit in Ottawa - Illinois, while in October 15, 1958 the first test was conducted, and in November 30, 1960 the last test. There was a huge number of vehicles with ten different types, while in periods of peak there were 185 employees and 400 soldiers. The main goal of the experiments was the investigation of behavior of paving and bridges under the action of a large number of dynamic loads, the conduct of measurements and the drawn of conclusions. The above experiments of dynamic investigation of bridges were carried out at the University of Illinois. In the works of Biggs, Suer & Louw in 1957 (theoretical research and a series of experiments), the bridge and the vehicle were replaced by mass oscillators without damping.

Also, in 1957 Oehler [21] published the results of experiments in 15 road bridges, where always the same experimental vehicle, a 3-axes track, was used.

Toledo Leyva & Veletsos, in 1958 [22], used the Wen program to study the influence of abnormalities of coating in the dynamic behavior of a simple beam. They studied bridges with openings 6 up to 27 meters and frequencies from 12,1 Hz up to 2,6 Hz, using a simple model vehicle, consisting from a mass oscillator without damping.

In 1962, Wright & Green [23] carried out a series of dynamic experiments in 47 road bridges, under regular traffic and speed of vehicles in Ontario, based on which the long Ontario tradition begun.

Varney & Galambos presented in 1965 a table with all the dynamical experiments in road bridges carried out from 1948 until 1965. The research at the University of Illinois was continued until the middle of the decade of 70s, when Fischer but also Veletsos, Walker, Nieto-Ramirez, Hwang, Eberhardt and Ruhl were mainly concerned with the improvement of bridge models. The work of Eberhardt [24] is worth mentioning, who in 1972 constructed a computer program, where three-dimensional models of vehicles and bridges were employed.

In 1979, the first code for bridge design is presented in Ontario, the "Ontario Highway Bridge Design Code". As the later code in 1983, this code also defines the dynamic factor as function of the fundamental frequency of the bridge.

In 1982, the A.S.C.E. committee for loads and forces on bridges, came to the conclusion that the studies relative to the phenomenon of impact, should have be collected and published, so that the basic bibliography for further promotion of the research exists.

Ghosn & Quingle [25] presented in 1989 a method, with which using the technique "Weigh-in-Motion" by Moses the determination of the dynamic amplitude of oscillation of bridges is possible.

In USA, a new regulation of loads on bridges is under preparation. Both this code and the Ontario code, will be publicly discussed, so that the dynamic factor is more precisely defined. Hwang and Novak [26], in 1989, described an analytical model with which something like that is possible.

The last years, and during the aforementioned experiments, certain interesting theoretical approaches have been presented in France. Palamas [27,28] in 1982 and 1985, dealed with simple models regarding the vehicles and the forms of abnormalities in paving. Coussy, under the supervision of whom Palamas was working, proceeded in 1989 with Said & Van Hoove to the theory of bi-axial vehicles with linear relation of spring - dampers and various forms of abnormalities.

Today, with the advanced development of technology, we are in position to deal more easily with such problems, as Bily [29] in Prague and Hymay in Ottawa, who employed three-dimensional models. However, both were not in position to examine the abnormalities of paving in a detailed manner.

A large number of theoretical and experimental studies on railway bridges, were conducted by the faculty in Dnepropetrovsk by Professor N.G. Bondar [30,31]. In the research centers of Moscow and Leningrad, extensive experimental studies in railway bridges are carried out. In USA, researchers have focused their efforts in experimental studies on railway bridges and road bridges, studying their behavior up to collapse. Research on moving loads is still carried out in the Universities of Illinois (Urbana), Michigan State (East Lansing), Northwestern (Evanston), Standard, M.I.T. and elsewhere. In Europe, similar research is carried out in Poland and Switzerland, mainly on railway bridges, in France on road bridges, while in Germany they are dealing with both types.

GENERAL THEORY OF MOVING MASS

Let us consider a beam, the axis of which passes through point O, which is also a support point, with cross-sectional area A, moment of inertia J_y and rotary inertia J_b, which is made from material with density ρ, and modulus of elasticity E.

A point mass is moving on the beam with speed $\overline{\upsilon}_S$, mass M and rotary inertia J_M.

Due to the excitation caused by the movement of the point mass M, the beam is vibrating and at time t, its deformation is w(x,t) and hence, point A has a velocity $\overline{\upsilon}_W$, parallel to axis z, which is engaged also by point mass M. It should be noted that in the displacement from A to A′, the horizontal component u of point A (due to the rotation of the beam about point 0) is neglected as being very small.

We shall consider the energy of the system.

a. Kinetic energy

The kinetic energy K of the system is constituted by the kinetic energy K_b of the beam and the kinetic energy K_M of the moving mass.

a1. For the beam, it is:

$$K_b = \frac{\rho \cdot A}{2}\int_0^L \overline{\upsilon}_w^2 + \frac{J_b}{2}\int_0^L \dot{w}'^2 \cdot dx$$

or finally: $K_b = \frac{\rho \cdot A}{2}\int_0^L \dot{w}^2 \cdot dx + \frac{J_b}{2}\int_0^L \dot{w}'^2 \cdot dx$ **(1)**

a2. For the moving mass, it is:

$$K_M = \frac{M}{2} \cdot \overline{\upsilon}^2 \big|_{x=S} + \frac{J_M}{2} \cdot \dot{w}'^2 \big|_{x=S} \quad \textbf{(2)}$$

Note here that the velocity $\overline{\upsilon}$ is a vectorial size and is the vectorial summation of two vectors $\overline{\upsilon}_w$ and $\overline{\upsilon}_S$, *i.e.*

$$\overline{\upsilon} = \overline{\upsilon}_w + \overline{\upsilon}_S \quad \textbf{(3)}$$

where the velocity component $\overline{\upsilon}_w$ has constant direction, while the component $\overline{\upsilon}_S$ has variable direction.

Hence, eq (2) can be written as:

$$K_M = \frac{M}{2}\overline{\upsilon}^2 \big|_{x=S} + \frac{J_M}{2} \cdot \dot{w}'^2 \big|_{x=S} = \frac{M}{2}(\overline{\upsilon}_w + \overline{\upsilon}_S)^2 + \frac{J_M}{2} \cdot \dot{w}'^2 \big|_{x=S} = \frac{M}{2}(\overline{\upsilon}_w^2 + \overline{\upsilon}_S^2 + 2 \cdot \overline{\upsilon}_M \cdot \overline{\upsilon}_S) \big|_{x=S} + \frac{J_M}{2} \cdot \dot{w}'^2 \big|_{x=S}$$

and since the product is: $\overline{\upsilon}_S \cdot \overline{\upsilon}_W = \upsilon_S \upsilon_W \cos\varphi$, and: $\cos\varphi \cong w'$ (since w′ is assumed very small), we will have:

$$K_M = \frac{M}{2}(\overline{\upsilon}_M^2 + \overline{\upsilon}_S^2 + 2\upsilon_M \upsilon_S w') \big|_{x=S} + \frac{J_M}{2} \cdot \dot{w}'^2 \big|_{x=S} \quad \textbf{(4)}$$

b. Potential energy

The potential energy of the system is given by the following expression:

$$V_b = \frac{EJ_y}{2} \cdot \int_2^L w''^2 \cdot dx \quad \textbf{(5)}$$

c. Finally, the load potential δΩ, of the external forces and the damping forces (that produce negative work) due to the infinitesimal displacements δw and ds will be:

$$\delta\Omega = \int_0^L [q(x,t) \cdot \delta w - c_y \cdot \dot{w} \cdot \delta w + u(t) \cdot \delta s \cdot \overline{\delta}(x-s)] \cdot dx \quad \textbf{(6)}$$

where the Dirac function will be denoted here with a bar, in contrast to the infinitesimal displacements δw and δs.

In relation (6), q(x,t) is the vertical load acting on the beam, while u(t) is the force applied on the mass M, which causes its movement.

Applying the Hamilton principle, we obtain:

$$\int_{t_1}^{t_2} \delta(L-\Omega)\, dt = \int_{t_1}^{t_2} (\delta K_b + \delta K_M - \delta V_b - \delta\Omega)\, dt = 0 \quad \textbf{(7)}$$

Then, we compute:

1. Variation of the kinetic energy of the beam:

$$\int_{t_1}^{t_2} \delta K_b dt = \int_{t_1}^{t_2}\int_0^L \left(\rho A \dot{w}\delta\dot{w} + J_b \dot{w}'\delta\dot{w}'\right) dx \cdot dt$$

which after integration by parts and taking into account the boundary and initial conditions takes the form:

$$\int_{t_1}^{t_2} \delta K_b dt = \int_{t_1}^{t_2}\int_0^L \left(-\rho A \ddot{w}\delta w + J_b \ddot{w}''\delta w\right) dx \cdot dt \quad \textbf{(8)}$$

2. Variation of the kinetic energy of the mass M:

$$\int_{t_1}^{t_2} \delta K_M dt = \int_{t_1}^{t_2}\int_0^L \{M\cdot[\overline{\upsilon}_W \delta\overline{\upsilon}_W + \overline{\upsilon}_S \delta\overline{\upsilon}_S + \delta(\overline{\upsilon}_W\overline{\upsilon}_S)] + J_M \dot{w}'\delta\dot{w}'\}\cdot\overline{\delta}(x-s)\, dx\cdot dt \quad \textbf{(9)}$$

If $\overline{\varepsilon}$ and $\overline{\tau}$ are the unit vectors on the tangent at point A and the parallel to axis z, we will have:

$$\overline{\upsilon}_W = \upsilon_W \cdot \overline{\tau} \quad , \quad \overline{\upsilon}_S = \upsilon_S \cdot \overline{\varepsilon} \quad \textbf{(10)}$$

Then, since the product of two vectors $\overline{\alpha}$ and $\overline{\beta}$ is: $\overline{\alpha}\cdot\overline{\beta} = \alpha\cdot\beta\cdot\cos(\alpha\hat{}\beta)$ and also since Fig. **1**: $\delta\overline{\varepsilon} = \overline{\varepsilon}\cdot\delta w' = \overline{\varepsilon}\cdot w''\cdot\delta w$, we will have:

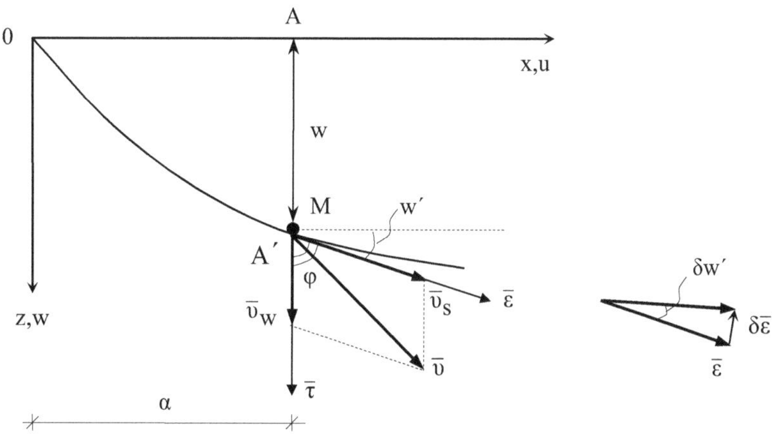

Figure 1: Point mass M moving on a vibrating beam

$$\begin{aligned}
\overline{\upsilon}_W\cdot\delta\overline{\upsilon}_W &= \upsilon_W\cdot\delta\upsilon_W\cdot\cos 0^0 = \upsilon_W\cdot\delta\upsilon_W = \dot{w}\cdot\delta\dot{w}\\
\overline{\upsilon}_S\cdot\delta\overline{\upsilon}_S &= \upsilon_S\cdot\overline{\varepsilon}\cdot\delta(\upsilon_S\cdot\overline{\varepsilon}) = \upsilon_S\cdot\overline{\varepsilon}\cdot(\upsilon_S\cdot\delta\overline{\varepsilon} + \overline{\varepsilon}\cdot\delta\upsilon_S) =\\
&= \upsilon_S^2\cdot\overline{\varepsilon}\cdot\delta\overline{\varepsilon} + \upsilon_S\cdot\delta\upsilon_S\cdot\overline{\varepsilon}\cdot\overline{\varepsilon} = \upsilon_S^2\cdot\overline{\varepsilon}^2\cdot w''\cdot\delta w + \upsilon_S\cdot\delta\upsilon_S\cdot\overline{\varepsilon}^2 =\\
&= \upsilon_S^2\cdot w''\cdot\delta w + \upsilon_S\cdot\delta\upsilon_S = \dot{s}^2\cdot w''\cdot\delta w + \dot{s}\cdot\delta\dot{s}
\end{aligned}$$

$$\begin{aligned}
\delta(\overline{\upsilon}_W\cdot\overline{\upsilon}_S) &= \delta(\upsilon_W\cdot\upsilon_S\cdot\cos\varphi) = \delta\cdot(\upsilon_W\cdot\upsilon_S\cdot w') =\\
&= \upsilon_S\cdot w'\cdot\delta\upsilon_W + \upsilon_W\cdot w'\cdot\delta\upsilon_S + \upsilon_W\cdot\upsilon_S\cdot\delta w' =\\
&= \dot{s}\cdot w'\cdot\delta\dot{w} + \dot{w}\cdot w'\cdot\delta\dot{s} + \dot{w}\cdot\dot{s}\cdot\delta w'
\end{aligned}$$

and hence, eq (9) is written as follows:

$$\int_{t_1}^{t_2} \delta K_M dt = \int_{t_1}^{t_2}\int_0^L \left\{ M(\dot{w}\delta\dot{w} + \dot{s}^2 w''\delta w + \dot{s}\delta\dot{s} + \dot{s}w'\delta\dot{w} + \dot{w}w'\delta\dot{s} + \dot{w}\dot{s}\delta w') + J_M\dot{w}'\delta\dot{w}'\right\}\cdot\overline{\delta}(x-s)dxdt$$

The above relation, after integration by parts and taking into account the boundary and initial conditions, is finally written as follows:

$$\left.\begin{aligned}\int_{t_1}^{t_2}\delta K_M dt &= \int_{t_1}^{t_2}\int_0^L\left[M(\ddot{w}+\ddot{s}w'+\dot{s}^2w''+2\dot{s}\dot{w}')\cdot\overline{\delta}(x-s)-\ J_M\ddot{w}'\overline{\delta}'(x-s)\right]\delta w dx dt\\ &+\int_{t_1}^{t_2}\int_0^L M(\ddot{s}+\ddot{w}w'+\dot{w}\dot{w}')\cdot\overline{\delta}(x-s)\ \delta s dx dt\end{aligned}\right\}\qquad\textbf{(11)}$$

3. Variation of the potential energy

$$\int_{t_1}^{t_2}\delta V_b dt = \int_{t_1}^{t_2}\int_0^L EI_y w''\delta w'' dx dt$$

The above relation, after integration by parts and taking into account the boundary and initial conditions, is finally written as follows:

$$\int_{t_1}^{t_2}\delta V_b dt = \int_{t_1}^{t_2}\int_0^L EI_y w''''\delta w dx dt \qquad\textbf{(12)}$$

Hence, eq (7), due to eqs (8), (11), (12) and (6), can be written as:

$$\int_{t_1}^{t_2}\int_0^L [EI_y w''''-J_b\ddot{w}''+M(\ddot{w}+\ddot{s}w'+\dot{s}^2w''+2\dot{s}\dot{w}')\cdot\overline{\delta}(x-s)-J_M\ddot{w}'\cdot\overline{\delta}'(x-s)-q(x,t)+c_y\dot{w}]\delta w dx dt +$$

$$+\int_{t_1}^{t_2}\int_0^L M\cdot[\ddot{s}+\ddot{w}w'+\dot{w}\dot{w}'-u(t)]\cdot\overline{\delta}(x-s)\ \delta s dx\cdot dt = 0$$

The above equation is valid only when the coefficients of δw and δs are zero, that is:

$$\left.\begin{aligned}&EI_y w''''+c_y\dot{w}-J_b\ddot{w}''+A\rho\ddot{w} = q(x,t)-M(\ddot{w}+\ddot{s}w'+\dot{s}^2w''+2\dot{s}\dot{w}')\cdot\delta(x-s)\ +J_M\ddot{w}'\delta'(x-s)\\ &M(\ddot{s}+\ddot{w}w'+\dot{w}\dot{w}') = u(t)\end{aligned}\right\}\qquad\textbf{(13)}$$

Note that the external load q(x,t) includes also the load of the moving mass, *i.e.*:

$$q(x,t) = p(x,t)+Mg\delta(x-s) \qquad\textbf{(14)}$$

Equations (13) are the governing equations for motion of a load $P = M\cdot g$, with mass M moving under the influence of a force u(t).

The first part of eq (13a) is the same as in a straight beam with one span regardless of the support conditions.

The second part of eq (13a) expresses the influence of the load q(x,t) and the moving mass on the beam motion. More analytically, the terms shown in this expression refer accordingly to the vertical component of the inertia force of mass M, the influence of the tangential at point A′ component of the mass M, the centrifugal force due to the curved motion of mass M on the curve with curvature w′′, the Coriolis force due to the rotation of the beam about point O and finally, the rotary inertia due to the rotation of mass M by an angle w′(x,t).

Finally, eq (13b) contains besides the moving force u(t) in the second part, the tangential at point A′ component of the mass M as well as the influence of dynamic displacement w and the vertical component of the inertia force of mass M. The influence of the centrifugal and Coriolis forces, as well as factors from which they are dependent, will be examined in the corresponding sections.

The solution of eq (13a), as well as other equations of motion that will be derived later on, is not possible to take a closed form, since these equation are very complex and also nonlinear when the above influences of mass M are taken into consideration.

For these reasons, we usually employ approximate iterative methods. Such a method is the one by Prof. A. Kounadis [32] applied in the Metal Structures Laboratory of NTUA. The procedure for obtaining such an approximate solution is as follows:

The governing equation is first written in the form:

$$F(f) = \Phi(f) \qquad \textbf{(a)}$$

where, function $F(f)$ contains all the linear terms, function $\Phi(f)$ contains all the nonlinear terms, while function $f(t)$ is an unknown function corresponding to the solution of eq(a) we are seeking.

We first consider a known function $f_o(t)$, the so-called starting function, which is introduced into the expression $\Phi(f)$, thus resulting:

$$F(f) = \Phi(f_0) \qquad \textbf{(b)}$$

Equation (b) is now a linear differential equation with known second part, which can be solved by known methods. The solution of eq(b) provides function $f_1(t)$, which is a first approximation, and which after introduction into the expression $\Phi(f)$ gives:

$$F(f) = \Phi(f_1) \qquad \textbf{(c)}$$

Equation (c) is again solved giving a second approximate solution f_2 etc.

It has been proven that after a number of iterations n, function $f_n - f_{n-1}$ tends to zero, and hence function is an f_n approximate solution of eq(a). The convergence speed of an iterative procedure, such the above, depends on various factors, the main important of which is the proper selection of the starting function. This function should be a very good approximation to the final solution of the problem, *i.e.* it should satisfy most of the conditions of the problem.

If for example, in the problem of the preceding paragraph we choose as starting function the solution of a vibrating beam under a uniformly applied harmonic load $q \cdot \sin \omega t$, we well have a very large number of iterations before we reach a satisfactory solution. Contrarily, if we choose as starting function the solution of a vibrating beam under a transverse load $P = M \cdot g$, moving on the beam with velocity υ (and neglecting the nonlinear terms in the second part of eq. 13a), the iteration procedure requires usually only one step to achieve a solution with sufficient accuracy.

In the first case, only the boundary conditions of the beam were satisfied, while the loading conditions were completely different.

In the second case, both the boundary conditions of the beam as well as the loading conditions of the structure (load value – velocity) were satisfied, while only secondary terms (influence terms regarding the mass M) with negligible influence have been omitted.

For the solution of problems with moving loads, we shall consider the following relatively simple cases.

Load P Moving on a Beam without Damping

Let us consider a load $P = M \cdot g$, moving on a beam with constant velocity υ, where damping is neglected as well as the secondary effects due to the mass M of the load P. Then, if P is the only dynamic load applied on the beam, eq (13a) is written as follows:

$$EI_y w''''(x,t) + m\ddot{w}(x,t) = P \cdot \delta(x-\alpha) \tag{15}$$

where α is the position of the load P on the beam at time t. We shall seek a solution in the form of separate variables such as:

$$w(x,t) = \sum_n W_n(x) \cdot T_n(t) \tag{16}$$

where $W_n(x)$ is the n-th shape function of the freely vibrating beam, while $T_n(t)$ is the corresponding time function, to be determined. Introducing the expression of w(x,t) from eq (16) into eq (15), we obtain:

$$EI_y \sum_n W''''T_n + m \cdot \sum_n W_n \ddot{T}_n = P \cdot \delta(x-\alpha) \tag{17}$$

For the freely vibrating beam, the following equation is valid:

$$EI_y W_n'''' - m\omega_n^2 W_n = 0 \tag{18}$$

where for a simply supported beam, for instance, it is:

$$W_n = \sin\frac{n\pi x}{\ell} \quad , \quad \omega_n^2 = \frac{n^4\pi^4 EI_y}{m \cdot \ell^4} \tag{19}$$

Equation (17) due to eq (18) becomes:

$$m\sum_n \omega_n^2 W_n T_n + m\sum_n W_n \ddot{T}_n = P \cdot \delta(x-\alpha) \tag{20}$$

Multiplying eq (20) by $W_\kappa(\kappa \neq n)$, integrating the outcome from 0 to ℓ and taking into account the orthogonality conditions, we will have:

$$\ddot{T}_\kappa + \omega_\kappa^2 T_\kappa = \frac{P}{m\int_0^\ell W_\kappa^2 dx} \cdot W_\kappa(\alpha) = \frac{P}{m\int_0^\ell W_\kappa^2 dx} \cdot W_\kappa(\upsilon t) \tag{21}$$

since $\alpha = \upsilon \cdot t$. The solution of eq (21) is given by the Duhamel's integral:

$$T_\kappa(t) = \frac{P}{m\omega_\kappa \int_0^\ell W_\kappa^2 dx} \cdot \int_0^t W_\kappa(\upsilon\tau) \cdot \sin\omega_\kappa(t-\tau)d\tau \tag{22}$$

For the simply supported case, eq (22) due to eqs (19), gives:

$$\left.T_\kappa(t) = \frac{2P}{m\ell} \cdot \frac{1}{\omega_\kappa^2 - \Omega_\kappa^2} \cdot \left(\sin\Omega_\kappa t - \frac{\Omega_\kappa}{\omega_\kappa} \cdot \sin\omega_\kappa t\right) \quad \text{where: } \Omega_\kappa = \frac{\kappa\pi\upsilon}{\ell} \quad \right\} \tag{23}$$

Load P Moving on a Beam with Damping

Under the conditions in the preceding paragraph including damping, the governing equation of motion for the problem is:

$$EI_y w'''' + c_y \dot{w} + m\ddot{w} = P\delta(x-\alpha) \quad \textbf{(24)}$$

Seeking again a solution in the form of separate variables eq (16) and following the same procedure, we arrive at the following equation, for the κth time function:

$$\ddot{T}_\kappa + \frac{c_y}{m}\dot{T}_\kappa + \omega_\kappa^2 T_\kappa = \frac{P}{m\cdot\int_0^\ell W_\kappa^2 dx}\cdot W_\kappa(\upsilon t) \quad \textbf{(25)}$$

The solution of eq (25), is given by the 2nd Duhamel's integral:

$$\left.T_\kappa(t) = \frac{P}{m\bar{\omega}_\kappa \int_0^\ell W_\kappa^2 dx}\cdot\int_0^t e^{-\beta_z(t-\tau)} W_\kappa(\upsilon\tau)\cdot\sin\bar{\omega}_\kappa(t-\tau)\cdot d\tau \quad \text{where:} \quad \beta_z = \frac{c_y}{2m}, \quad \bar{\omega}_\kappa = \sqrt{\omega_\kappa^2 - \beta_z^2}\right\} \quad \textbf{(26)}$$

For the simply supported case, the above equation due to eqs (19), gives:

$$T_\kappa(t) = \frac{2P}{m\ell}\cdot\frac{1}{\bar{\omega}_\kappa^4 + 2(\beta_z^2 - \bar{\omega}_\kappa^2)\Omega_\kappa^2 + \Omega_\kappa^4}\cdot \left\{ e^{-\beta_z \cdot t}\left[2\beta_z\Omega_\kappa \cos\bar{\omega}_\kappa t + \frac{\Omega_\kappa}{\bar{\omega}_\kappa}(\beta_z^2 - \bar{\omega}_\kappa^2 + \Omega_\kappa^2)\cdot\sin\bar{\omega}_\kappa t\right] + \left[-2\beta_z\Omega_\kappa \cos\Omega_\kappa t + (\bar{\omega}_\kappa^2 - \Omega_\kappa^2)\cdot\sin\Omega_\kappa t\right]\right\} \quad \textbf{(27)}$$

where: $\Omega_\kappa = \frac{\kappa\pi\upsilon}{\ell}$ and hence:

$$w(x,t) = \sum_n W_n(x) T_n(t) \quad \textbf{(27a)}$$

The influence of damping on the time function T_κ, is shown in the diagram in Fig. **2**, where c=0,02 and κ =1, and for time required for the load to cross the beam length, *i.e.* for time interval 0 to $\frac{\ell}{\upsilon}$.

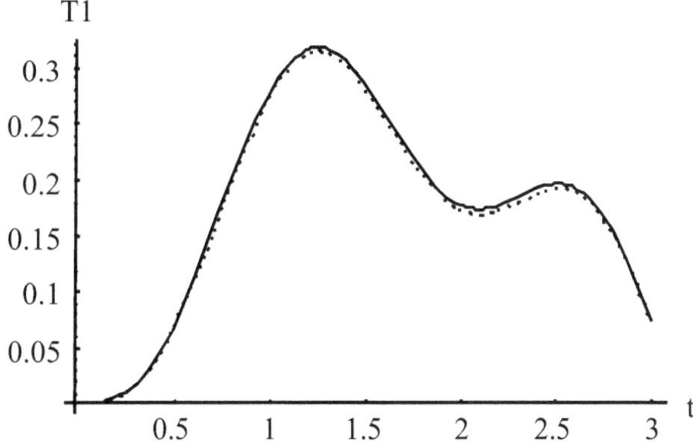

Figure 2: The influence of damping on the time function T_κ

General Solution of the Equation of Motion

In the general case of a beam under moving load, considering constant velocity and neglecting the rotary inertia of the beam, eq (13a) becomes:

$$EI_y w'''' + c_y \dot{w} + m\ddot{w} = M(g - \ddot{w} - \upsilon^2 w'' - 2\upsilon\dot{w}')\cdot\delta(x-\alpha) + J_M \ddot{w}'\delta'(x-\alpha) \quad \textbf{(28)}$$

where υ is the constant velocity: $\dot{s} = \upsilon = \text{const.}$, and $\ddot{s} = 0$

We shall seek a solution in the form:

$$w(x,t) = \sum_n W_n(x)\cdot T_n(t) \quad \textbf{(29)}$$

where $W_n(x)$ is the nth shape function of the freely vibrating beam, and $T_n(t)$ is the corresponding time function, to be determined. Introducing the expression for w(x,t) from eq (29) into eq (28), we obtain:

$$\left.\begin{aligned} & EI_y \sum_n W_n'''' \; T_n + c_y \sum_n W_n \dot{T}_n + m\sum_n W_n \ddot{T}_n = \\ & = M\left[g - \sum_n W_n \ddot{T}_n - \; \upsilon^2 \sum_n W_n'' T_n - 2\upsilon \sum_n W_n' \dot{T}_n \right]\cdot\delta(x-\alpha) + J_M \sum_n W_n' \ddot{T}_n \cdot \delta'(x-\alpha) \end{aligned}\right\} \quad \textbf{(30)}$$

and since eq (18) is valid for the freely vibrating beam, eq (30) becomes:

$$\left.\begin{aligned} & m\sum_n \omega_n^2 W_n T_n + c_y \sum_n W_n \dot{T}_n + m\sum_n W_n \ddot{T}_n = \\ & = Mg\cdot\delta(x-\alpha) - M\left[\sum_n W_n \; \ddot{T}_n + \; \upsilon^2 \sum_n W_n'' T_n + 2\upsilon \sum_n W_n' \; \dot{T}_n \right]\cdot\delta(x-\alpha) + J_M \sum_n W_n' \ddot{T}_n \cdot \delta'(x-\alpha) \end{aligned}\right\} \quad \textbf{(31)}$$

Multiplying eq (31) by $W_\kappa\left(\kappa \neq n\right)$, integrating the outcome from 0 to ℓ and taking into account the orthogonality conditions, we obtain:

$$\left.\begin{aligned} & \ddot{T}_\kappa + \frac{c_y}{m}\dot{T}_\kappa + \omega_n^2 T_\kappa = \frac{1}{m\cdot\int_0^\ell W_\kappa^2 \cdot dx}\cdot\Big\{ MgW_\kappa(\alpha) - \\ & -MW_\kappa(\alpha)\left[\sum_{n=1}^{\infty} W_n(\alpha)\ddot{T}_n + \upsilon^2 \sum_{n=1}^{\infty} W_n''(\alpha)T_n + 2\upsilon\sum_{n=1}^{\infty} W_n'(\alpha)\dot{T}_n \right] - J_M\left[W_\kappa'(\alpha)\sum_{n=1}^{\infty} W_n'(\alpha)\ddot{T}_n + W_\kappa(\alpha)\sum_{n=1}^{\infty} W_n''(\alpha)\ddot{T}_n \right]\Big\} \end{aligned}\right\} \quad \textbf{(32)}$$

It is obvious that a closed form solution of eq (32) is not possible to obtain. For the case of a simply supported beam, and given that:

$$W_n(\alpha) = \sin\frac{n\pi\alpha}{\ell} = \sin\frac{n\pi\upsilon t}{\ell} = \sin\Omega_n t \quad , \quad \Omega_n = \frac{n\pi\upsilon}{\ell}$$

eq (32), can be written as:

$$\left.\begin{aligned} & \ddot{T}_\kappa + \frac{c_y}{m}\dot{T}_\kappa + \omega_\kappa^2 T_\kappa = \frac{2Mg}{m\ell}\cdot\sin\Omega_\kappa t - \frac{2M}{m\ell}\cdot\sin\Omega_\kappa t\cdot\left[\sum_n \sin\Omega_\kappa t\cdot\ddot{T}_n - \upsilon^2\sum_n \Omega_n^2\cdot\sin\Omega_n t\cdot T_n + \right. \\ & \left. + \; 2\upsilon\sum_n \Omega_n\cdot\cos\Omega_n t\cdot\dot{T}_n \right] - \frac{2J_M}{m\ell}\cdot\left[\Omega_\kappa\cdot\cos\Omega_\kappa t\cdot\sum_n \Omega_n\cdot\cos\Omega_n t\cdot\ddot{T}_n - \sin\Omega_\kappa t\cdot\sum_n \Omega_n^2\cdot\sin\Omega_n t\cdot\ddot{T}_n \right] \end{aligned}\right\} \quad \textbf{(33)}$$

We shall employ an approximate technique. We choose as starting function the expression given by eq (23). Then eq (33) becomes:

$$\left.\begin{aligned}
&\ddot{T}_\kappa + \frac{c_y}{m}\dot{T}_\kappa + \omega_n^2\ T_\kappa = \frac{2Mg}{m\ell}\cdot\sin\Omega_\kappa t - \\
&-\left(\frac{2M}{m\ell}\right)^2 g\cdot\sin\Omega_\kappa t\cdot\sum_n \frac{\Omega_n^2}{\omega_n^2-\Omega_n^2}\cdot\sin\Omega_n t\cdot\left(-\sin\Omega_n t+\frac{\omega_n}{\Omega_n}\cdot\sin\omega_n t\right)+ \\
&+\left(\frac{2M}{m\ell}\right)^2 g\cdot\sin\Omega_\kappa t\cdot\sum_n \frac{\Omega_n^2}{\omega_n^2-\Omega_n^2}\cdot\sin\Omega_n t\cdot\left(\sin\Omega_n t-\frac{\Omega_n}{\omega_n}\cdot\sin\omega_n t\right) \\
&-\left(\frac{2M}{m\ell}\right)^2 g\cdot\sin\Omega_\kappa t\cdot\sum_n \frac{\Omega_n^2}{\omega_n^2-\Omega_n^2}\cdot\cos\Omega_n t\cdot(\cos\Omega_n t-\cos\omega_n t) \\
&-\frac{4MJ_M g}{(m\ell)^2}\Omega_\kappa\cos\Omega_\kappa t\cdot\sum_n \frac{\Omega_n^3}{\omega_n^2-\Omega_n^2}\cdot\cos\Omega_n t\left(-\sin\Omega_n t+\frac{\omega_n}{\Omega_n}\sin\omega_n t\right)+ \\
&+\frac{2MJ_M g}{(m\ell)^2}\cdot\sin\Omega_\kappa t\cdot\sum_n \frac{\Omega_n^4}{\omega_n^2-\Omega_n^2}\cdot\sin\Omega_n t\cdot\left(-\sin\Omega_n t+\frac{\omega_n}{\Omega_n}\sin\omega_n t\right)
\end{aligned}\right\} \tag{34}$$

The solution of eq (34), due to the initial conditions, is given by Duhamel's integral:

$$T_\kappa(t) = \frac{g}{\overline{\omega}_\kappa}\cdot\int_0^t\left[F_M(\tau)+F_J(\tau)\right]\cdot e^{-\beta_z\cdot(t-\tau)}\cdot\cos\overline{\omega}_\kappa\left(t-\tau\right)\cdot d\tau \tag{34a}$$

with β_z and $\overline{\omega}_\kappa$ from eq (26b), where:

$$\left.\begin{aligned}
F_M(t) &= \frac{2M}{m\ell}\cdot\sin\Omega_\kappa t + \\
&+\left(\frac{2M}{m\ell}\right)^2\sum_n A_n\left[\sin(\Omega_\kappa-2\Omega_n)t+\sin(\Omega_\kappa+2\Omega_n)t\right]+ \\
&+\left(\frac{2M}{m\ell}\right)^2\sum_n B_n\left[\sin(\Omega_\kappa-\Omega_n+\omega_n)t+\sin(\Omega_\kappa+\Omega_n-\omega_\kappa)t\right]+ \\
&+\left(\frac{2M}{m\ell}\right)^2\sum_n \Gamma_n\left[\sin(\Omega_\kappa-\Omega_n-\omega_n)t+\sin(\Omega_\kappa+\Omega_n+\omega_n)t\right] \\
\text{and:}& \\
A_n &= -\frac{\Omega_n^2}{\omega_n^2-\Omega_n^2}\ ,\quad B_n = -\frac{\Omega_n(\omega_n-\Omega_n)}{4\omega_n(\omega_n+\Omega_n)}\ ,\quad \Gamma_n = \frac{\Omega_n(\omega_n+\Omega_n)}{4\omega_n(\omega_n-\Omega_n)}
\end{aligned}\right\} \tag{34b}$$

and also the following terms, which correspond to the rotary inertia of the moving load and are usually omitted:

$$\left.\begin{aligned}
F_J(t) &= \frac{2MJ_M}{(m\ell)^2}\cdot\sum_n\frac{A_n}{2}\cdot\sin\Omega_\kappa t + \\
&+\frac{2MJ_M}{(m\ell)^2}\cdot\sum_n\left[\Delta_n\sin(\Omega_\kappa-2\Omega_n)t+E_n\sin(\Omega_\kappa+2\Omega_n)t\right]+ \\
&+\frac{2MJ_M}{(m\ell)^2}\cdot\sum_n Z_n\left[\sin(\Omega_\kappa+\Omega_n-\omega_n)t-\sin(\Omega_\kappa+\Omega_n+\omega_n)t\right]+ \\
&+\frac{2MJ_M}{(m\ell)^2}\cdot\sum_n H_n\left[\sin(\Omega_\kappa-\Omega_n-\omega_n)t+\sin(\Omega_\kappa-\Omega_n+\omega_n)t\right] \\
\text{and:}\quad \Delta_n &= \frac{\Omega_n^3(\Omega_\kappa+\Omega_n)}{4(\omega_n^2-\Omega_n^2)}\ ,\quad E_n = \frac{\Omega_n^3(\Omega_n-\Omega_\kappa)}{4(\omega_n^2-\Omega_n^2)} \\
Z_n &= \frac{\omega_n\Omega_n^2(\Omega_\kappa+\Omega_n)}{4(\omega_n^2-\Omega_n^2)}\ ,\quad H_n = \frac{\omega_n\Omega_n^2(\Omega_n-\Omega_\kappa)}{4(\omega_n^2-\Omega_n^2)}
\end{aligned}\right\} \tag{34c}$$

DYNAMIC INFLUENCE LINES

Nondimensionalization of the Governing Equations

The motion of a beam due to the passage of a load $P = M \cdot g$, moving with constant velocity υ, is given by:

$$w(x,t) = \sum_n W_n(x) \cdot T_n(t) \tag{35}$$

where, if we currently neglect the influence of mass forces of the load, the time function $T_n(t)$ is given by eq (27), while for the simply supported beam the expressions for W_n and ω_n are given by eq (19).

In order to be able to compare the deflections of a beam due to a moving load, we shall consider the static deflection of the same beam due to a load P applied at the midspan. If w_0 is the maximum static deflection, we know that:

$$w_0 = \frac{P\ell^3}{48EI_y} = \frac{2,02936 \cdot P\ell^3}{\pi^4 EI_y}$$

and if ω_1 is the fundamental frequency of the beam, we obtain from eq (19b) for n=1

$$\omega_1^2 = \frac{\pi^4 EI_y}{m\ell^4}$$

and hence:

$$w_0 = \frac{2,02936 \cdot P}{m\ell\omega_1^2} \tag{36}$$

We notice also that, the coefficient β_z has frequency dimensions ($\sec^{-1}$). Then, we set:

$$\left.\begin{aligned} &\kappa = \frac{\beta_z}{\omega_1} \quad \text{and hence: } \beta_z = \kappa \cdot \omega_1 \\ &\lambda = \frac{\omega_\upsilon}{\omega_1} \quad \text{and hence: } \omega_\upsilon = \lambda \cdot \omega_1 \\ &\text{where: } \omega_\upsilon = \frac{\pi \cdot \upsilon}{\ell} = \frac{\Omega_n}{n} \quad \text{and hence: } \Omega_n = n\lambda\omega_1 \\ &\text{and finally: } \bar{\omega}_n^2 = \omega_n^2 - \beta_z^2 = n^4\omega_1^2 - \kappa^2\omega_1^2 = (n^4 - \kappa^2)\,\omega_1^2 \end{aligned}\right\} \tag{37a}$$

Also, for nondimensionalization of eq (27) we set:

$$\xi = \frac{x}{\ell} \;, \quad \tau = \frac{t}{T_1} \;, \quad \text{where: } T_1 = \frac{2}{\pi} \cdot \sqrt{\frac{m\ell^4}{EI_y}} \;, \quad \bar{\upsilon} = \frac{\upsilon \cdot T_1}{\ell} \tag{37b}$$

We define the critical velocity υ_{cr}, which is the velocity required for the load P, to cross the beam length in time equal to the fundamental half-period. That is:

$$\upsilon_{cr} = \frac{2\ell}{T_1} \tag{37c}$$

Then, we shall have:

$$\left.\begin{aligned}
&\lambda=\frac{\omega_{\upsilon}}{\omega_1}=\frac{\pi\upsilon}{\ell\omega_1}=\frac{\pi\upsilon}{\ell\dfrac{2\pi}{T_1}}=\frac{\upsilon}{\dfrac{2\ell}{T_1}}=\frac{\upsilon}{\upsilon_{cr}}\\
&\overline{\upsilon}=\upsilon\frac{T_1}{\ell}=\frac{2\upsilon}{\upsilon_{cr}}=2\lambda\ ,\quad \beta_z t=\kappa\omega_1\tau T_1=2\kappa\pi\tau\\
&\overline{\omega}_n\cdot t=\omega_1\sqrt{n^4-\kappa^2}\cdot\tau\ T_1=2\pi\tau\sqrt{n^4-\kappa^2}\ ,\quad \tau=\frac{\xi}{\overline{\upsilon}}=\frac{\xi}{2\lambda}\\
&\frac{\Omega_n}{\overline{\omega}_n}=\frac{n\lambda\omega_1}{\omega_1\sqrt{n^4-\kappa^2}}=\frac{n\lambda}{\sqrt{n^4-\kappa^2}}\\
&\Omega_n t=n\lambda\omega_1\tau\ T_1=2n\pi\lambda\tau
\end{aligned}\right\}\tag{37d}$$

and hence, eq (27) can be written in a nondimensional form:

$$\left.\begin{aligned}
w(\xi,\tau)=\frac{w(x,t)}{w_0}&=0,9855\sum_n\frac{1}{n^2[n^2(n^2-\lambda^2)^2+4\lambda^2\kappa^2]}\cdot\\
&\cdot\left\{e^{-2\kappa\pi\tau}\left[2n\kappa\lambda\cos\left[2\pi\tau\sqrt{n^4-\kappa^2}\right]+\right.\right.\\
&\left.+\frac{n\lambda(2\kappa^2-n^4+n^2\lambda^2)}{\sqrt{n^4-\kappa^2}}\cdot\sin\left[2\pi\tau\sqrt{n^4-\kappa^2}\right]\right]+\\
&\left.\left[-2n\kappa\lambda\cos\left[2n\pi\lambda\tau\right]+n^2(n^2-\lambda^2)\cdot\sin[2n\pi\lambda\tau]\right]\right\}\cdot\sin(n\pi\xi)
\end{aligned}\right\}\tag{38}$$

In the case where damping is neglected, *i.e.* $\beta_z=0$, $\kappa=0$, eq (38) gives:

$$\left.\begin{aligned}
w(\xi,\tau)=0,9855\sum_n\frac{1}{n^4(n^2-\lambda^2)}\cdot&\left\{\ n\lambda(n^2+\lambda^2)\cdot\sin[2n^2\pi\tau]+\right.\\
&\left.+n^2(n^2-\lambda^2)\cdot\sin[2n\pi\lambda\tau]\ \right\}\cdot\sin(n\pi\xi)
\end{aligned}\right\}\tag{39}$$

Finally, for the case of the static influence line with $\overline{\upsilon}=0$ and given that: $2\lambda\tau=\xi_P$, we will have:

$$w(\xi)=\frac{w(x)}{w_0}=0,9855\sum_n\frac{1}{n^4}\cdot\sin(n\pi\xi)\cdot\sin(n\pi\xi_P)\tag{40}$$

where ξ_P denotes the load position, while ξ is the position to which the influence line is referred to.

When damping is very small ($\beta<<1$), then the terms κ and κ^2 can be omitted since their influence is negligible. In this case, some summation terms for $\lambda=j$ in Equations (38) or (39), may take indeterminate form.

If the above equations are determined though Laplace transformation, it can be shown that these terms acquire real value that tends to zero and hence, they can be omitted and finally eq (38) can be written as:

$$\left.\begin{aligned}
w(\xi,\tau)=0,9855\sum_{\substack{n=1\\ n\neq j}}^{\infty}\frac{1}{n^4(n^2-\lambda^2)}\cdot&\left\{\ n\lambda(n^2+\lambda^2)\cdot\sin[2n^2\pi\tau]+\right.\\
&\left.+n^2(n^2-\lambda^2)\cdot\sin[2n\pi\lambda\tau]\ \right\}\cdot\sin(n\pi\xi)
\end{aligned}\right\}\tag{39a}$$

Finally, note that when the moving load leaves the beam, this will continue vibrating freely according to the formula:

$$\left.\begin{aligned}
&w(\xi,\tau) = 0{,}9855\sum_n e^{-2\kappa\pi\cdot(\tau-\tau_0)}\cdot\Big(A_n \sin\Big[2\pi(\tau-\tau_0)\sqrt{n^4-\kappa^2}\Big] + \\
&\qquad + B_n \cos\Big[2\pi(\tau-\tau_0)\sqrt{n^4-\kappa^2}\Big]\Big)\cdot\sin(n\pi\xi) \\
&\text{where } A_n \text{ and } B_n \text{ are:} \\
&A_n = \frac{\left.\dfrac{dw}{d\tau}\right|_{\tau=\tau_0} + \beta_z\cdot B_n}{2\pi\sqrt{n^4-\kappa^2}} \\
&B_n = w(\xi,\tau_0) \text{ and } \tau_0 = \frac{1}{2\lambda}
\end{aligned}\right\} \quad \textbf{(41)}$$

The above constants A_n, B_n have been obtained by the deformation conformity conditions, and velocities at time $\tau=\tau_0$. Next, we shall examine the influence of the velocity and damping on the beam deformation, but always for values $\kappa<1$, *i.e.* for feeble damping, since for $\kappa>1$ the dynamic behavior of a bridge depends also from other factors (see "Critical Damping" in page 71 and Chapter 7).

The Influence of Velocity

In Fig. **3** one can see the dynamic influence lines of the displacement at the midspan of a simply supported bridge for various values of λ, with damping factor $\beta=0{,}02$ (feeble damping), and $\kappa=0{,}60$. It is obvious that:

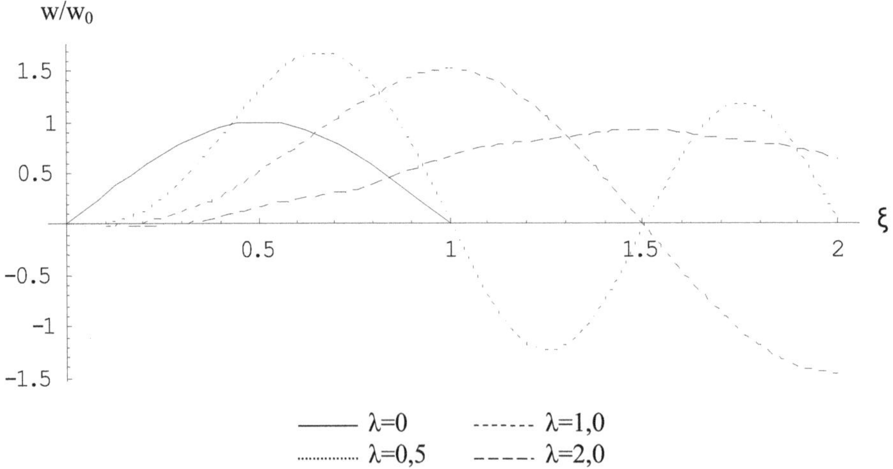

Figure 3: Dynamic influence lines of a simply supported bridge for various λ

a. For velocity values less than the critical velocity $\lambda<1$, the maximum dynamic displacement at the midspan of the bridge occurs when the load is within the span of the bridge.

b. For velocity equal to the critical one $\lambda=1$, the maximum dynamic displacement occurs at the instant when the load is exiting the bridge.

c. For supercritical velocities $\lambda>1$, the maximum dynamic displacement occurs after the load has left the bridge which is vibrating freely.

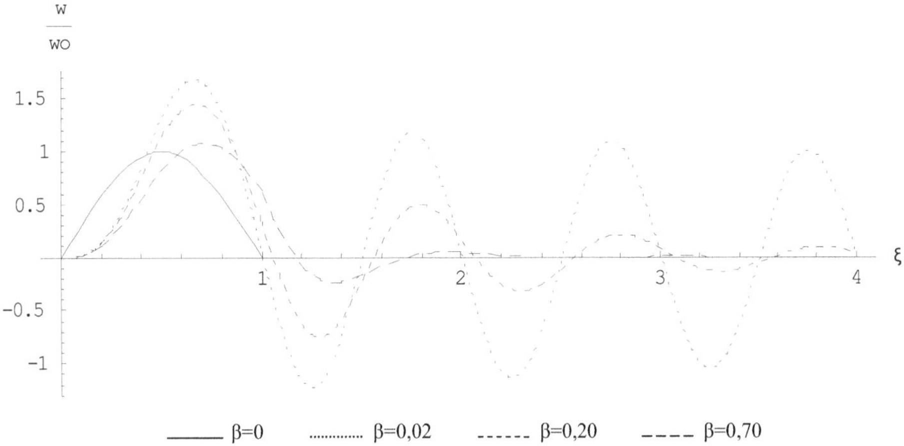

Figure 4: Dynamic influence lines of a simply supported bridge for various β

The Influence of Damping

In Fig. **4**, one can see the dynamic influence lines of the midspan deflection of a simply supported bridge due to a load moving with velocity λ=0,5 and for damping factors: β=0,02, β=0,20 and β=0,70.

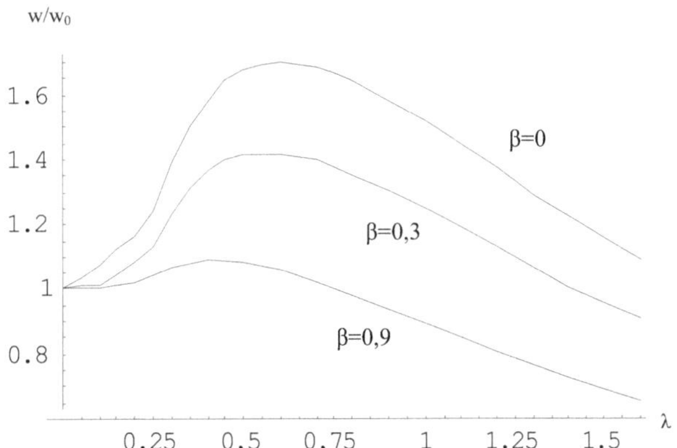

Figure 5: Dynamic influence lines of a simply supported bridge for various β

We initially observe the effect of damping on the beam deformation which is decreasing when the damping factor β increases. We also observe the influence of damping on the final half-period value: $\overset{*}{T} = \frac{2 \cdot \pi}{\overline{\omega}}$, which increases as the damping factor increases.

Finally, the diagrams in Fig. **5**, show the characteristic influence of the velocity λ on the nondimensional midspan deflection of the beam for three sub-critical damping values : β=0,2, β=0,3, β=0,9.

Eccentric Moving Loads [33], [34]

In this case, the basic relations in paragraphs of pages 61 and 64 are valid, which for an eccentric load P_z, moving at distance e from the bridge axis of symmetry (see Fig. **6**), can be written as:

$$\left.\begin{aligned} &EI_y w'''' + c_y \dot{w} + m\ddot{w} = P_z \delta(x-\alpha) \\ &EI_z \upsilon'''' + EI_z z_M \theta'''' + c_\theta \dot{\upsilon} + m\ddot{\upsilon} = 0 \\ &EC_S \theta'''' - EI_z z_M \upsilon'''' + c_\theta \dot{\theta} - GJ_d \theta'' + \Theta_x \ddot{\theta} = eP_z \delta(x-\alpha) \end{aligned}\right\} \qquad \textbf{(42a,b,c)}$$

Seeking solutions in the form:

$$\left.\begin{aligned} w(x,t) &= \sum_n W_n(x)\cdot T_n(t) \\ \upsilon(x,t) &= \sum_n V_n(x)\cdot P_n(t) \\ \theta(x,t) &= \sum_n \Phi_n(x)\cdot P_n(t) \end{aligned}\right\} \quad \textbf{(43a,b,c)}$$

where W_n, V_n, Φ_n are the shape functions of the beam in Fig. **6**, already determined in Chapter 3, while $T_n(t)$ and $P_n(t)$ are the time functions to be determined.

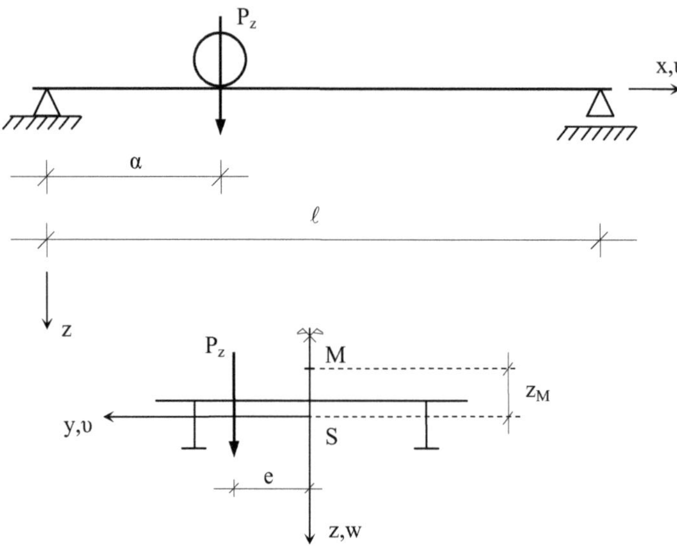

Figure 6: Eccentric moving load

We observe that deformations $\upsilon(x,t)$ and $\theta(x,t)$ will have a common time function, since they depend to each other. Introducing eqs (43) into eqs (42), and following a procedure similar to the one in paragraph of page 88 and taking into account the orthogonality conditions in relations (115) and (120) of chapter 3, we finally obtain:

$$\left.\begin{aligned} T_n(t) &= \frac{P_z}{m\overline{\omega}_{yn}\int_0^\ell W_n^2 dx}\int_0^t e^{-\beta_z\cdot(t-\tau)}\cdot W_n(\upsilon\tau)\ \ \sin\overline{\omega}_{yn}(t-\tau)d\tau \\ P_n(t) &= \frac{eP_z}{\overline{\omega}_{\theta n}\left(m\int_0^\ell V_n^2 dx+\Theta_x\int_0^\ell \Phi_n^2 dx\right)}\cdot\int_0^t e^{-\beta_\theta\cdot(t-\tau)}\Phi_n(\upsilon\tau)\cdot\sin\overline{\omega}_{\theta_n}(t-\tau)d\tau \\ \text{where: } \overline{\omega}_{y_n} &= \sqrt{\omega_{y_n}^2-\beta_z^2}\ ,\quad \beta_z=\frac{c_y}{2m} \\ \overline{\omega}_{\theta_n} &= \sqrt{\omega_{\theta_n}^2-\beta_\theta^2}\ ,\quad \beta_\theta=\frac{c_z}{2m} \end{aligned}\right\} \quad \textbf{(44a,b,c,d,e,f)}$$

In Figs. **7** and **8** one can see the dynamic influence lines of a bridge with $\ell = 50\text{m}$ subjected to an eccentric load P_z=1500kp moving at e=1,00 and e=2,00m respectively. The material and structural properties of the bridge are: J_y=0,227m^4, J_z=1,654m^4, z_M=1,376m, C_M=0,88m^6, $m = 2.512\text{kp}\cdot\text{m}/\sec^2$, and $\Theta_M = 1.476{,}6\text{kp}\cdot\sec^2$.

In order to avoid the cumbersome procedure for solution and determination of the shape functions described in chapter 3, we can follow the Galerkin method setting:

$$\left.\begin{aligned} V_n(x) &= \sum_{\kappa} c_{\kappa} \cdot \overline{Y}_{\kappa}(x) \\ \Phi_n(x) &= \sum_{\kappa} d_{\kappa} \cdot \overline{\theta}_{\kappa}(x) \end{aligned}\right\} \quad \textbf{(45a,b)}$$

where $\overline{Y}_{\kappa}$ and $\overline{\theta}_{\kappa}$ are taken from paragraph "**The Special Case z$_M$=0**" on page 69 of chapter 3 with z_M=0, that are:

$$\left.\begin{aligned} \overline{Y}_{\kappa}(x) &= \sin\frac{n\pi x}{\ell} \\ \overline{\theta}_{\kappa}(x) &= \sin\lambda_{1\kappa}x - \frac{\sin\lambda_{1\kappa}\ell}{\sinh\lambda_{2\kappa}\ell}\cdot\sinh\lambda_{2\kappa}x \end{aligned}\right\} \quad \textbf{(46a,b)}$$

while: λ_{1n}, λ_{2n} are taken from eqs 96 of chapter 3.

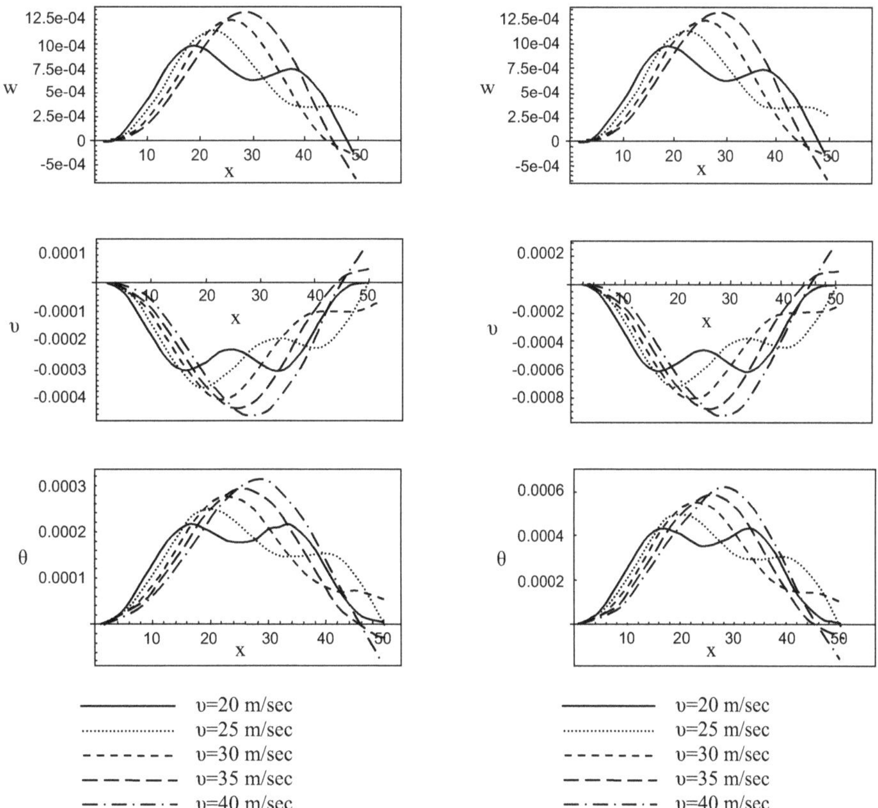

Figure 7: Dynamic influence lines

Figure 8: Dynamic influence lines

Introducing eqs (46) into the equations for free vibration (*i.e.* eqs.42b,c with zero second parts and c_0=0), we multiply successively the first equation by the $\overline{Y}_1, \overline{Y}_2, ..., \overline{Y}_{\kappa}$, and the second by $\overline{\theta}_1, \overline{\theta}_2, ..., \overline{\theta}_{\kappa}$ and we integrate the outcomes from 0 to ℓ. This way, we obtain a linear system of $2 \cdot \kappa$ equations, from which we can determine the eigenfrequencies $\omega_{\theta n}$ and the corresponding shape functions V_n, Φ_n.

Moving Uniform Load

Let us consider the uniform load in Fig. **9**, moving on a bridge, the length L of which is sufficiently longer than the length ℓ of the bridge.

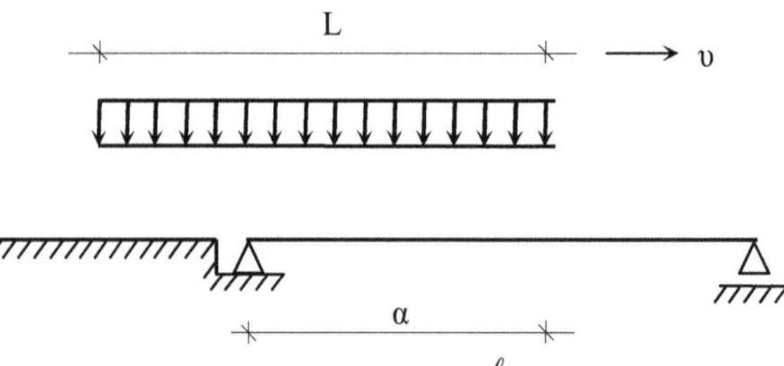

Figure 9: Uniform load moving on a bridge

Assuming that the bridge at time t=0, when the load is entering the span, is at rest, we will have the following equation of motion:

$$EI_y w'''' + c_y \dot{w} + m\ddot{w} = p_z(x) \tag{47}$$

Seeking a solution in the form:

$$w(x,t) = \sum_n W_n(x) \cdot T_n(t) \tag{48}$$

where: $W_n(x) = \sin\frac{n\pi x}{\ell}$. Following the well-known procedure, we arrive finally at the following differential equation:

$$\ddot{T}_n + \frac{c_y}{m}\dot{T}_n + \omega_n^2 T_n = \frac{\int_0^{\alpha} p_z(x) W_n(x) dx}{m\int_0^{\ell} W_n^2 dx}$$

which for the usual case where $p_z(x)=p=$ const., gives:

$$\ddot{T}_n + \frac{c_y}{m}\dot{T}_n + \omega_n^2 T_n = \frac{2p}{nm\pi}\cdot\left(1-\cos\frac{n\pi\alpha}{\ell}\right) = \frac{2p}{nm\pi}\cdot\left(1-\cos\Omega_n t\right) \tag{49}$$

where: $\Omega_n = \frac{n\pi\upsilon}{\ell}$ and $\alpha = \upsilon\cdot t$.

The solution to the above Equation (49), with initial conditions: $w(x,0) = \dot{w}(x,0) = 0$, is given by the Duhamel's integral:

$$\left.\begin{aligned} T_n(t) = \frac{2p}{nm\pi\bar{\omega}_n}\cdot\Bigg\{ & \frac{\bar{\omega}_n - e^{-\beta_z\cdot t}\cdot\left(\bar{\omega}_n\cos\bar{\omega}_n t + \beta_z\cdot\sin\bar{\omega}_n t\right)}{\beta_z^2+\bar{\omega}_n^2} + \\ & + \frac{e^{-\beta_z\cdot t}\left[\bar{\omega}_n(\beta_z^2+\bar{\omega}_n^2-\Omega_n^2)\cos\bar{\omega}_n t + \beta_z(\beta^2+\bar{\omega}_n^2+\Omega_n^2)\sin\bar{\omega}_n t\right]}{(\beta_z^2+\bar{\omega}_n^2)^2+2(\beta_z^2-\bar{\omega}_n^2)\cdot\Omega_n^2+\Omega_n^4} - \\ & - \frac{2\beta_z\bar{\omega}_n\Omega_n\sin\Omega_n t + \bar{\omega}_n\cdot(\beta_z^2+\bar{\omega}_n^2-\Omega_n^2)\cdot\cos\Omega_n t}{(\beta_z^2+\bar{\omega}_n^2)^2+2(\beta_z^2-\bar{\omega}_n^2)\Omega_n^2+\Omega_n^4}\Bigg\} \\ \text{with: } \beta_z = \frac{c_y}{2m}\ ,\ \ & \bar{\omega}_n = \sqrt{\omega_n^2-\beta_z^2} \end{aligned}\right\} \tag{50}$$

Moving Harmonic Load

Let us consider again the simply supported bridge shown in Fig. **10**, on which a harmonic load $P_0 \cos\Omega_p t$ is moving with constant velocity. This problem applies to the case of railway bridges and in combination with the case analyzed in the previous paragraph, can be considered as a train model moving on a bridge with constant velocity υ.

The equation of motion of the bridge is:

$$EI_y w'''' + c_y \dot{w} + m\ddot{w} = P_0 \cdot \cos\Omega_p t \cdot \delta(x-\alpha) \qquad \textbf{(51)}$$

Seeking a solution in the form of eq (48) and following the well-known procedure, we arrive at the following differential equation:

$$\ddot{T}_n + \frac{c_y}{m}\dot{T}_n + \omega_n^2 T_n = \frac{2P_0}{m\ell} \cdot \cos\Omega_p t \cdot \sin\Omega_n t \qquad \textbf{(52)}$$

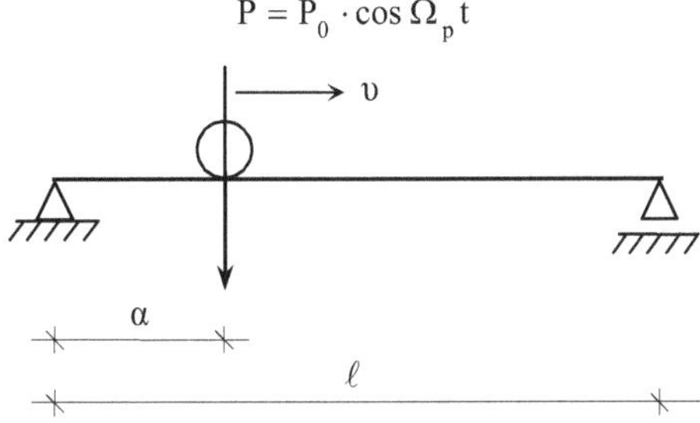

Figure 10: Harmonic load moving on a bridge

The solution to the above equation is given by Duhamel's integral:

$$\left.\begin{aligned}
T_n(t) = {} & \frac{P_0}{m\ell\overline{\omega}_n} \cdot \frac{1}{A^4 + 2A^2(\beta_z^2 - \overline{\omega}_n^2) + (\beta_z^2 + \overline{\omega}_n^2)^2} \cdot \\
& \cdot \Big\{ Ae^{-\beta_z \cdot t}\left[(A^2 + \beta_z^2 - \overline{\omega}_n^2) \cdot \sin\overline{\omega}_n t + 2\beta_z\overline{\omega}_n \cos\overline{\omega}_n t\right] + \\
& + \overline{\omega}_n \cdot \left[(-A^2 + \beta_z^2 + \overline{\omega}_n^2) \cdot \sin At - 2A\beta_z \cdot \cos At\right] \Big\} + \\
& + \frac{P_0}{m\ell\overline{\omega}_n} \cdot \frac{1}{B^4 + 2B^2 \cdot (\beta_z^2 - \overline{\omega}_n^2) + (\beta_z^2 + \overline{\omega}_n^2)^2} \cdot \\
& \cdot \Big\{ Be^{-\beta_z \cdot t}\left[(B^2 + \beta_z^2 - \overline{\omega}_n^2) \cdot \sin\overline{\omega}_n t + 2\beta_z\overline{\omega}_n \cos\overline{\omega}_n t\right] + \\
& + \overline{\omega}_n \cdot \left[(-B^2 + \beta_z^2 + \overline{\omega}_n^2) \cdot \sin Bt - 2B\beta_z \cos Bt\right] \Big\} \\
\text{where: } & A = \frac{\Omega_n + \Omega_p}{2} \quad \text{and} \quad B = \frac{\Omega_n - \Omega_p}{2}
\end{aligned}\right\} \qquad \textbf{(53)}$$

THE INFLUENCE OF SEVERAL FACTORS

The Influence of the Real Vehicle

Let us consider a simply supported bridge, as the one shown in Fig. **11**, on which a vehicle with mass M is moving with constant velocity υ. The position of the vehicle is defined by the distance α of the front wheel axis from the left end of the bridge, which is also the load entrance position in the bridge. Thus, at time t the distance α from the left end of the bridge will be $\upsilon \cdot t (= \alpha)$, where time t starts at the instance when the front wheels enter the bridge.

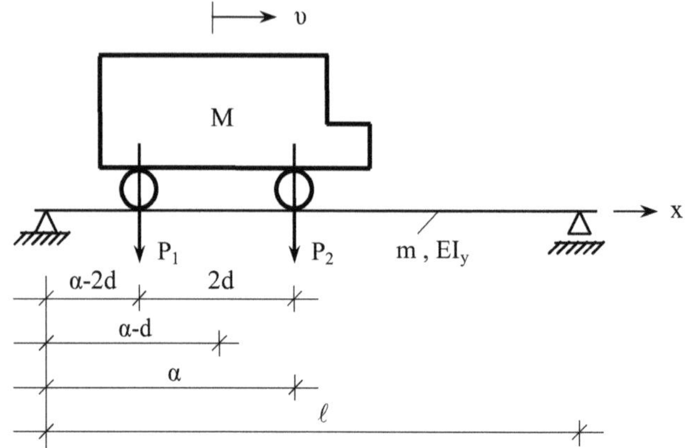

Figure 11: Model of a real vehicle moving on a bridge

Neglecting the effects due to the mass of the vehicle on the bridge vibration, the corresponding equation of motion is:

$$EI_y w'''' + c_y \dot{w} + m\ddot{w} = P_1 \cdot \delta(x-\alpha) + P_2 \cdot \delta(x-\alpha+2d) \tag{54}$$

Seeking a solution in the form:

$$w(x,t) = \sum_n W_n(x) T_n(t) \tag{55}$$

and following the usual procedure, we arrive at the following differential equation for the time function:

$$\ddot{T}_n + \frac{c_y}{m}\dot{T}_n + \omega_n^2 T_n = \frac{2P_1}{m\ell} W_n(\upsilon t) + \frac{2P_2}{m\ell} W_n(\upsilon t - 2d) \tag{56}$$

The solution to the above equation is given by the 2nd Duhamel's integral and is as follows:

$$\left.\begin{aligned} T_n(t) &= \frac{2P_1}{m\ell\bar{\omega}_n} \cdot \int_0^t e^{-\beta_z(t-\tau)} W_n(\upsilon\tau) \sin\bar{\omega}_n(t-\tau) d\tau + \\ &+ \frac{2P_2}{m\ell\bar{\omega}_n} \cdot \int_{t_1}^t e^{-\beta_z(t-\tau)} W_n(\upsilon\tau - 2d) \cdot \sin\bar{\omega}_n(t-\tau) d\tau \\ \text{where: } & \beta_z = \frac{c_y}{2m} \;,\; \bar{\omega}_n = \sqrt{\omega_n^2 - \beta_z^2} \;,\; t_1 = \frac{2d}{\upsilon} \end{aligned}\right\} \tag{57}$$

Performing the integrations in the above equation, we get:

$$\left.\begin{aligned} T_n(t) &= \frac{2P_1}{m\ell} \cdot \frac{1}{\omega_n^2 + 2(\beta_z^2 - \bar{\omega}_n^2)\Omega_n^2 + \Omega_n^4} \cdot \\ &\cdot \left\{ e^{-\beta_z \cdot t} \left[2\beta_z \Omega_n \cdot \cos\bar{\omega}_n t + \frac{\Omega_n}{\bar{\omega}_n} \cdot (\beta_z^2 - \bar{\omega}_n^2 + \Omega_n^2) \cdot \sin\bar{\omega}_n t \right] + \right. \\ &\left. + \left[-2\beta_z \Omega_n \cos\Omega_n t + (\omega_n^2 - \Omega_n^2) \cdot \sin\Omega_n t \right] \right\} \\ \text{for } & 0 \le t \le \frac{2d}{\upsilon} \end{aligned}\right\} \tag{58}$$

$$
\left.
\begin{aligned}
&T_n(t)=\frac{2P_1}{m\ell}\cdot\frac{1}{\omega_n^2+2(\beta_z^2-\overline{\omega}_n^2)\Omega_n^2+\Omega_n^4}\cdot\\
&\quad\cdot\left\{e^{-\beta_z\cdot t}\left[2\beta_z\Omega_n\cos\overline{\omega}_n t+\frac{\Omega_n}{\overline{\omega}_n}\cdot(\beta_z^2-\overline{\omega}_n^2+\Omega_n^2)\cdot\sin\overline{\omega}_n t\right]+\right.\\
&\quad\left.+\left[-2\beta_z\Omega_n\cos\Omega_n t+(\omega_n^2-\Omega_n^2)\cdot\sin\Omega_n t\right]\right\}+\\
&\quad+\frac{2P_2}{m\ell}\cdot\frac{1}{\omega_n^2+2(\beta_z^2-\overline{\omega}_n^2)\Omega_n^2+\Omega_n^4}\cdot\left\{e^{-\beta_z\cdot(t-t_1)}\left[2\beta_z\Omega_n\cos\overline{\omega}_n(t-t_1)+\right.\right.\\
&\quad\left.+\frac{\Omega_n}{\overline{\omega}_n}\cdot(\beta_z^2-\overline{\omega}_n^2+\Omega_n^2)\sin\overline{\omega}_n(t-t_1)\right]+\\
&\quad\left.+\left[-2\beta_z\Omega_n\cos\Omega_n(t-t_1)+(\omega_n^2-\Omega_n^2)\sin\Omega_n(t-t_1)\right]\right\}\\
&\text{for }\ \frac{2d}{\upsilon}\le t\le\frac{\ell}{\upsilon}\ ,\quad\text{where: }\ \Omega_n=\frac{n\pi\upsilon}{\ell}\ ,\quad t_1=\frac{2d}{\upsilon}
\end{aligned}
\right\}\qquad\textbf{(58)}
$$

For the usual case where: $P_1=P_2=\frac{Mg}{2}=\frac{P}{2}$ and following the nondimensionalization procedure of paragraph "Nondimensionalization of the governing equations" in page 92, the solution to eq (47) is:

$$
\left.
\begin{aligned}
&w_1(\xi,\tau)=\frac{w(x,t)}{w_0}=\sum_n\frac{0{,}4928}{n^2[n^2(n^2-\lambda^2)+4\lambda^2\kappa^2]}\cdot\\
&\quad\cdot\left\{e^{-2\kappa\pi\tau}\cdot\left[2n\kappa\lambda\cos\left(2\pi\tau\sqrt{n^4-\kappa^2}\right)+\right.\right.\\
&\quad\left.+\frac{n\lambda(2\kappa^2-n^4+n^2\lambda^2)}{\sqrt{n^4-\kappa^2}}\cdot\sin\left(2\pi\tau\sqrt{n^4-\kappa^2}\right)\right]+\\
&\quad\left.+\left[-2n\kappa\lambda\cdot\cos(2n\pi\lambda\tau)+n^2(n^2-\lambda^2)\cdot\sin(2n\pi\lambda\tau)\right]\right\}\cdot\sin(n\pi\xi)\\
&\text{for: }\ 0\le\tau\le\overset{*}{\tau}\quad\text{where: }\ \overset{*}{\tau}=\frac{\overline{d}}{2\lambda}\ ,\quad\overline{d}=\frac{2d}{\ell}
\end{aligned}
\right\}\qquad\textbf{(59a)}
$$

$$
\left.
\begin{aligned}
&w(\xi,\tau)=w_1(\xi,\tau)+\sum_n\frac{0{,}4928}{n^2[n^2(n^2-\lambda^2)+4\lambda^2\kappa^2]}\cdot\\
&\quad\cdot\left\{e^{-2\kappa\pi\cdot(\tau-\overset{*}{\tau})}\cdot\left[2n\kappa\lambda\cdot\cos\left(2\pi(\tau-\overset{*}{\tau})\sqrt{n^4-\kappa^2}\right)+\right.\right.\\
&\quad\left.+\frac{n\lambda(2\kappa^2-n^4+n^2\lambda^2)}{\sqrt{n^4-\kappa^2}}\cdot\sin\left(2\pi(\tau-\overset{*}{\tau})\sqrt{n^4-\kappa^2}\right)\right]+\\
&\quad+\left[-2n\kappa\lambda\cdot\cos\left[2n\pi\lambda(\tau-\overset{*}{\tau})\right]+\right.\\
&\quad\left.\left.+n^2(n^2-\lambda^2)\cdot\sin\left[2n\pi\lambda(\tau-\overset{*}{\tau})\right]\right]\right\}\sin(n\pi\xi)\\
&\text{for: }\ \overset{*}{\tau}\le\tau\le\frac{\xi}{2\lambda}\quad\text{where: }\ \overset{*}{\tau}=\frac{\overline{d}}{2\lambda}\ ,\quad\overline{d}=\frac{2d}{\ell}
\end{aligned}
\right\}\qquad\textbf{(59b)}
$$

The diagrams shown in Fig. **12**, are the influence lines for $P_1 = P_2 = \frac{Mg}{2} = \frac{P}{2}$ and wheel axes to bridge length ratios d/ ℓ = 0,10, 0,20, 0,30, while with solid line is shown the influence line for a concentrated load P corresponding to d/ ℓ = 0.

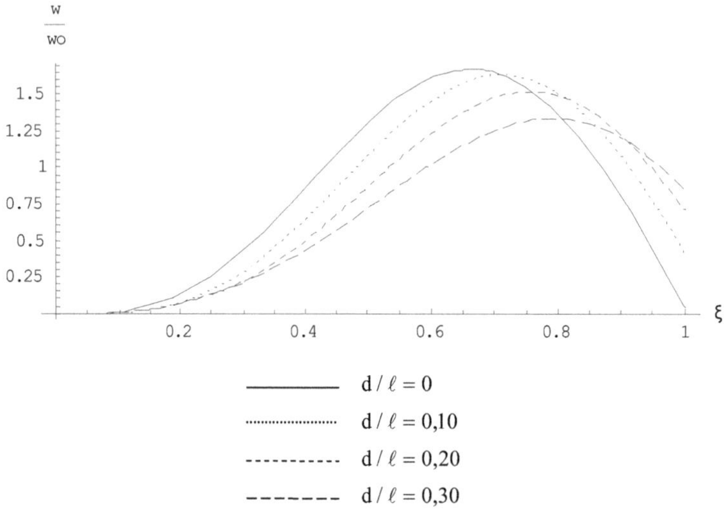

Figure 12: Influence lines for various values of d / ℓ = 0,10

The Influence of the Vehicle's Mass [35]

Considering only the inertia force due to the mass M of a moving vehicle, eq (28) is written as:

$$EI_y w'''' + \frac{c_y}{m}\dot{w} + m\ddot{w} = M\left(g - \ddot{w}\right)\cdot\delta\left(x - \alpha\right) \tag{60}$$

Seeking a solution in the form:

$$w\left(x,t\right) = \sum_n W_n(x)T_n(t) \tag{61}$$

Introducing eq (61) into eq (60) and following the usual procedure, we arrive at the following differential equation for the time function T_n.

$$\ddot{T}_n + \frac{c_y}{m}\dot{T}_n + \omega_n^2 T_n = \frac{2M}{m\ell}(g\sin\Omega_n t - \sin\Omega_n t\cdot\sum_\rho \sin\Omega_\rho t\cdot\ddot{T}_\rho) \tag{62}$$

In order to solve the above equation, we set as a starting function the expression given in eq (23). Then, we will have:

$$\ddot{T}_n + \frac{c_y}{m}\dot{T}_n + \omega_n^2 T_n = \frac{2Mg}{m\ell}\cdot\sin\Omega_n t - \left(\frac{2M}{m\ell}\right)^2 g\sin\Omega_n t\cdot\sum_\rho \frac{\Omega_\rho^2}{\omega_\rho^2 - \Omega_\rho^2}\cdot\left(-\sin\Omega_\rho t + \frac{\omega_\rho}{\Omega_\rho}\cdot\sin\omega_\rho t\right)\cdot\sin\Omega_\rho t$$

or after some manipulation:

$$\left.\begin{aligned}
&\ddot{T}_n+\frac{c_y}{m}\dot{T}_n+\omega_n^2 T_n=\frac{2Mg}{m\ell}\cdot Z_n(t)\quad \text{where:}\\
&Z_n(t)=\frac{2Mg}{m\ell}\cdot\sin\Omega_n t+\left(\frac{2M}{m\ell}\right)^2 g\cdot\sum_\rho\frac{\Omega_\rho^2}{2(\omega_\rho^2-\Omega_\rho^2)}\cdot\sin\Omega_n t\\
&-\left(\frac{2M}{m\ell}\right)^2 g\cdot\sum_\rho\frac{\Omega_\rho^2}{4(\omega_\rho^2-\Omega_\rho^2)}\cdot\left[\sin(\Omega_n+2\Omega_\rho)t+\sin(\Omega_n-2\Omega_\rho)t\right]\\
&-\left(\frac{2M}{m\ell}\right)^2 g\cdot\sum_\rho\frac{\omega_\rho\Omega_\rho}{4(\omega_\rho^2-\Omega_\rho^2)}\cdot\left[\sin(\Omega_n+\omega_\rho-\Omega_\rho)t+\sin(\Omega_n-\omega_\rho+\Omega_\rho)t\right]\\
&+\left(\frac{2M}{m\ell}\right)^2 g\cdot\sum_\rho\frac{\omega_\rho\Omega_\rho}{4(\omega_\rho^2-\Omega_\rho^2)}\cdot\left[\sin(\Omega_n+\omega_\rho+\Omega_\rho)t+\sin(\Omega_n-\omega_\rho-\Omega_\rho)t\right]
\end{aligned}\right\}\quad \textbf{(63a,b)}$$

The solution to the above is given by the Duhamel's integral:

$$T_n(t)=\frac{2Mg}{m\ell}\cdot\frac{1}{\bar{\omega}_n}\int_0^t e^{-\beta_z\cdot(t-\overset{*}{t})}Z_n(\overset{*}{t})\cdot\sin\bar{\omega}(t-\overset{*}{t})d\overset{*}{t}\quad \textbf{(64)}$$

Due to eq (36) and the nondimensionalization procedure of paragraph "Nondimensionalization of the governing equations" in page 92 we get:

$$w_0=\frac{2,02936}{\omega_1^2}\cdot\frac{Mg}{m\ell}\ \text{and hence:}\ \bar{w}=\frac{w}{w_0}=\frac{1}{w_0}\sum_n W_n T_n=\frac{1}{2,02936\cdot\frac{Mg}{m\ell}}\cdot\omega_1^2\cdot\sum_n W_n\cdot T_n$$

Given that:

$$\frac{\omega_1}{\bar{\omega}_n}\omega_1 dt=\frac{\omega_1}{\omega_1\sqrt{n^4-\kappa^2}}\cdot\left(\omega_1 T_1 dt\right)=\frac{2\pi}{\sqrt{n^4-\kappa^2}}d\tau$$

we have:

$$\bar{w}(\xi,\tau)=$$

$$0,9855\cdot\sum_{\substack{n\\ n\neq\kappa}}\left\{\left(\int_0^\tau\frac{2\pi}{\sqrt{n^4-\kappa^2}}\cdot e^{-2\pi\kappa(\tau-\overset{*}{\tau})}\bar{Z}_n(\overset{*}{\tau})\cdot\sin\left[2\pi(\tau-\overset{*}{\tau})\sqrt{n^4-\kappa^2}\right]d\overset{*}{\tau}\right)\cdot\sin(n\pi\xi)\right\}\quad \textbf{(65a)}$$

with:

$$\left.\begin{aligned}
&\bar{Z}_n(\tau)=\sin(2n\pi\lambda\tau)+2\frac{M}{m\ell}\cdot\sum_{\substack{\rho=1\\ \rho\neq\lambda}}^{\infty}\frac{\lambda^2}{2(\rho^2-\lambda^2)}\sin(2n\pi\lambda\tau)-\\
&-2\frac{M}{m\ell}\cdot\sum_{\substack{\rho=1\\ \rho\neq\lambda}}^{\infty}\frac{\lambda^2}{4(\rho^2-\lambda^2)}\Big[\sin\left[2(n+2\rho)\pi\lambda\tau\right]+\sin\left[2(n-2\rho)\pi\lambda\tau\right]\Big]-\\
&-2\frac{M}{m\ell}\cdot\sum_{\substack{\rho=1\\ \rho\neq\lambda}}^{\infty}\frac{\rho^3\lambda}{4\rho^2(\rho^2-\lambda^2)}\Big[\sin[2(n\lambda+n^2-\rho\lambda)\pi\tau]+\\
&+\sin[2(n\lambda-n^2+\rho\lambda)\pi\tau]\Big]+\\
&+2\frac{M}{m\ell}\cdot\sum_{\substack{\rho=1\\ \rho\neq\lambda}}^{\infty}\frac{\rho^3\lambda}{4\rho^2(\rho^2-\lambda^2)}\Big[[\sin 2(n\lambda+n^2+\rho\lambda)\pi\tau]+\\
&+\sin[2(n\lambda-n^2-\rho\lambda)\pi\tau]\Big]
\end{aligned}\right\}\quad \textbf{(65b)}$$

In Fig. **13**, one can see the displacement ratio $\frac{w_M}{w_P}$ versus the mass ratio $\frac{M}{m\ell}$ (mass of the moving load to the total mass of the bridge), for various values of the velocity parameter $\lambda = \frac{\upsilon}{\upsilon_{cr}}$. In the above, w_M is the midspan deflection, taking into account the mass M of the moving load P and w_P is the same deflection without the influence of mass M.

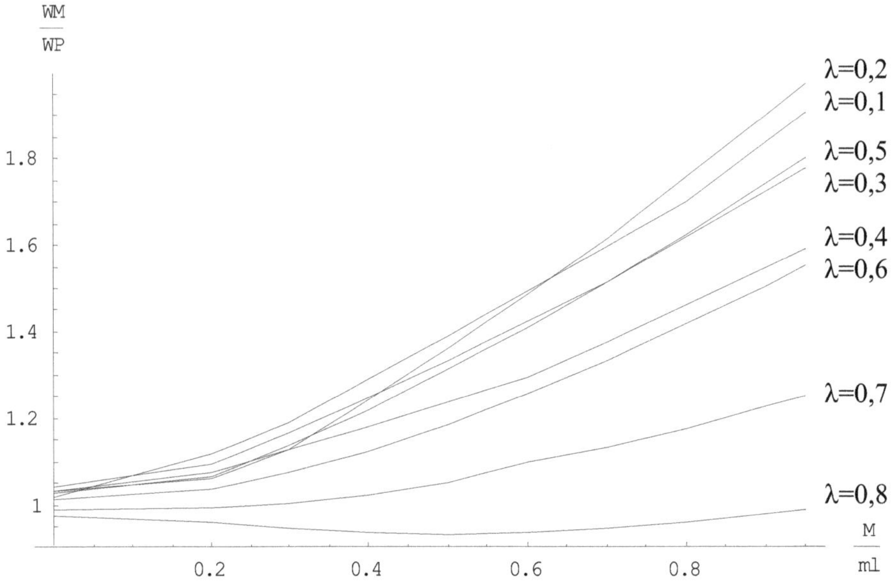

Figure 13: Deflection vs mass ratios for various values of λ

In Fig. **14**, one can see the ratio $\frac{w_M}{w_P}$ versus the velocity parameter λ for various values of the mass ratio $\bar{M} = \frac{M}{m\ell}$.

It is worthy to observe the disturbance near the value $\lambda \cong 0,40$ since for values $\lambda > 0,50$, the influence of mass M is gradually diminished. This is due to the fact that the maximum deflection occurs when the load has exited the bridge.

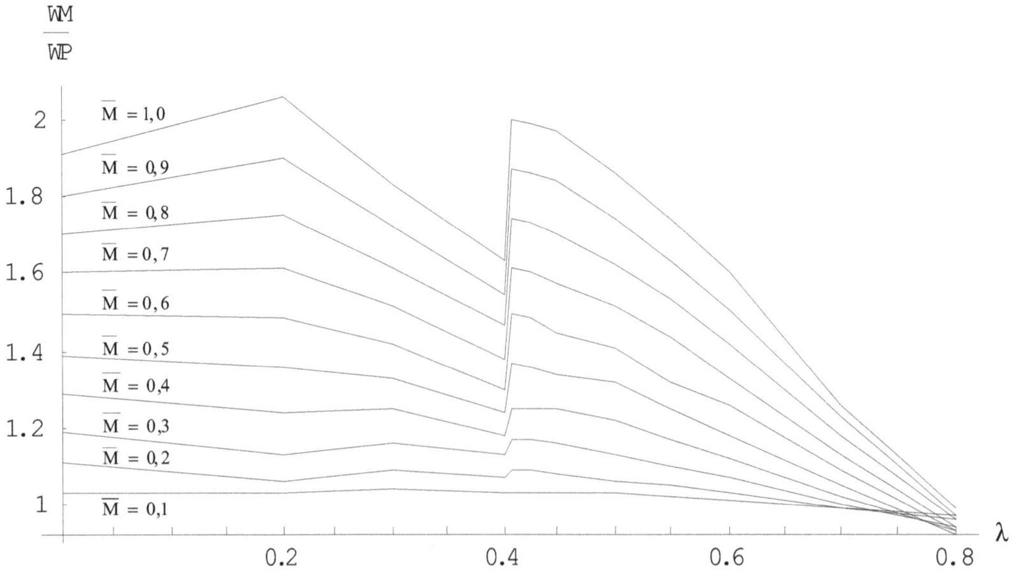

Figure 14: Deflection vs λ for various values of the mass ratio

Secondary Effects of the Vehicle Mass [36,37,38,39]

In this section, we shall determine and study the additional influence of the mass inertia secondary forces, *i.e.* the Coriolis force and the centrifugal force.

In contrast to the classical theory, we assume that the inertia forces due to the mass of the vehicle are negligible (as shown in page 102). as well as the mass inertia secondary forces, such as the Coriolis and the centrifugal force.

Setting J_M=0 into eq (13), we arrive at the following equation:

$$EI_y w'''' + c_y \dot{w} + m\ddot{w} = M(g - \ddot{w} - \upsilon^2 w'' + 2\upsilon \dot{w}') \cdot \delta(x - \alpha) \qquad \textbf{(66)}$$

Following the previously described procedure in the preceding sections, we obtain the following:

$$\left.\begin{aligned} &\ddot{T}_n + \frac{c_y}{m}\dot{T}_n + \omega_n^2 T_n = \frac{2Mg}{m\ell}\cdot Z_n(t) \qquad \text{which gives:} \\ &T_n(t) = \frac{2M_g}{m\ell}\cdot\frac{1}{\overline{\omega}}\int_0^t e^{-\beta_z(t-\overset{*}{t})}\cdot Z_n(\overset{*}{t})\cdot \sin\overline{\omega}_n(t-\overset{*}{t})\cdot d\overset{*}{t} \end{aligned}\right\} \qquad \textbf{(67a,b)}$$

and $Z_n(t)$:

$$\left.\begin{aligned} Z_n(t) = & \\ = \sin\Omega_n t &- \frac{2M}{m\ell}\cdot\sum_\rho \frac{\Omega_\rho}{\omega_\rho^2 - \Omega_\rho^2}\left[\sin(\Omega_n - 2\Omega_\rho)t + \sin(\Omega_n - 2\Omega_\rho)t\right] \\ &- \frac{2M}{m\ell}\cdot\sum_\rho \frac{\Omega_\rho(\omega_\rho - \Omega_\rho)}{4\omega_\rho(\omega_\rho + \Omega_\rho)}\left[\sin(\Omega_n - \Omega_\rho + \omega_\rho)t + \sin(\Omega_n + \Omega_\rho - \omega_\rho)t\right] \\ &+ \frac{2M}{m\ell}\cdot\sum_\rho \frac{\Omega_\rho(\omega_\rho + \Omega_\rho)}{4\omega_\rho(\omega_\rho - \Omega_\rho)}\left[\sin(\Omega_n - \Omega_\rho - \omega_\rho)t + \sin(\Omega_n + \Omega_\rho + \omega_\rho)t\right] \end{aligned}\right\} \qquad \textbf{(67c)}$$

Performing nondimensionalization in the above expression, we finally obtain:

$$\overline{w}(\xi,\tau) = 0{,}9855 \sum_{\substack{n \\ n\neq\kappa}}\left\{\left(\int_0^\tau \frac{2\pi}{\sqrt{n^4-\kappa^2}} e^{-2\pi\kappa(\tau-\overset{*}{\tau})}\overline{Z}_n(\overset{*}{\tau})\sin\left[2\pi(\tau-\overset{*}{\tau})\sqrt{n^4-\kappa^2}\right]d\overset{*}{\tau}\right)\sin(n\pi\xi)\right\} \qquad \textbf{(68a)}$$

$$\left.\begin{aligned} \text{with}\quad \overline{Z}_n(\tau) = &\sin(2n\pi\lambda\tau) \\ &- \frac{2M}{m\ell}\cdot\sum_{\substack{\rho \\ \rho\neq\lambda}}\frac{\lambda^2}{4(\rho^2-\lambda^2)}\cdot\left[\sin\left[2(n+2\rho)\pi\lambda\tau\right] + \sin\left[2(n-2\rho)\pi\lambda\tau\right]\right] \\ &- \frac{2M}{m\ell}\cdot\sum_\rho \frac{\lambda(\rho-\lambda)}{4\rho(\rho+\lambda)}\cdot\left[\sin\left[2(n\lambda - n^2 + \rho\lambda)\pi\tau\right] + \sin\left[2(n\lambda + n^2 - \rho\lambda)\pi\tau\right]\right] \\ &+ \frac{2M}{m\ell}\cdot\sum_{\substack{\rho \\ \rho\neq\lambda}}\frac{\lambda(\rho+\lambda)}{4\rho(\rho-\lambda)}\cdot\left[\sin\left[2(n\lambda - n^2 - \rho\lambda)\pi\tau\right] + \sin\left[2(n\lambda + n^2 + \rho\lambda)\pi\tau\right]\right] \end{aligned}\right\} \qquad \textbf{(68b)}$$

In Fig. **15**, one can see the ratio $\frac{w_c}{w_M}$, versus the parameter λ for various values of the ratio $\overline{M} = \frac{M}{m\ell}$. In the above ratio, w_c is the midspan nondimensional deflection of the bridge, when the influence of inertia forces due to mass M as well as the Coriolis and centrifugal forces, while w_M is the same deflection taking into account only the inertia forces due to mass M according to the paragraph "The influence of the vehicle's mass" in page 102.

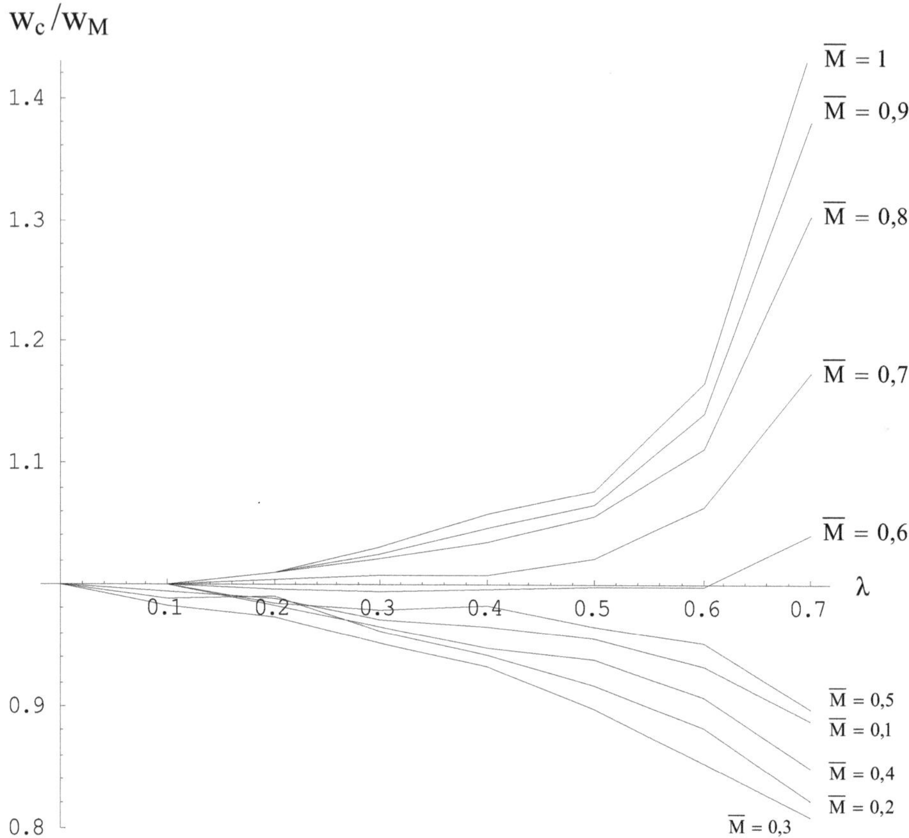

Figure 15: Deflection ratio versus λ for various values of the mass ratio

By comparing Figs. **14** and **15** we observe that for $\lambda < 0{,}5$, the influence of the inertia forces due to the mass is significantly bigger than that the one due to Coriolis and centrifugal forces. For $\lambda > 0{,}7$ the Coriolis and centrifugal forces become important, while the influence of the mass M forces is practically negligible.

Finally, one can also note that for $\frac{M}{m\ell} < 0{,}5$, the Coriolis and centrifugal forces are acting beneficially, while for $\frac{M}{m\ell} > 0{,}5$ this action is reversed.

The Real Conditions for Mass Suspension

In practice, any moving mass M, is supported through a suspension system on wheels with total mass m_0 (which can be neglected). This system is composed from the following two basic parts: the elastic springs with constant k_p and the dampers with constant c_p. The mathematical model adopted in this section is shown in Fig. **16**.

The basic assumptions for the mathematical bridge – moving mass model shown in Fig. **16** are the following:

1. We can study the bridge behavior by considering a single beam with equivalent bending stiffness, inertia, mass and damping.
2. The real vehicle, with two or more wheel axes, is modeled by the equivalent single degree-of-freedom system shown in Fig. **16**.
3. As the vehicle is moving on the bridge, the wheel with mass m_0, is always in contact with the road surface (without bouncing or impact on surface anomalies etc.), which is assumed completely smooth

without irregularities or roughness.

4. The load (vehicle) is moving with constant velocity υ.

At time t=0, the load enters the bridge from the left, while the bridge is initially at rest, *i.e.*:

$$w(x,0) = \dot{w}(x,0) = 0 \qquad \textbf{(69)}$$

When the load, *i.e.* the system of masses M and m_0, is on the left end of the bridge (t = 0), the spring is statically deformed by $\frac{P}{k_p} = \frac{Mg}{k_p}$. Taking axis ε as the reference system for measuring the motion of the mass M (Fig. **16**), which passes through the static equilibrium position for masses M, m_0 and applying the Newton's second Law, we obtain:

$$M\ddot{z} = P - k_p\left(\frac{P}{k_p} + z - w(\alpha)\right) - c_p\left(\dot{z} - \dot{w}(\alpha)\right) \text{ or } M\ddot{z} + c_p\dot{z} + k_p z = k_p w(\alpha) + c_p\dot{w}(\alpha) \qquad \textbf{(70)}$$

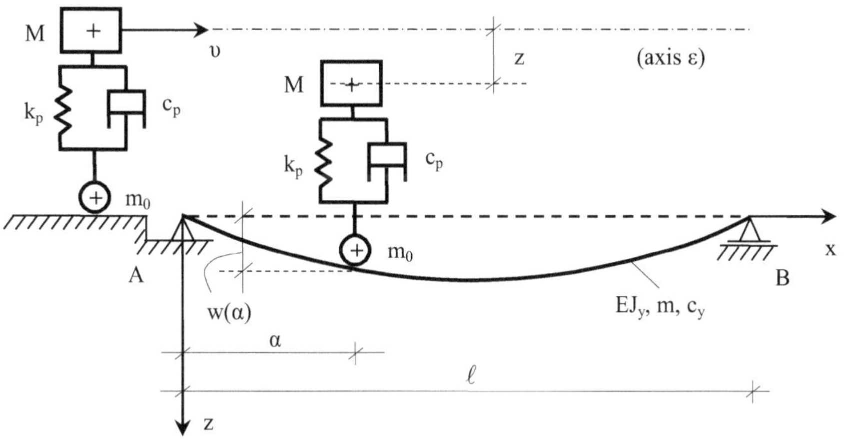

Figure 16: Suspended mass M moving on a bridge

The above relation can be also written as follows:

$$\left.\begin{aligned} &\ddot{z} + 2\beta_p\dot{z} + \omega_p^2 z = \Phi(t) \\ \text{where:}\quad &\Phi(t) = \omega_p^2 w(\alpha) + 2\beta_p\dot{w}(\alpha) \\ &\beta_p = \frac{c_p}{2M} \quad , \quad \omega_p^2 = \frac{k_p}{M} \end{aligned}\right\} \qquad \textbf{(71a,b,c,d)}$$

The solution to the above differential Equation (71a) is given by the following relation:

$$\left.\begin{aligned} &z(t) = \frac{1}{\overline{\omega}_p}\cdot\int_0^t e^{-\beta_p(t-\tau)}\Phi(\tau)\cdot\sin\overline{\omega}_p(t-\tau)d\tau \\ &\text{with :}\qquad \overline{\omega}_p = \sqrt{\omega_p^2 - \beta_p^2} \end{aligned}\right\} \qquad \textbf{(72a,b)}$$

Applying the Leibnitz formula, according to which the following function: $f(x) = \int_{\alpha(x)}^{\beta(x)} F(x,t)\cdot dt$ can be differentiated as follows:

$$\frac{df(x)}{dx} = F\big(x,\beta(x)\big)\cdot\frac{d\beta(x)}{dx} - F\big(x,\alpha(x)\big)\cdot\frac{d\alpha(x)}{dx} + \int\limits_{\alpha(x)}^{\beta(x)} \frac{\partial F(x,t)}{\partial x}\cdot dt$$

eq (72a) gives successively:

$$\left.\begin{aligned} &\dot{z}(t) = \frac{1}{\overline{\omega}_p}\cdot\int\limits_0^t e^{-\beta_p\cdot(t-\tau)}\Phi(\tau)\left[\overline{\omega}_p\cos\overline{\omega}_p(t-\tau) - \beta_p\sin\overline{\omega}_p(t-\tau)\right]\cdot d\tau \\ &\ddot{z}(t) = \\ &\quad = \Phi(t) + \frac{1}{\overline{\omega}_p}\cdot\int\limits_0^t e^{-\beta_p\cdot(t-\tau)}\Phi(\tau)\left[(\beta_p^2 - \overline{\omega}_p^2)\sin\overline{\omega}_p(t-\tau) - 2\beta_p\overline{\omega}_p\cos\overline{\omega}_p(t-\tau)\right]\cdot d\tau \end{aligned}\right\} \tag{73a,b}$$

The total load applied on the bridge at position x=α, is: $M(g-\ddot{z}) + m_0\big(g - \ddot{w}(\alpha)\big)$ and hence, the equation of motion for the bridge will be :

$$EI_y w'''' + c_y\dot{w} + m\ddot{w} = \left\{(M+m_0)g - M\ddot{z} - m_0\ddot{w}(\alpha)\right\}\cdot\delta(x-\alpha) \tag{74}$$

We seek a solution in the form:

$$\left.\begin{aligned} &w(x,t) = \sum_n W_n(x)T_n(t) \\ &\text{where:}\quad W_n(x) = \sin\frac{n\pi x}{\ell} \end{aligned}\right\} \tag{75a,b}$$

Following the procedure described in the preceding sections, we easily conclude to the following differential equation for the time function:

$$\ddot{T}_n + \frac{c_y}{m}\dot{T}_n + \omega_n^2 T_n = \frac{2M}{m\ell}\cdot\left\{\frac{M+m_0}{M}g - \ddot{z} - \frac{m_0}{M}\ddot{w}(\alpha)\right\}\cdot\sin\Omega_n t \quad \text{where:}\quad \left.\Omega_n = \frac{n\pi\upsilon}{\ell}\right\} \tag{76a,b}$$

The solution to eq (76a) is given by the Duhamel's integral:

$$\left.\begin{aligned} &T_n(t) = \frac{2M}{m\ell}\cdot\frac{1}{\overline{\omega}_n}\cdot\int\limits_0^t e^{-\beta_z(t-\tau)}Z_n(\tau)\sin\overline{\omega}_n(t-\tau)\cdot d\tau \\ &\text{with:}\quad Z_n(t) = \left\{\frac{M+m_0}{M}\cdot g - \ddot{z}(t) - \frac{m_0}{M}\cdot\ddot{w}(\alpha,t)\right\}\cdot\sin\Omega_n t \\ &\qquad \beta_z = \frac{c_y}{2m} \quad , \quad \overline{\omega}_n = \sqrt{\omega_n^2 - \beta_z^2} \end{aligned}\right\} \tag{77a,b,c,d}$$

and hence:

$$w(x,t) = \sum_n W_n(x)T_n(t) \tag{78}$$

Using the nondimensional sizes in eq (37a,b,c,d), as well as the following nondimensional quantities:

$$\sigma = \frac{\beta_\rho}{\omega_1} \quad , \quad \rho = \frac{\omega_p}{\omega_1} \quad , \quad \vartheta = \frac{m_0}{M} \quad , \quad \bar{M} = \frac{M}{m\ell} \tag{79}$$

and as starting function the one given by eq (23), we can obtain the nondimensional expression of eq (78).

The diagrams in Figs **17** and **18**, show the influence of springs and dampers on the deflection of the bridge, respectively. More specifically, in Fig. **17** one can see the effect of springs for ρ=0, 1.0, 1.5 and velocity parameter value λ=0.5 and σ=0.3.

In Fig. **18**, one can see the effect of dampers for σ=0.3 και σ=0.9 and for λ=0.5 and ρ=1.

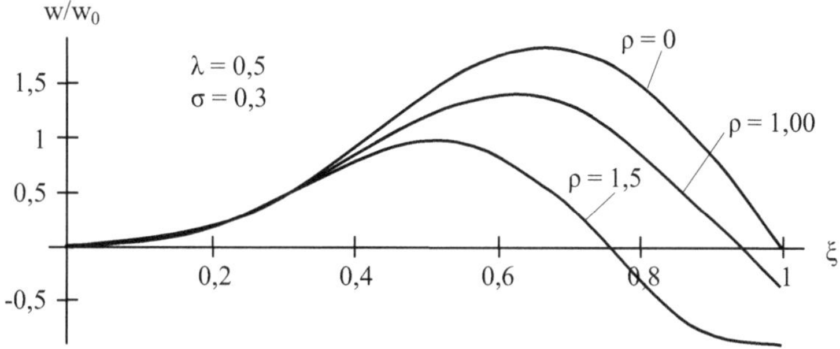

Figure 17: The effect of springs for λ=0.5 and σ=0.3 and various values of ρ

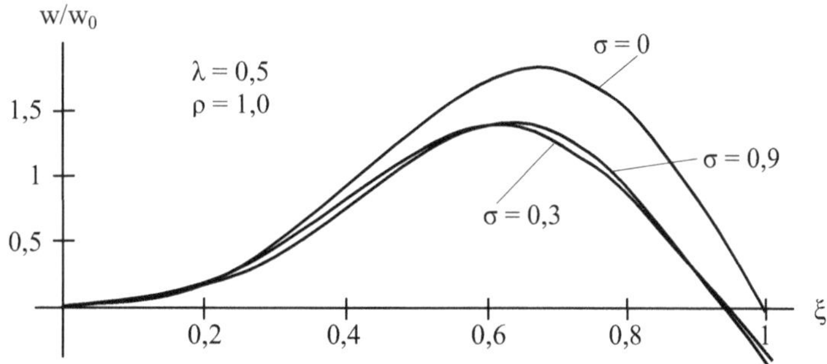

Figure 18: The effect of dampers for λ=0.5 and ρ=1 and various values of σ

The Influence of Deck-Surface Irregularities [40]

The influence of a rough deck-surface on the dynamic response of a bridge depends on various factors. The irregularities on a deck-surface, may be due the roughness of the deck-surface, but also due to random or on purpose existing anomalies for traffic reasons.

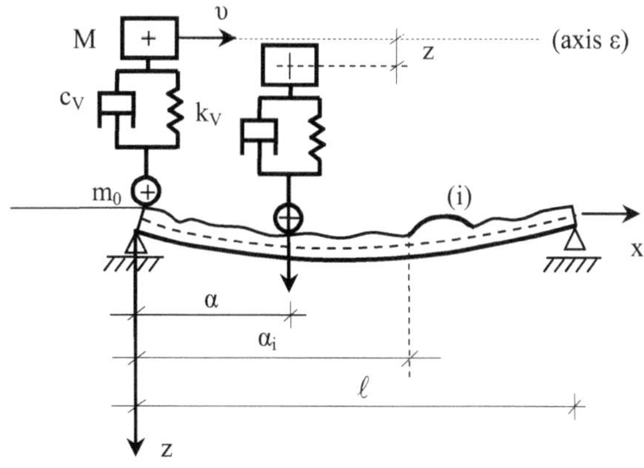

Figure 19: Model of a vehicle moving on a bridge with irregular surface

The adopted bridge model as well as the one of the moving vehicle, used for the analysis, is shown in Fig. **19**. The corresponding equations of motion of the bridge are:

$$\left.\begin{aligned} & EI_y w'''' + c_b \dot{w} + m\ddot{w} = \\ & = [(M+m_0)g - M\ddot{z} - m_0\ddot{w}(\alpha)]\cdot\delta(x-\alpha) + P_{imp}\cdot\delta(x-\alpha_i) \\ & \\ & M\ddot{z} + c_p\dot{z} + k_p z = k_p[w(\alpha)+r(\alpha)] + c_p\dot{w}(\alpha) \end{aligned}\right\} \quad \textbf{(80a,b)}$$

where:

M	is the mass of the moving vehicle
k_v	is the spring constant of the moving vehicle
c_v	is the damping constant of the moving vehicle
z	is the distance of mass M from axis (ε)
r(x)	is a function which expresses the surface of the irregularity type on the deck
α_i	is the position of the irregularity (i) starting point
P_{imp}	is the impact force generated by the impact of the m_0 on the irregularity or the surface roughness.

The Roughness of the Deck-Surface

It is obvious that the roughness of the bridge-deck can be considered as a special case of a deck with irregularities and hence, it can be studied using the set of eqs (80).

The roughness of the deck consists actually from numerous small irregularities of the coating (usually asphalt – or even concrete), which are of course randomly distributed. Their modeling *via* ordinarily repeated bumps etc. leads to more unfavorable (conservative) results from dynamical point of view.

Many important experimental investigations have been conducted by numerous researchers such as Wang and Huang [41], who presented the response spectrum of the roughness intensity, or the modeling of a harmonically varied surface presented by Cheng et al [42], as well as the studies by Fafard [43] and others.

The term "roughness" usually covers also the deck surface anomalies in the form of repeated waves with small amplitudes (in the range between 5 to 40cm).

For the last case, the most commonly used method [49, 50] is the one that is presented next, besides some deficiencies.

The random of surface anomalies in a bridge, can be modeled by employing a random process, in which we consider a random dynamical spectrum of the intensity (dynamic spectral density PSD), that is given by the following relation:

$$S_r(\omega_s) = A_r(\omega_{s\theta})\left(\frac{\omega_s}{\omega_{s\theta}}\right)^{-2} \quad \textbf{(81)}$$

where $S_r(\omega_s)$ is the dynamic spectral density of the roughness, A_r is the roughness coefficient given by the standards ISO [50], $\omega_{s\theta}$ is the discontinuity frequency that is usually taken equal to $1/2\pi$ or according to ISO is taken equal to 0.10 and ω_s is the domain frequency.

From the above dynamic spectral density of the roughness and using the Fourier transformation, we obtain a mathematical expression for the surface shape given by:

$$r(x) = \sum_{k=1}^{N} \sqrt{4A_r \left(\frac{2k\pi}{L_c \omega_{s0}} \right)^{-2} \frac{2\pi}{L_c}} \cdot \cos(\omega_{sk} x - \varphi_k) \tag{82}$$

where $\omega_{sk} = k \cdot \Delta\omega_s$, $\Delta\omega_s = 2\pi / L_c$, L_c is double the length of the bridge and φ_k is a random number between 0 and 2π, that can be obtained from the Monte-Carlo method (estimation).

The above function r(x) is introduced into eq (80b) in order to proceed with the analysis.

The main deficiency of the above method is that the wheel of the vehicle is considered to follow the whole irregularity curve (since it is modeled as a point) (see also Fig. **20**), which is not correct since the dimensions of the wheel (Fig. **22b**) do not allow the modeling of the above case for irregularity waves between 5 to 40cm (Fig. **20**). Finally, the roughness is modeled by a cosine wave surface, which results to a more favorable and actually non-applicable dynamic response of the system.

In order to construct a fully analytical model [44] for the roughness of the deck surface and the response of the bridge under moving loads, we shall model the roughness using a saw-type surface, such as the one shown in Fig. **21**. Even though this model is very simplified in comparison to the dynamic spectral density model (PSD functions) used widely in the relative studies, it is very realistic and representative for a big number of surface coatings in use. For an ideally smooth surface coating it is d=θ=0.

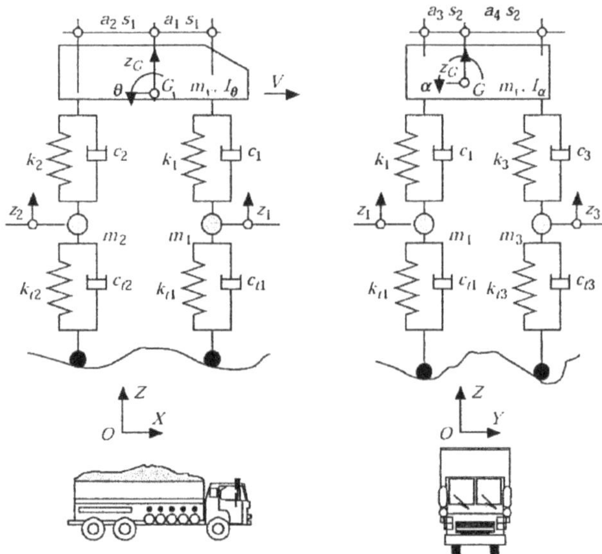

Figure 20: Model vehicle moving on a deck with surface irregularities

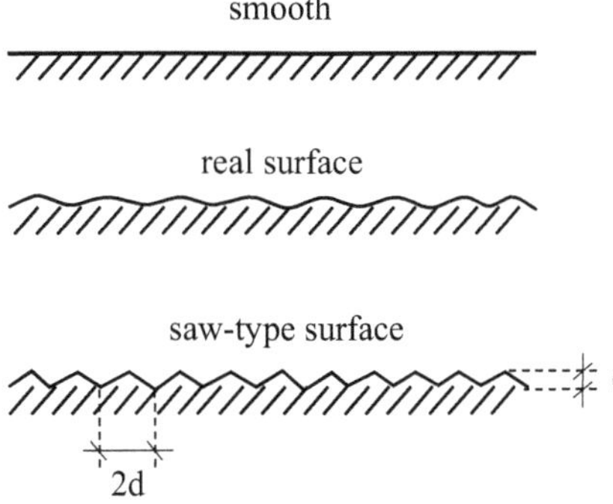

Figure 21: Modeling of the deck surface roughness

On a perfectly smooth deck surface, the mass M in Fig. **19**, is moving on a straight line (ε). Also, the wheel with mass m_0 moves depending on its behavior, *i.e.* on the developing forces and reactions between the wheel and the deck surface. We first consider that the wheel comes in contact with the deck through an orthogonal contact area with length f (Fig. **22**) and width equal to the width of the wheel. The contact length f, depends on the wheel diameter, the internal air-pressure and the carried load (*i.e.* the mass M). The usual range of values is from 5 to 10cm for small cars and from 10 to 20cm for larger vehicles.

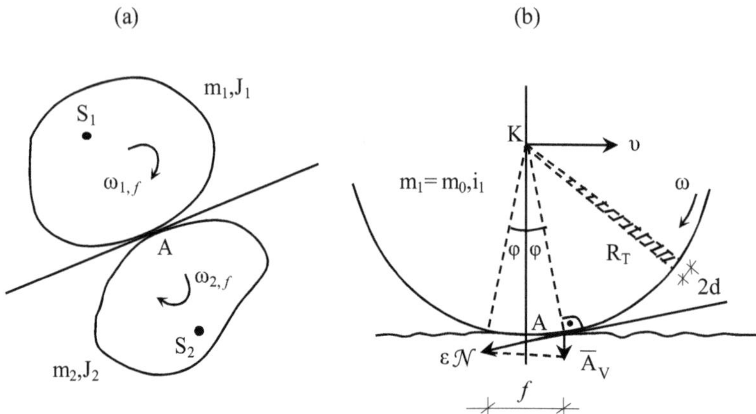

Figure 22: Vehicle wheel in contact with the deck-surface

The developing forces depend on:

- changes of the surface on which the wheel is moving. This factor is negligible for roughness cases where: f>>2d, while it becomes important for surface irregularities (see the next paragraph).
- impact forces.

It is obvious that the actual problem is to assess these impact forces, since the developing impact forces due to rotational impact do not attribute to the total mass m_0.

We shall employ the rotational impact theory [45] that is as follows:

When the solid bodies shown in Fig. **22a**, with rotational velocities $\omega_{i,f}$ and rotational inertias J_i (i=1,2) are colliding, after their bouncing, the new rotational velocity, of the first body after impact is given by:

$$\omega_{1,\alpha} = \frac{(J_1 - \varepsilon J_2)\omega_{1,f} + (1+\varepsilon)J_2\omega_{2,f}}{J_1 + J_2} \tag{83}$$

where ε is the percentage of the energy remaining after impact, which for regular wheel types is between 0,70 and 0,95.

Let us now consider the case of the wheel shown in Fig. **22**, with mass m_0, rotary inertia J_1 and inertia radius i_1 that is moving with velocity υ on a deck surface.

The horizontal movement with velocity υ does not produce any impact forces, since the wheel is moving on a direction parallel to the contact surface.

Setting $J_2 = \infty$ and $\omega_{2,f} = 0$ into eq (83), we obtain:

$$\left.\begin{aligned} &\omega_{1,\alpha} = -\varepsilon \cdot \omega_{1,f} = -\varepsilon \cdot \omega \\ \text{and hence:}\quad &\mathcal{M} = J_1 \cdot \omega = \mathcal{N} \cdot R \end{aligned}\right\} \tag{84a,b}$$

We assume arbitrarily that every circular sector with width 2d (Fig. **22b**) produces a corresponding impact force, or in other words that every projection (or peek) in the saw-model corresponds to $\frac{2d}{2\pi R}\cdot m_0$, that produces the corresponding impact force. Then, we will have: $J_1 = m_0 i_1^2 = m_0 \frac{R^2}{2}$ and since $\omega = \frac{\upsilon}{R}$, eq (84b) becomes:

$$\mathcal{N} = \frac{d}{\pi R}\cdot\frac{w_0 \upsilon}{2} \tag{85}$$

Then, the actual impact force acting at point A on the deck will be:

$$\bar{A}_v = \frac{\varepsilon \cdot d}{2\pi R}\cdot m_0 \upsilon \sin\varphi \tag{86}$$

and the impact forces in eq (80a) will be:

$$\left.\begin{aligned} &P_{imp} = \bar{A}_v \sum_{i=1}^{S} \delta(t - t_i) \\ \text{where:} \quad &t_i = t_{i-1} + \frac{2d}{\upsilon} \end{aligned}\right\} \tag{87a,b}$$

The produced deformations and oscillations of the bridge are not important and they can consequently be omitted. On the contrary, the forces due to the deck roughness present an important influence on the vehicle developing small-vibrations and the resulting sense of nausea.

In the diagram shown in Fig. **23**, one can see the deformation at the mid point of the bridge due to the deck roughness only.

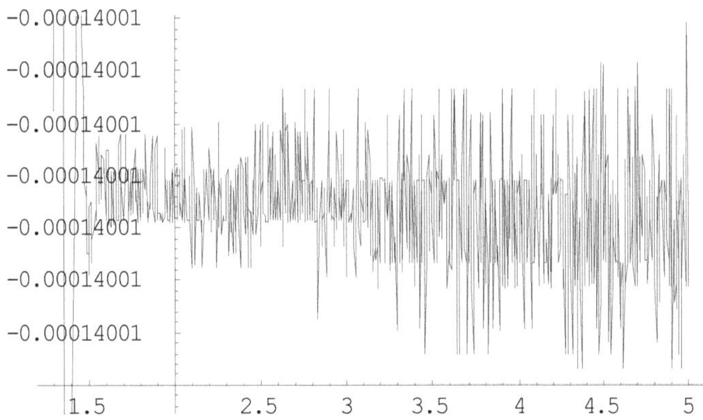

Figure 23: Midpoint deformation of a bridge with deck roughness

Irregularities of the Deck [46]

If the irregularities of the deck are random, they will have also a random surface, which is out of the scope of this section.

We are mostly concerned with irregularities, constructed on purpose on the deck, mainly for reasons of reducing the speed of vehicles entering the bridge.

With the term irregularity, we describe geometrical changes (bumps) in the deck surface with a function given by an expression r(x). The most common types of such geometrical anomalies are two, as shown in Fig. **24**.

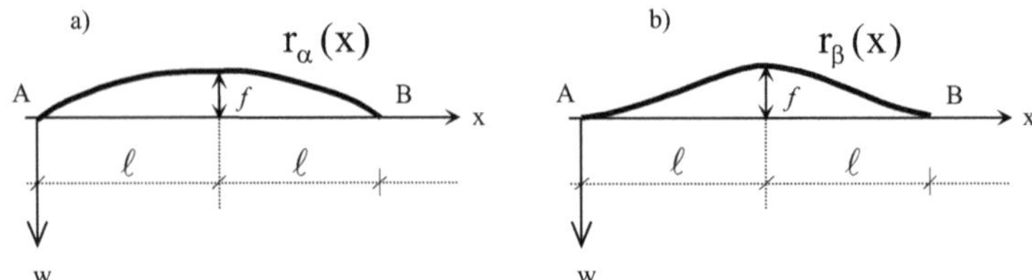

Figure 24: Surface irregularities models

The main characteristic of the irregularity type shown in Fig. **24a** is that the moving load enters and exits the irregularity in an abrupt manner due to the impact forces developing at both entering and exiting instances. This type of irregularity is modeled as follows:

$$r_\alpha(x) = \frac{f}{\ell^2}x^2 - \frac{2f}{\ell}x \tag{88}$$

The second type of irregularity shown in Fig. **24b** is characterized by a smooth entering and exiting of the wheel, since the tangents at the starting and ending points A and B are horizontal. This type of irregularity is modeled as follows:

$$r_\beta(x) = -\frac{f}{\ell^4}(x-\ell)^4 + \frac{2f}{\ell^2}(x-\ell)^2 - f \tag{89}$$

For the second irregularity type it is $P_{imp}=0$. The impact forces P_{imp} for type "a" irregularity are developing when the mass m_0 (wheel's mass) collides (since there is no smooth entrance like in type "b") with velocity υ onto the irregularity at both entering and exiting instances (Fig. **25**). The last case is considered with ∞ mass and 0 velocity. Mass m_0, after impact, changes direction to $\overline{K_1 w}$ (when entering) or to $\overline{K_2 w}$ (when exiting), as shown in Fig. **25**.

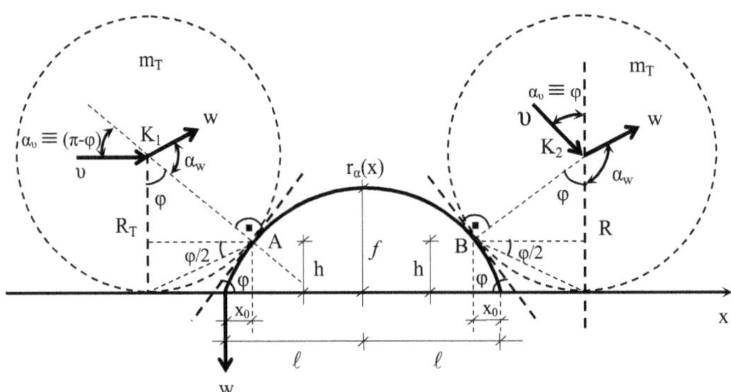

Figure 25: Wheel model entering and exiting an irregularity surface

According to the general theory of impact, we will have:

$$\left.\begin{aligned}
&\text{when entering:}\quad P_{impact}^{in} = \varepsilon m_0 \upsilon \cdot \frac{\sin\varphi \cdot \cos(\alpha_w + \varphi)}{\cos\alpha_w} \\
&\text{when exiting:}\quad P_{impact}^{out} = \varepsilon m_0 \upsilon \cdot \sin\varphi \\
&\text{where:}\quad \sin\frac{\varphi}{2} = \sqrt{\frac{h}{2R_T}} \quad , \quad \tan\alpha_w = -\frac{\cot\varphi}{\varepsilon} \, , \\
&\qquad \tan\varphi = r'_\alpha(x_0) \quad , \quad h = r_\alpha(x_0) \\
&\qquad \varepsilon \;\; : \text{ impact coefficient} \\
&\qquad x_0 \;\; : \text{ the positive root of equation:} \\
&\qquad r'_\alpha(x_0)\cdot[R_T - r_\alpha(x_0)] - \sqrt{r_\alpha(x_0) - [2R_T - r_\alpha(x_0)]} = 0
\end{aligned}\right\} \tag{90}$$

Introducing the above expressions into eq (80) and after some manipulations, we can obtain nondimensional solutions in order to study the influence of each irregularity type on the dynamic response of the system.

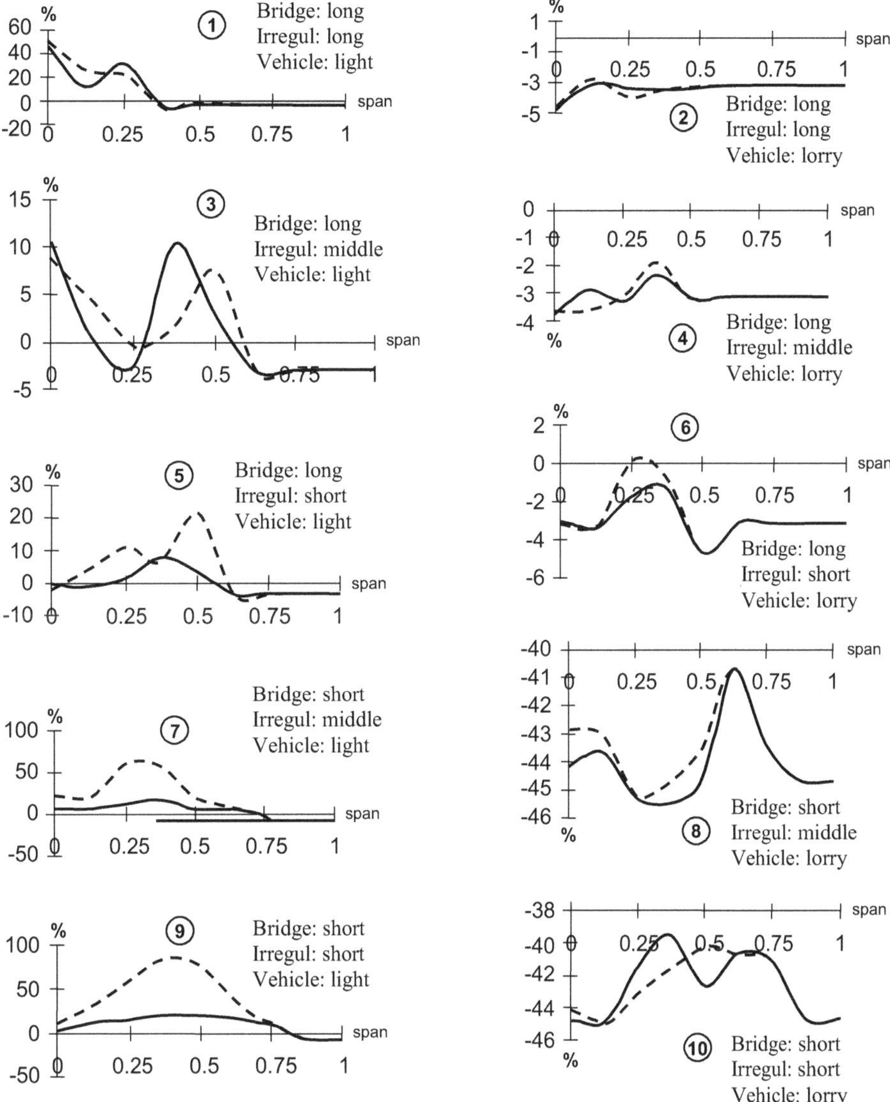

Figure 26: Influence of type "a" irregularities on the dynamic response of bridges

From this study, we can obtain the diagrams shown in Figs. **26** and **27** and arrive at the following basic conclusions:

1. **Regarding moving loads.** The use of a detailed two-axes model vehicle (instead of the single wheel vehicle) is necessary for light vehicles, where the deviation between the results from the two models reaches 44%, while the use of a single axis vehicle without springs and dampers gives results that deviate up to 85%. For short span bridges, the deviation is 25% and 60%, respectively. For heavy vehicles (two-axes trucks), the corresponding values are 40% for short span bridges, while for long bridges the difference is up to 5%.
2. **Regarding the irregularity type (type "a" or "b").** Type "a" irregularity causes increase to the dynamic deflections of the bridge from 10% up to 50% for light vehicles and from 0 up to 5% for heavy vehicles. Type "b" irregularity, causes small increased values from 2% up to 12% for light vehicles and up to 2% for heavy vehicles.
3. **Regarding the irregularity position.** The existence of an irregularity at the entering position corresponds to the most unfavorable results with significant differences compared to the cases when

the irregularity is at $\frac{L}{4}$ or at $\frac{L}{2}$ on the bridge (the deflections are almost double). Therefore, if irregularities are used in order to reduce the traffic speed, it is recommended to place outside the bridge and close to the entrance point. Moreover, we should note that small length irregularities, up to 1m, have the most unfavorable influence. Therefore, the recommended length for irregularities is between 2m and 4m. An irregularity with longer length, acts as two independent irregularities at the entrance and exit points of the wheel, especially for irregularities of type "b". Finally, we should note the optimum length of an irregularity depends on the characteristic properties of the bridge and more specifically on the eigenfrequency spectrum.

The diagrams in Fig. **26** with the solid line show the percentage increases of the midpoint deflections due to the use of two-axes models in comparison with the one-axis model shown in Fig. **19** for several positions of the irregularity on the bridge span. The same increases are shown with dotted lines comparing two-axes and single axis models without springs and dampers. The diagrams in Fig. **26** correspond to irregularities of type "a". In Fig. **27**, one can see the same results for irregularities of type "b".

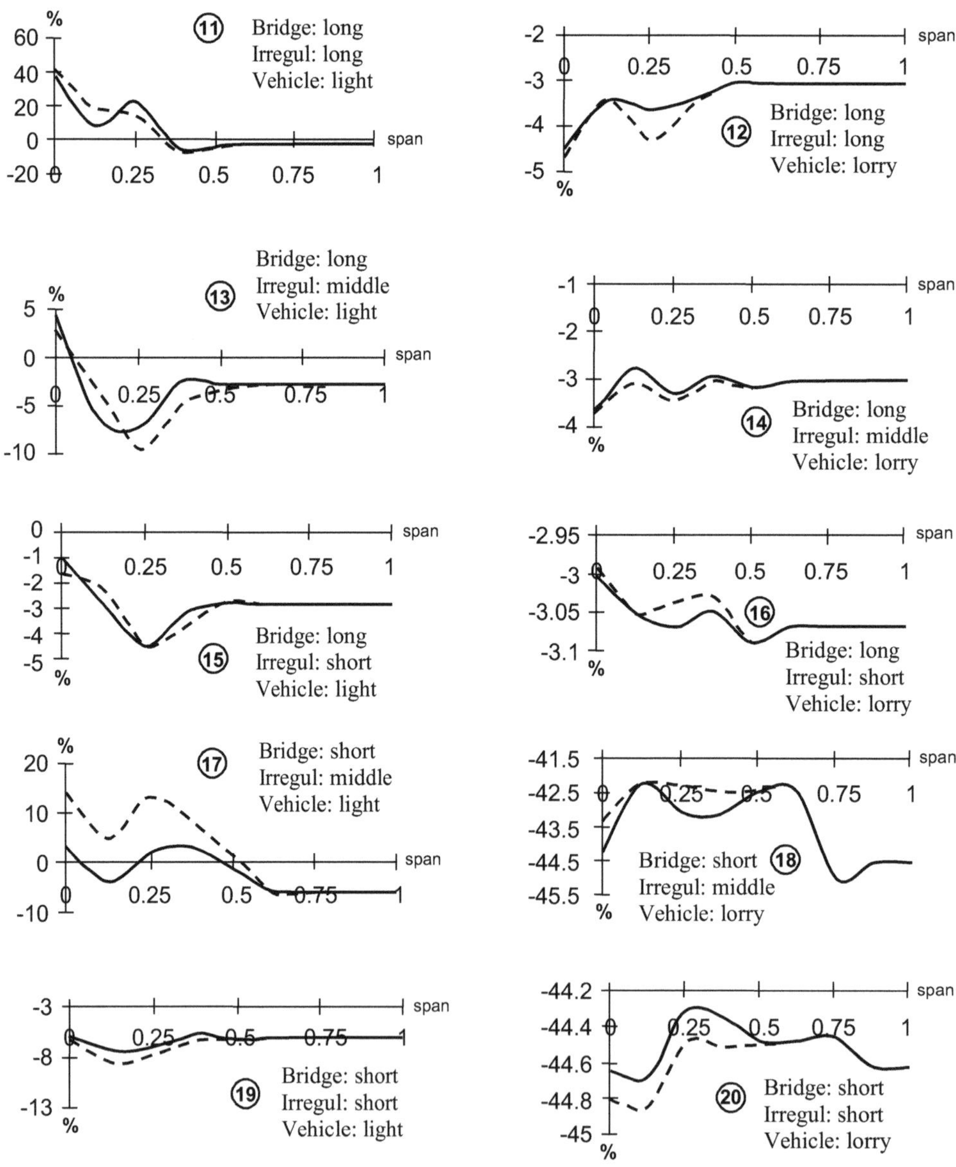

Figure 27: Influence of type "b" irregularities on the dynamic response of bridges

REFERENCES

[1] R. Willis *et al.* "Preliminary Essay to the Appendix B. Experiments for determining the effects produced by causing weights to travel over bars with different velocities", In: G. Grey *et al.* "Report of the Commissioners Appointed to inquire into the application of Iron to Railway Structures", 1849, W. Gloves and Sons, London. Reprinted in: P. Barlow "Treatise on the strength of Timber", 1851, Cast iron and Malleable Iron, London.

[2] G.G. Stokes "Discussion of a Differential Equation relating to the Breaking of Railway Bridges", 1849, Transactions, Cambridge Philosoph. Soc., Part 5.

[3] H. Zimmermann "Die Schwingungen eines Trägers mit bewegter Last", Centralblatt der Bauverwaltung, 1896, Vol.16, No 23, No 23A, No 24, No 26.

[4] A.N. Krýlov "Mathematical Collection of Papers of the Academy of Sciences", 1905, Vol.61, Petersburg, A.N. Krýlov "Über die erzwungenen Schwingungen von gleichförmigen elastische Stäben", 1905, Mathematischen Annalen, Vol. 61.

[5] S.P. Timoshenko "On the Forced Vibration of Bridges", 1922, Philosoph. Magazine, Ser. 6, Vol. 43.

[6] N.G. Bondar "Dynamic Calculations of Beams Subjected to a moving load", 1954, (in Russian) Issledoviniya po theorii sooruzhenii, Vol. 6, Moscow.

[7] H. Saller "Eifluss bewegter Last auf Eisenbahnoberbau und Brücken", 1921, Kreidels Verlag, Berlin.

[8] H.H. Jeffcott "On the Vibration of beams under the action of moving loads", 1929, Philosoph. Magazin, ser. 7, Vol. 8, No 48.

[9] C.E. Inglis "A Mathematical Treatise on Vibration in Railway Bridges", 1934, The University Press, Cambridge.

[10] K.V. Kolousêk "Structural Dynamics of continuous Beams and Frame Systems", (in Czech), 1950, German translation: "Baudynamik der Durchlaufträger und Rahmen", 1953, Fachbuchverlag, Leipzig.

[11] K.V. Kolousêk "Schwingungen der Brücken aus Stahl und Stahlbeton", 1956, Publication IABSE, XVI, Zürich.

[12] A. Hillerborg "Dynamic Influences of Smoothly Running Loads on Simply Supported Girderrs", 1951, Kungl. Tkn. Högskolan, Stockholm.

[13] J.M. Biggs, H.S. Suer, J.M. Louw "Vibration of simple-span highway bridges", 1959, Trans. ASCE, Vol. 124.

[14] R. Willis "An Essay on the Effects Produced by Causing Weights to travel over Elastic Bars", P. Barlow "A treatise on the Strength of Timber, Cast and Malleable Iron", 1851, John Weal, London.

[15] M. Deslandres "Action des chocs rythės sur les traveės metalliques", 1892, Annales, Tome IV, No 43.

[16] J. Melan "Uber die Dynamische Wirkung bewegter Lasten auf Brücken", 1893, Z. des Österr. Ing.- und Arch. - Vereines XLX. Nr 20.

[17] N. Van Eaman "Live Load Stress Measurements on Fort Loubon Bridge", 1952, Final Report, HRB Proc. 31.

[18] R.C. Edgerton, G.W. Beecroft "Dynamic Studies of two Continuous Plate-Girder Bridges", 1956, HRB Bulletin 124.

[19] AASHO "Road Test, History and Description of Project", 1961, HRB Special Report 61A, NAS-NRC Publication 816.

[20] AASHO "Road Test, Bridge Research", 1962, HRB Special Report 61D, NAS-NRC Publication 953.

[21] L.T. Oehler "Vibration Susceptibilities of Various Highway Bridge Type", 1957, J. Struct. Div., 83.

[22] A.S. Veletsos, J. Toledo Layva "Effects of Roadway Uneveness on Dynamic Response of Simple Span Highway Bridges", 1958, University of Illinois, Urbane, Civ. Eng. Studies, Struct. Research Series 168.

[23] D.T. Wright, R. Green "Highway Bridge Vibrations Part II", 1962, Ontario Test Pragramme, Report 5, Queen's University, Kingston, Ontario.

[24] A.C. Eberhardt, W.H. Walker "A Finite Element Approach to the Dynamic Analysis of Continuous highway Bridges", 1972, University of Illinois, Civ. Eng. Studies, Struct. Research Series 394.

[25] M. Ghosh, X. Quingle "Estimating Bridge Dynamic Using the Weigh-In-Motion Algorithm", 1989, TRB 68th Annual Meeting.

[26] E.S. Hwang, A.S. Novak "Dynamic Analysis of Girder Bridges", 1989, TRR 1223.

[27] J. Palamas "Imperfections du profil d' un pont et vibrations sous traffic", 1982, Thése Ecole Nat. des Ponts et Chaussées et Univ. Pierre et Marie Curie, Paris.

[28] J. Palamas, O. Coussy, Y. Bamberger "Effect of Surface Irregularities upon the Dynamic Response of Bridges under Suspended Moving Loads", 1985, J. of Sound and Vibration, 99 (2).

[29] V. Bily "Analysis of System Consisting of Bridge Structure and Moving Vehicle by the Component Element Method", 1991, Proc. 2nd International Conference on Traffic Effects on Structures and Environment.

[30] N.G. Bondar "Dynamic Calculations of Beams Subjected to a Moving Load" (in Russian), 1954, Issledovaniya po teorii sooruzhenii, Vol. 6, Moscow.

[31] N.G. Bondar, I.I. Kazei, B.F. Lesokhin, Yu.G. Kozmin "Dynamic of Railway Bridges", 1956, Transport, Moscow.

[32] A. Kounadis "A very efficient approximate method for Solving non-linear boundary-value problems", 1985, Scientific

papers of N.T.U.A., 9(3,4).
[33] G.T. Michaltsos, A.G. Sarandithou, D. Sophianopoulos "Flexural-torsional vibration of simply supported open cross-section steel beams under moving loads", 2005, Journal of Sound and Vibration, 280.
[34] D.S. Sophianopoulos, G.T. Michaltsos "Combined torsional-lateral vibrations of beams under vehicular loading", 1999, Facta Universitatis, Vol.2, No 9.
[35] G.T. Michaltsos, A.N. Kounadis, D.S.Sophianopoulos "The effect of a moving mass and other parameters on the dynamic response of a simply supported beam", 1996, Journal of Sound and Vibration, 191 (3).
[36] G.T. Michaltsos, A.N. Kounadis "The effect of Centripetal and Coriolis forces on the dynamic response of light (steel) bridges under moving loads", 2001, Journal of Vibrations and Control, Vol.7, No 3.
[37] G.T. Michaltsos "The influense of Centripetal and Coriolis forces on the dynamic response of light bridges under moving vehicles", 2001, Journal of Sound and Vibration, 247 (2).
[38] G.T. Michaltsos, D.S. Sophianopoulos "The dynamic effect of distributed loads and other parameters on the behaviour of light railway bridges", 2002, Proc. of Eurodyn 2002, Munich.
[39] F. Khalily, M.F. Golnaraghi, G.R. Heppler "On the dynamic behaviour of a flexible beam carrying a moving Mass", 1994, Nonlinear Dynamics, 5.
[40] G.T. Michaltsos "Parameters affecting the dynamic response of light (steel) Bridges", 2000, Facta Universitatis-Series: Mechanics, Automatic Control and Robotics, Vol.2, No 10.
[41] T.L. Wang, P. Huang "Cable stayed bridge vibration due to random surface roughness", 1995, Journal of Structural Engineering, 118 (5).
[42] Y.S. Cheng, F.T.K. Au, Y.K. Cheng and D.Y. Zherg "On the separation between moving vehicles and bridges", 1999, Journal of Sound and Vibration, 222 (50).
[43] M. Fafard, M. Laflamme, M. Savarad, M. Bennur "Dynamic analysis of existing continuous bridge", 1998, Journal of Bridge Engineering, 3(1).
[44] G.T. Michaltsos, D.S. Sophianopoulos "The effect of roughness in conjunction with other parameters on the dynamic response of steel highway bridges under vehicular loading", 2001, 6th National Conference of Mechanics, Salonica.
[45] W. Goldsmith "Impact", 1960, Adward Arnold, London.
[46] G.T. Michaltsos, T.G. Konstantakopoulos "Dynamic response of a bridge with surface deck irregularities", 2000, Journal of Vibration and Control, 6.
[47] P. Chang, H. Lee "Impact factors for simple-span highway girder bridges", 1994, Journal of Structural Engineering, 120 (3).
[48] D.P. Hang, T.L. Wang "Vibration of highway steel bridges with longitudinal grades", 1998, Computers and Structures, 69 (2).
[49] K. Henchi, M. Farard, M. Talbot, G. Dhatt "An efficient algorithm for dynamic analysis of bridges under moving vehicles using a coupled modal und physical components approach", 1998, Journal of Sound and Vibration, Vol. 212, No 4.
[50] ISO 8608 "Mechanical vibration - Road surface profiles - reporting of measured data", 1995 (E).

CHAPTER 5

Motion of Supports

Abstract: In this chapter, the special problem of support settlement is exclusively treated. The corresponding shape functions for settlement of one or more supports of simple beams with up to three spans with various boundary conditions are derived and presented in detail. Among the examined cases are supports vibrating along the longitudinal and the transverse of the bridge or the vertical direction as well.

INTRODUCTION

The excitation cause of the dynamic strain of a bridge may be acting on the bridge structure (for example wind pressure, moving loads, etc) or may be acting externally. In this case, the dynamic excitation is transferred to the bridge structure through its supports.

The mechanism is as following: the excitation cause is on the ground and it acts on the bridge's supports imposing a temporary or a continuing for some time motion.

The motion of the supports evidently sets the bridge in motion, which reacts with the developing inertia forces due to its mass. The latter may be considered as a virtual loading on the bridge.

The excitation cause may be distinguished into:

a. sudden settlement or ground displacement, that is usually caused by geological reasons and is generally a permanent ground displacement, and

b. ground motions passing through the soil, which are harmonic or not functions of time, caused mainly by earthquakes and rarely result to a permanent ground displacement.

The type of the arising forced motion of a bridge, subjected to the above described instantaneous or not ground motions depends on a number of factors.

The most important factors are:

a. The in-plane direction of the excitation or ground displacement (Fig. **1**) and the possible existence of a significant or not vertical component that is depended on the epicenter distance.

b. The height and the slenderness of the piers.

c. The distance z_E of the supports from the gravity center 0 of the bridge's cross-section.

d. The distance z_M of the shear center M of the bridge's cross-section from the gravity center 0.

The above are only a few factors from which the type of strain that will be developed depends on (namely axial or vertical or lateral-torsional bending or all together).

Finally, we must note that possible foundation of supports on different soil type (Fig. **1**), but also a possible long distance between the supports (mainly in long span bridges), or the existence of a fault between supports may result to support motions described by different excitation time functions or different amplitudes of movement. Thus, the excitation cause will be completely different for each support.

In this case, the following assumption will be valid further on:

"In a system with multiple supports and connections to the ground, we accept that the motion of each support may be different. Although these movements may be big, we accept that their differences at each time instant t will be small, so that linear behaviour of the beam is retained."

George T. Michaltsos and Ioannis G. Raftoyiannis

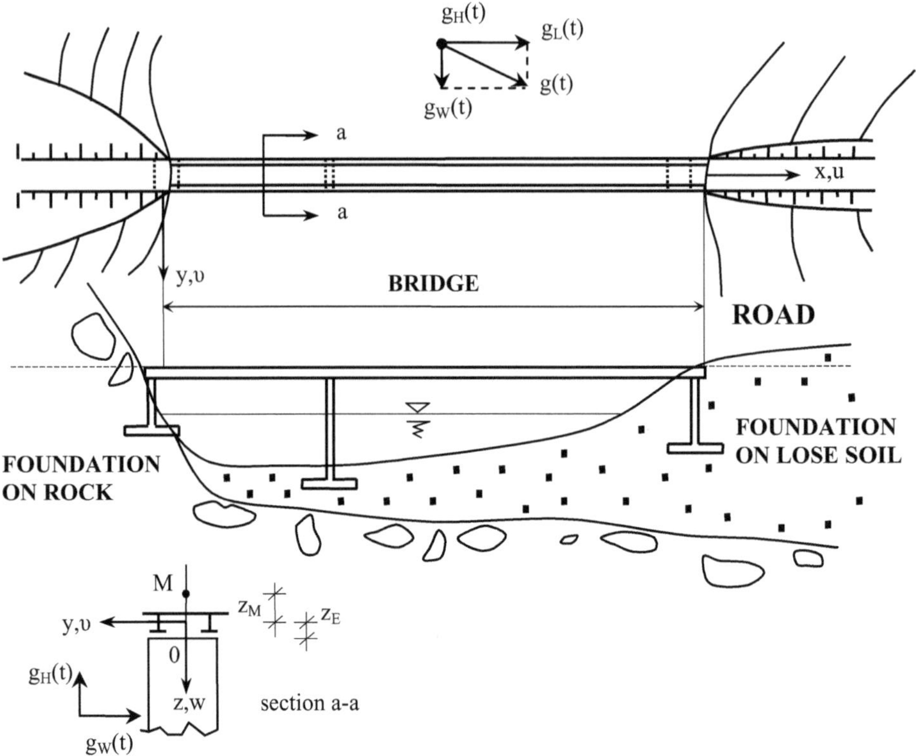

Figure 1: Top and side view of a 2-span bridge

THE LONGITUDINAL MOTION OF SUPPORTS [1]

We will study the longitudinal motion of a beam with prismatic cross-section caused by the movement of one or more of its supports. We will study at first the simple case of a cantilever beam and then a beam resting on two supports (simply supported, fixed-fixed or fixed-pinned).

The Cantilever Beam

We consider a cantilever beam with prismatic cross-section Fig. **2**, whose support (ground) moves along x direction according to the following relation:

$$u_g = f(t) \tag{1}$$

Let us consider now the point A on the beam, which oscillates in direction parallel to x-axis and which at instant t goes to point A′ that is at distance u from its initial position at time t=0.

This total movement u is analyzed into the movement u_g which undergoes the beam moving like a rigid body, and into the movement u_o which is caused by the elastic deformation of the beam. Therefore, it is:

$$u = u_g + u_0 \tag{2}$$

Since no load is acting on the span of the beam, the equation of motion is given by (see 63a of chapter 3):

$$EA(u_g + u_o)'' - c_x(u_g + u_o)^{\bullet} - m(u_g + u_o)^{\bullet\bullet} = 0$$

and since: $u_g'' = f''(t) = 0$, the above equation is finally written as:

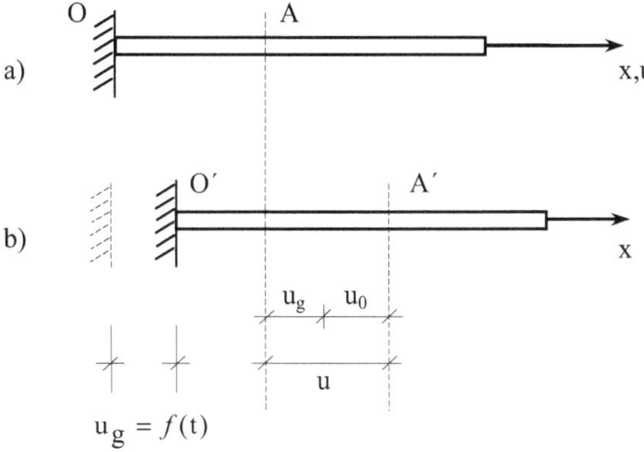

Figure 2: Axial deformation of a beam subjected to support movement

$$EAu_0'' - c_x\dot{u}_0 - m\ddot{u}_0 = m\ddot{f} + c_x\dot{f} \tag{3}$$

Comparing eq (3) to eq(51a) of chapter 3, we see that the corresponding equivalent distributed load is $q_x(x,t) = m\ddot{f}(t) + c_x\dot{f}(t)$.

In order to solve eq (3), we are searching for a solution in the form:

$$u_o(x,t) = \sum_n U_n(x)T_n(t) \tag{4}$$

where $U_x(x)$ is the shape function of a cantilever beam in free longitudinal vibration and $T_n(t)$ is the corresponding time function to be determined. Thus, the differential equation giving the time function $T_n(t)$ (see also eq(126) of chapter 3), will be:

$$\ddot{T}_n + \frac{c_x}{m}\dot{T}_n + \omega_{xn}^2 T_n = -\frac{\int_0^\ell q_x(x,t)U_n(x)dx}{m\cdot\int_0^\ell U_n^2 dx} \quad \text{or}$$

$$\left.\begin{aligned} &\ddot{T}_n + 2\beta_x\dot{T}_n + \omega_{xn}^2 T_n = -\frac{\int_0^\ell U_n(x)dx}{\int_0^\ell U_n^2(x)dx}\cdot[\ddot{f}(t) + 2\beta_x\dot{f}(t)] \\ &\text{where: } \beta_x = \frac{c_x}{2m} \text{ and } \omega_{xn} \text{ are the eigenfrequencies} \end{aligned}\right\} \tag{5}$$

The solution of eq (5) is given by the Duhamel's integral:

$$\left.\begin{aligned} &T_n(t) = -\frac{1}{\overline{\omega}_{xn}}\cdot\frac{\int_0^\ell U_n(x)dx}{\int_0^\ell U_n^2(x)dx}\cdot\int_0^t e^{-\beta_x(t-\tau)}\,[\ddot{f}(\tau) + 2\beta_x\dot{f}(\tau)]\cdot\sin\overline{\omega}_{xn}(t-\tau)d\tau \\ &\text{where: } \overline{\omega}_{xn} = \sqrt{\omega_{xn}^2 - \beta_x^2} \end{aligned}\right\} \tag{6}$$

Finally the longitudinal motion of the beam is given by the equation:

$$U(x,t) = f(t) + \sum_n U_n(x)T_n(t) \tag{7}$$

Beam on Two Supports

We consider the beam $\overline{12}$ shown in Fig. **3**, the supports 1 and 2 of which may be of any type: namely, they may be fixed or pinned supports. Thus, the supports in Fig. **3** are only indicative. We suppose, in addition, that supports 1 and 2 may move independently to each other according to the following laws:

$$\left.\begin{array}{l} \text{support 1: } u_{g_1} = f_1(t) \\ \text{support 2: } u_{g_2} = f_2(t) \end{array}\right\} \tag{8a,b}$$

We point out that the assumption of introduction is always valid, according to which the movements u_{g1} and u_{g2} may be big but their differences at any time instant t are considered small.

We assume that the beam moves from position $\overline{12}$ to position $\overline{1'2'}$. Thus, the deformation is caused only by the gradually (non-dynamic) movement of the supports. Then, edges 1 and 2 will go to their new positions 1′ and 2′, while the arbitrary point A will go to position A′. Namely, both supports will have undergone the movement:

$$u_g = \frac{\ell - x}{\ell} \cdot u_{g_1} + \frac{x}{\ell} \cdot u_{g_2} = \frac{\ell - x}{\ell} \cdot f_1(t) + \frac{x}{\ell} \cdot f_2(t) \tag{9}$$

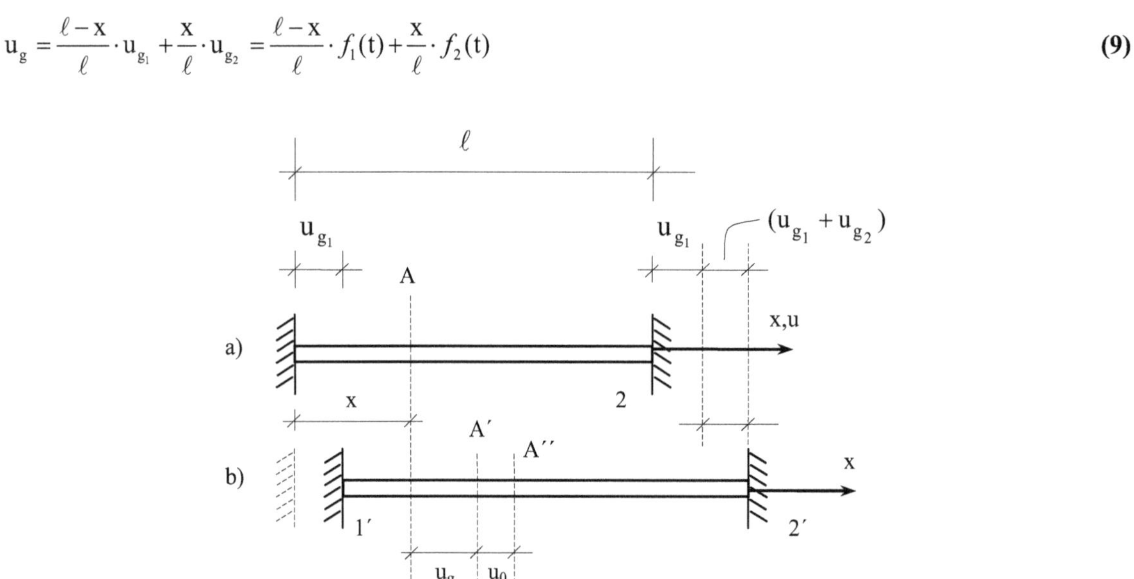

Figure 3: Axial deformation of a beam subjected to support movement

The movement from $\overline{12}$ to $\overline{1'2'}$ has also a dynamic character and thus, the beam will perform an axial vibration as well. The beam, at instant t, will be exactly at point A′′ (because of the axial vibration), with $\overline{A'A''} = u_0$, where u_o is the additional deformation caused by the vibration only. Then, the total deformation of point A will be:

$$u = u_g + u_0 = \frac{\ell - x}{\ell} \cdot f_1(t) + \frac{x}{\ell} \cdot f_2(t) + u_0 \tag{10}$$

Introducing the above relation into eq(63a) of chapter 3, we get:

$$EA(u_g + u_0)'' - c_x(u_g + u_0)^{\bullet} - m(u_g + u_0)^{\bullet\bullet} = 0$$

or since $u_g'' = 0$, the above equation is finally written as:

$$EAu_0'' - c_x \dot{u}_0 - m\ddot{u}_0 = m\ddot{u}_g + c_x \dot{u}_g \tag{11}$$

We are searching for a solution in the form:

$$u_0(x,t) = \sum_n U_n(x)T_n(t) \tag{12}$$

Thus, the differential equation for $T_n(t)$ will be:

$$\left.\begin{aligned} \ddot{T}_n + 2\beta_x \dot{T}_n + \omega_{xn}^2 T_n = & -\frac{\int_0^\ell \frac{\ell - x}{\ell} \cdot U_n dx}{\int_0^\ell U_n^2 dx} \cdot [\ddot{f}_1(t) + 2\beta_x \dot{f}_1(t)] \\ & -\frac{\int_0^\ell \frac{x}{\ell} \cdot U_n dx}{\int_0^\ell U_n^2 dx} \cdot [\ddot{f}_2(t) + 2\beta_x \dot{f}_2(t)] \end{aligned}\right\} \tag{13}$$

The solution of eq (13) is given by the Duhamel's integral:

$$\left.\begin{aligned} T_n(t) = & -\frac{1}{\overline{\omega}_{xn}} \cdot \frac{\int_0^\ell \frac{\ell - x}{\ell} \cdot U_n dx}{\int_0^\ell U_n^2 dx} \cdot \int_0^t e^{-\beta_x (t-\tau)} [\ddot{f}_1(\tau) + 2\beta_x \dot{f}_1(\tau)] \cdot \sin \overline{\omega}_n (t-\tau) d\tau \\ & -\frac{1}{\overline{\omega}_{xn}} \cdot \frac{\int_0^\ell \frac{x}{\ell} \cdot U_n dx}{\int_0^\ell U_n^2 dx} \cdot \int_0^t e^{-\beta_x (t-\tau)} [\ddot{f}_2(\tau) + 2\beta_x \dot{f}_2(\tau)] \cdot \sin \overline{\omega}_n (t-\tau) d\tau \end{aligned}\right\} \tag{14}$$

Thus, the axial deformation of the beam is given by the equation:

$$u(x,t) = f_1(t) + f_2(t) + \sum_n U_n(x)T_n(t) \tag{15}$$

When $u_{g_1} = u_{g_2} = f(t)$, eq (9) gives $u_g = f(t)$, while eq (14) coincides with eq (5).

THE VERTICAL MOTION OF SUPPORTS

In the previous paragraph, we have studied the dynamic strain of a beam caused by the movement of its supports along x-axis.

In the present paragraph, we will study the dynamic strain of a beam whose supports are moving vertically, namely along z-axis, which is an axis of symmetry for the beam's cross-section.

We will distinguish again the cases of beams with a single support and beams with more than one supports.

The Cantilever Beam

We consider the cantilever beam $\overline{01}$ shown in Fig. **4**, where the support 0 (ground) moves along the vertical direction z, according to the relation:

$$w_g = f(t) \tag{16a}$$

If the beam is moved slowly from the position $\overline{01}$ to $\overline{0'1'}$ (static case), it would undergo deformations due to the gradually (non-dynamic) movement of its supports only and the ends 0 and 1 would be displaced to their new positions 0′ and 1′ while the arbitrary point A would be displaced to A′: namely, the beam undergoes the displacement $w_g = f(t)$. But the movement of the beam from $\overline{01}$ to $\overline{0'1'}$ has a dynamic nature as well and therefore, it will perform a bending vibration also. At instant t, the beam will go to the position 0′1′′ (exactly because of the above bending vibration), and point A′ to position A′′, with $\overline{A'A''} = w_0$, where w_0 is the additional deformation caused by the vibration only. Thus, the total movement of point A will be:

$$w = w_g + w_0 = f(t) + w_0 \tag{16b}$$

Introducing the expression of w into eq(63b) of chapter 3, we have:

$$EI_y[f(t)+w_0]'''' + c_y[f(t)+w_0]^{\bullet} + m[f(t)+w_0]^{\bullet\bullet} = 0$$

or after differentiations:

$$EI_y w'''' + c_y \dot{w}_0 + m\ddot{w}_0 = -m\ddot{f}(t) - c_y \dot{f}(t) \tag{17}$$

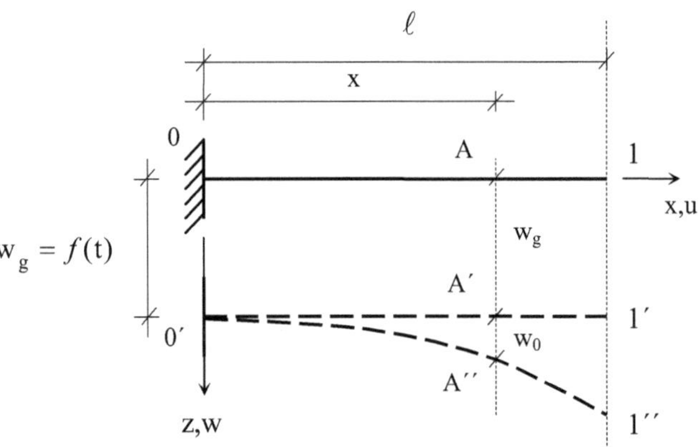

Figure 4: Bending deformation of a beam subjected to support movement

We are searching for a solution in the form:

$$w_0(x,t) = \sum_n W_n(x) T_n(t) \tag{18}$$

Then, the differential equation for $T_n(t)$ will be:

$$\left.\begin{aligned} &\ddot{T}_n(t) + 2\beta_y \dot{T}_n(t) + \omega_{yn}^2 T_n(t) = -\frac{\int_0^\ell W_n(x)dx}{\int_0^\ell W_n^2(x)dx} \cdot [\ddot{f}(t) + 2\beta_y \dot{f}(t)] \\ &\text{where: } \beta_y = \frac{c_y}{2m} \text{ and } \omega_{yn} \text{ the eigenfrequencies} \end{aligned}\right\} \tag{19}$$

The solution of eq (19) is given by the Duhamel's integral:

$$T_n(t) = -\frac{1}{\overline{\omega}_{yn}} \cdot \frac{\int_0^\ell W_n(x)dx}{\int_0^\ell W_n^2(x)dx} \cdot \int_0^t e^{-\beta_y(t-\tau)}[\ddot{f}(\tau) + 2\beta_y \dot{f}(\tau)] \cdot \sin\overline{\omega}_{yn}(t-\tau) \cdot d\tau$$

$$\text{where: } \overline{\omega}_{yn} = \sqrt{\omega_{yn}^2 - \beta_y^2} \qquad \textbf{(20)}$$

Thus, the motion of the beam due to the movement of support 0, is given by the following equation:

$$w(x,t) = f(t) + \sum_n W_n(x)T_n(t) \qquad \textbf{(21)}$$

The Simply Supported Beam

Let us consider now the simply supported beam shown in Fig. **5**, which is subjected to a dynamic motion of either the left support 1 or the right support 2 or of both supports simultaneously.

It is obvious that, for all the above cases, the beam will move firstly as a rigid body and then it will perform a vibration. According to Fig. **5**, we may write:

$$\begin{aligned} &\text{left edge:} && w_g = w_{g_\alpha}\left(1-\frac{x}{\ell}\right) = \left(1-\frac{x}{\ell}\right)f_\alpha(t) = g_\alpha(x)f_\alpha(t) \\ &\text{right edge:} && w_g = w_{g_\delta}\frac{x}{\ell} = \frac{x}{\ell}f_\delta(t) = g_\delta(x)f_\delta(t) \end{aligned} \qquad \textbf{(22)}$$

The above functions $g_\alpha(x)$ and $g_\delta(x)$ are called **influence functions** of the movement and correspond to unit settlement of the supports.

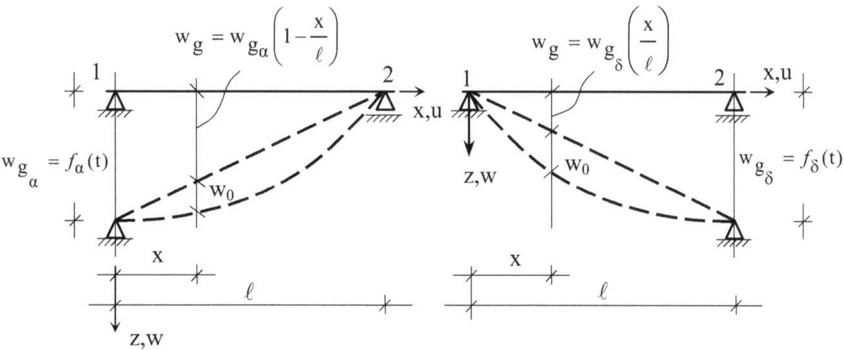

Figure 5: Displacements of a simply supported beam due to support settlements

Therefore, we can write, always assuming that movements $w_{g_\alpha}, w_{g_\delta}$ may be big but their difference small:

$$w(x,t) = g_\alpha(x)f_\alpha(t) + g_\delta(x)f_\delta(t) + w_0(x,t) \qquad \textbf{(23)}$$

Introducing eq (23) into eq(63b) of chapter 3 we have:

$$EI_y[g_\alpha f_\alpha + g_\delta f_\delta + w_0]'''' + c_y[g_\alpha f_\alpha + g_\delta f_\delta + w_0]^{\bullet} + m[g_\alpha f_\alpha + g_\delta f_\delta + w_0]^{\bullet\bullet} = 0$$

or after differentiations:

$$EI_y w'''' + c_y \dot{w}_0 + m\ddot{w}_0 = -m[g_\alpha \ddot{f}_\alpha + g_\delta \ddot{f}_\delta] - c_y[g_\alpha \dot{f}_\alpha + g_\delta \dot{f}_\delta] \tag{24}$$

We are searching for a solution in the form:

$$w_0(x,t) = \sum_n W_n(x) T_n(t) \tag{25}$$

Then, $T_n(t)$ is given by the following differential equation:

$$\left.\begin{array}{l} \ddot{T}_n(t) + 2\beta_y \dot{T}_n(t) + \omega_{y_n}^2 T_n(t) = -\dfrac{\int_0^\ell g_\alpha W_n dx}{\int_0^\ell W_n^2 dx} \cdot [\ddot{f}_\alpha + 2\beta_y \dot{f}_\alpha] - \dfrac{\int_0^\ell g_\delta W_n dx}{\int_0^\ell W_n^2 dx} \cdot [\ddot{f}_\delta + 2\beta_y \dot{f}_\delta] \\ \text{where: } \beta_y = \dfrac{c_y}{2m}, \text{ and } \omega_{y_n} \text{ the eigenfrequencies} \end{array}\right\} \tag{26}$$

The solution of eq (26) is given by the Duhamel's integral:

$$\left.\begin{array}{l} T_n(t) = -\dfrac{\int_0^\ell g_\alpha W_n dx}{\bar{\omega}_n \int_0^\ell W_n^2 dx} \cdot \int_0^\ell e^{-\beta_y (t-\tau)} [\ddot{f}_\alpha(\tau) + 2\beta_y \dot{f}_\alpha(\tau)] \cdot \sin \bar{\omega}_{yn}(t-\tau) \cdot d\tau - \\ \qquad -\dfrac{\int_0^\ell g_\delta W_n dx}{\bar{\omega}_n \int_0^\ell W_n^2 dx} \cdot \int_0^\ell e^{-\beta_y (t-\tau)} [\ddot{f}_\delta(\tau) + 2\beta_y \dot{f}_\delta(\tau)] \cdot \sin \bar{\omega}_{yn}(t-\tau) \cdot d\tau \\ \text{where: } \bar{\omega}_{yn} = \sqrt{\omega_{yn}^2 - \beta_y^2} \end{array}\right\} \tag{27}$$

The Influence Functions

The study of the dynamic behavior of beams subjected to time depended movements of their supports becomes significantly easier by using the so-called influence functions. As an influence function, we determine the one expressing the static (non-dynamic) deformation of a beam caused by a unit settlement of one of its supports. It is obvious that, due to a support settlement, a statically determined beam will move as a rigid body. But in statically undetermined systems a possible support settlement will induce movement to the beam in the form of an elastic deformable body.

In the following, the influence functions are given for unit settlement or rotation of a support, for the most usual cases in bridge structures.

The Cantilever Beam

1. Unit settlement of support 1:

$$g(x) = 1 \tag{28}$$

2. Unit rotation of support 1:

$$g(x) = x \tag{29}$$

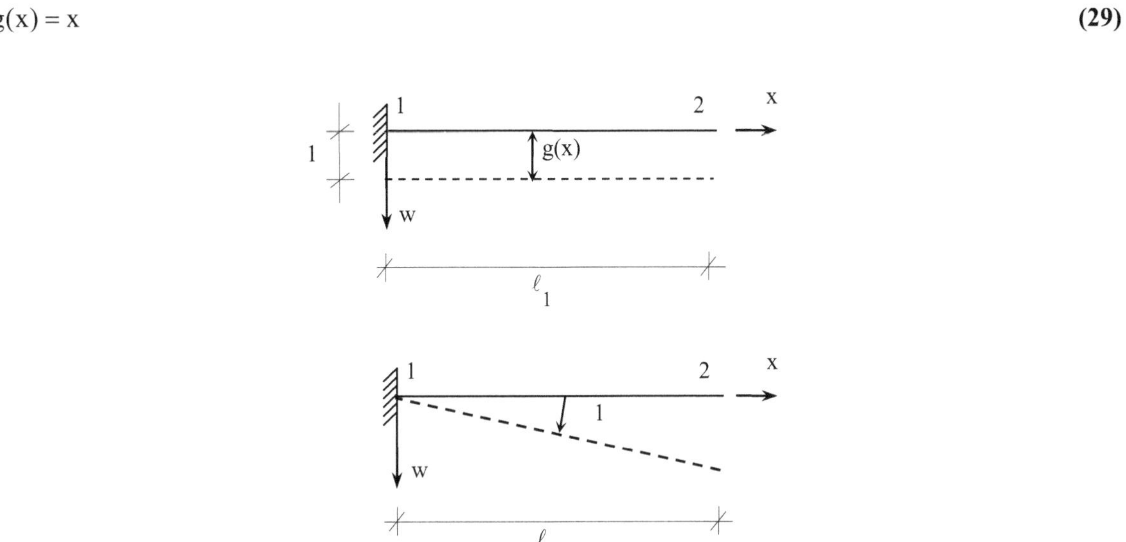

Figure 6: Influence functions of a cantilever due to support settlement and rotation

The Simply Supported Beam

1. Unit settlement of support 1:

$$g(x) = 1 - \frac{x}{\ell} \tag{30}$$

2. Unit settlement of support 2:

$$g(x) = \frac{x}{\ell} \tag{31}$$

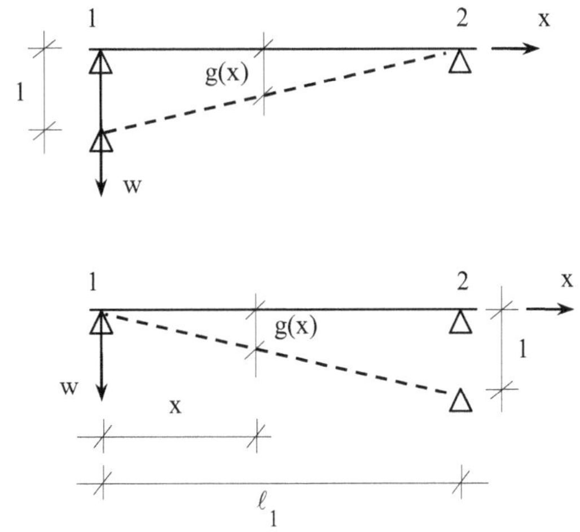

Figure 7: Influence function of a simply-supported beam due to support settlements

The Fixed-Fixed Beam

1. Unit settlement of support 1:

$$g(x) = 1 - 3\frac{x^2}{\ell^2} + 2\frac{x^3}{\ell^3} \tag{32}$$

2. Unit settlement of support 2:

$$g(x) = 3\frac{x^2}{\ell^2} - 2\frac{x^3}{\ell^3} \tag{33}$$

3. Unit rotation of support 1:

$$g(x) = x - 2\frac{x^2}{\ell} + \frac{x^3}{\ell^2} \tag{34}$$

4. Unit rotation of support 2:

$$g(x) = -\frac{x^2}{\ell} + \frac{x^3}{\ell^2} \tag{35}$$

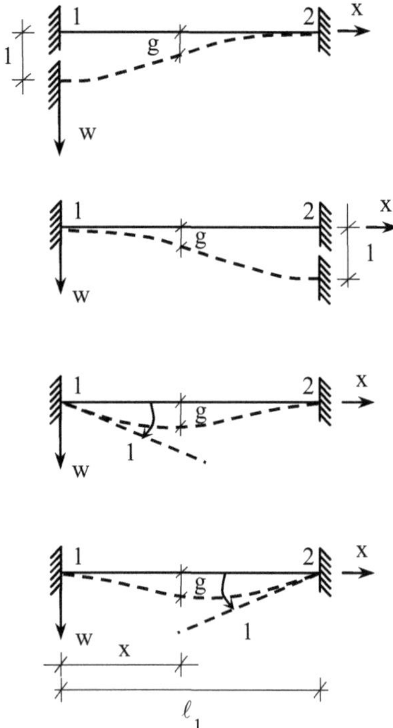

Figure 8: Influence function of a fixed-fixed beam due to support settlements and rotations

The Fixed-Pinned Beam

1. Unit settlement of support 1:

$$g(x) = 1 - \frac{3x^2}{2\ell^2} + \frac{x^3}{2\ell^3} \tag{36}$$

2. Unit settlement of support 2:

$$g(x) = \frac{3x^2}{2\ell^2} - \frac{x^3}{2\ell^3} \quad (37)$$

3. Unit rotation of support 1:

$$g(x) = x - \frac{3x^2}{2\ell} + \frac{x^3}{2\ell^2} \quad (38)$$

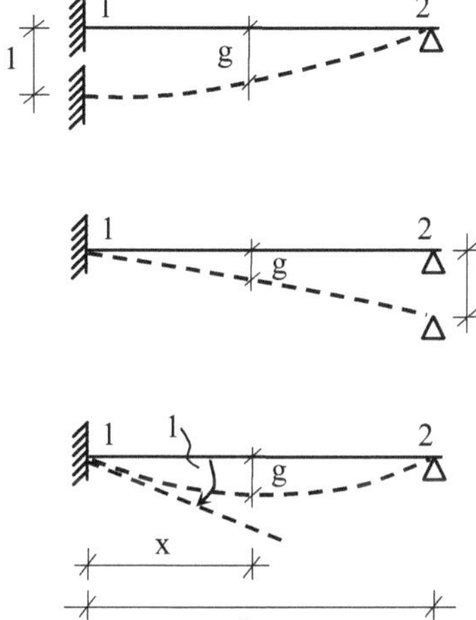

Figure 9: Influence function of a fixed-simply supported beam due to support settlement and rotation

The Two-Span Beam

1. Unit settlement of support 1:

$$\left.\begin{aligned} g_1(x_1) &= \frac{\ell_1 - x_1}{\ell_1} - \frac{1}{2\ell_1(\ell_1+\ell_2)}\cdot\left(-\frac{x_1^3}{\ell_1} + \ell_1 x_1\right) \\ g_2(x_2) &= -\frac{1}{2\ell_1(\ell_1+\ell_2)}\cdot\left(\frac{x_2^3}{\ell_2} - 3x_2^2 + 2\ell_2 x_2\right) \end{aligned}\right\} \quad (39)$$

2. Unit settlement of support 2:

$$\left.\begin{aligned} g_1(x_1) &= \frac{x_1}{\ell_1} + \frac{1}{2\ell_1\ell_2}\cdot\left(-\frac{x_1^3}{\ell_1} + \ell_1 x_1\right) \\ g_2(x_2) &= \frac{\ell_2 - x_2}{\ell_2} + \frac{1}{2\ell_1\ell_2}\cdot\left(\frac{x_2^3}{\ell_2} - 3x_2^2 + 2\ell_2 x_2\right) \end{aligned}\right\} \quad (40)$$

3. Unit settlement of support 3:

$$\left.\begin{aligned} g_1(x_1) &= -\frac{1}{2\ell_2(\ell_1+\ell_2)}\cdot\left(-\frac{x_1^3}{\ell_1} + \ell_1 x_1\right) \\ g_2(x_2) &= \frac{x_2}{\ell_2} - \frac{1}{2\ell_2\cdot(\ell_1+\ell_2)}\cdot\left(\frac{x_2^3}{\ell_2} - 3x_2^2 + 2\ell_2 x_2\right) \end{aligned}\right\} \quad (41)$$

4. Unit rotation of support 2:

$$\left.\begin{aligned} g_1(x_1) &= -\frac{1}{2\ell_1}\cdot\left(-\frac{x_1^3}{\ell_1}+\ell_1 x_1\right) \\ g_2(x_2) &= -\frac{1}{2\ell_2}\cdot\left(\frac{x_2^3}{\ell_2}-3x_2^2+2\ell_2 x_2\right) \end{aligned}\right\} \tag{42}$$

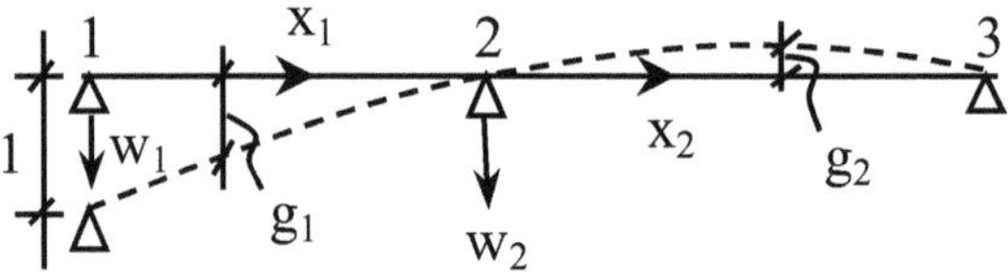

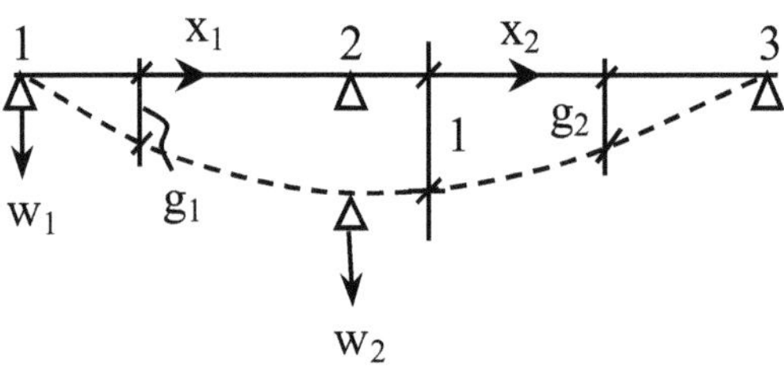

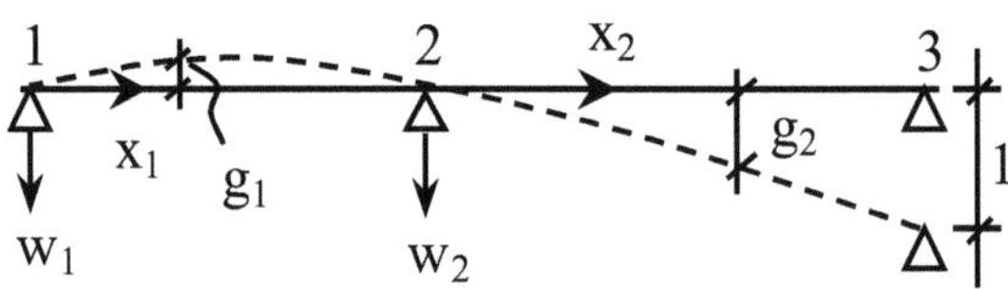

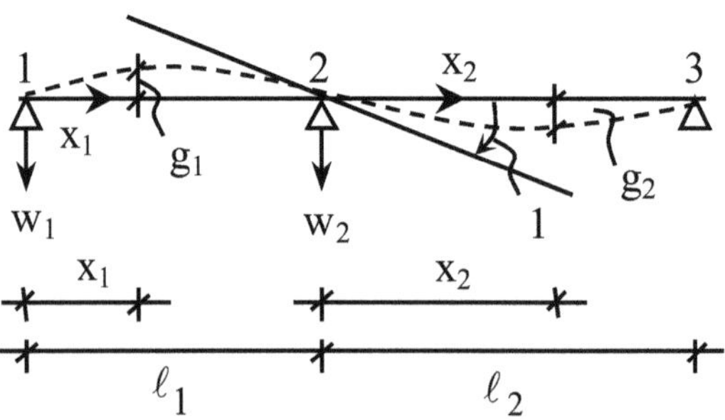

Figure 10: Influence function of a 2-span beam due to support settlements and rotations

The Three-Span Beam

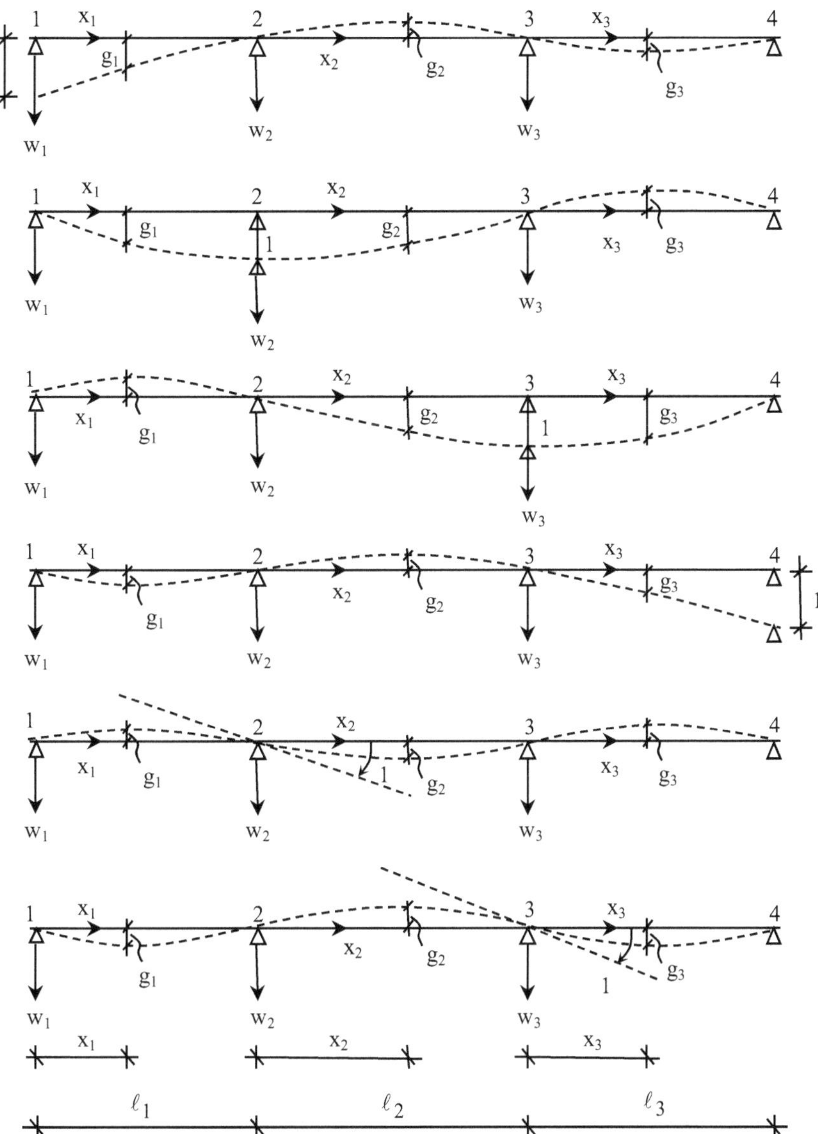

Figure 11: Influence function of a 3-span beam due to support settlements and rotations

1. Unit settlement of support 1:

$$\left.\begin{aligned}
&g_1(x_1)=\frac{\ell_1-x_1}{\ell_1}-2\kappa_1\left(\ell_2+\ell_3\right)\left(-\frac{x_1^3}{\ell_1}+\ell_1 x_1\right)\\
&g_2(x_2)=-2\kappa_1\left(\ell_2+\ell_3\right)\left(\frac{x_2^3}{\ell_2}-3x_2^2+2\ell_2 x_2\right)-\ell_2\kappa_1\left(\frac{x_2^3}{\ell_2}-\ell_2 x_2\right)\\
&g_3(x_3)=\ell_2\kappa_1\left(\frac{x_3^3}{\ell_3}-3x_3^2+2x_3\ell_3\right)\\
&\text{where:}\quad \kappa_1=\frac{1}{\ell_1(3\ell_2^2+4\ell_1\ell_2+4\ell_1\ell_3+4\ell_2\ell_3)}
\end{aligned}\right\}\qquad \textbf{(43)}$$

2. Unit settlement of support 2:

$$\left.\begin{array}{l} g_1(x_1)=\dfrac{x_1}{\ell_1}+\kappa_2\left[2\ell_2(\ell_2+\ell_3)+\ell_1(3\ell_2+2\ell_3)\right]\left(-\dfrac{x_1^3}{\ell_1}+\ell_1x_1\right) \\ g_2(x_2)=\dfrac{\ell_2-x_2}{\ell_2}+\kappa_2\left[2\ell_2(\ell_2+\ell_3)+\ell_1(3\ell_2+2\ell_3)\right]\left(\dfrac{x_2^3}{\ell_2}-3x_2^2+2\ell_2x_2\right)- \\ \qquad -\kappa_2(\ell_1+\ell_2)\ (2\ell_1+\ell_2)\left(-\dfrac{x_2^3}{\ell_2}+\ell_2x_2\right) \\ g_3(x_3)=-\kappa_2(\ell_1+\ell_2)\ (2\ell_1+\ell_2)\left(\dfrac{x_3^3}{\ell_3}-3x_3^2+2x_3\ell_3\right) \\ \text{where:}\quad \kappa_2=\dfrac{1}{\ell_1\ell_2(3\ell_2^2+4\ell_1\ell_2+4\ell_1\ell_3+4\ell_2\ell_3)} \end{array}\right\} \tag{44}$$

3. Unit settlement of support 3:

$$\left.\begin{array}{l} g_1(x_1)=\kappa_3\left[2\ell_1(\ell_2+\ell_3)+\ell_2(2\ell_2+3\ell_3)\right]\left(-\dfrac{x_1^3}{\ell_1}+\ell_1x_1\right) \\ g_2(x_2)=-\dfrac{x_2}{\ell_2}+\kappa_3(\ell_2+\ell_3)\ (\ell_2+2\ell_3)\left(\dfrac{x_2^3}{\ell_2}-3x_2^2+2\ell_2x_2\right)+ \\ \qquad -\kappa_3\left[2\ell_1(\ell_2+\ell_3)+\ell_2(2\ell_2+3\ell_3)\right]\left(-\dfrac{x_2^3}{\ell_2}+\ell_2x_2\right) \\ g_3(x_3)=-\dfrac{\ell_3-x_3}{\ell_3}+\kappa_3\left[2\ell_1(\ell_2+\ell_3)+\ell_2(2\ell_2+3\ell_3)\right]\left(\dfrac{x_3^3}{\ell_3}-3x_3^2+2x_3\ell_3\right) \\ \text{where:}\quad \kappa_3=\dfrac{1}{\ell_2\ell_3(4\ell_1\ell_2+3\ell_2^2+4\ell_1\ell_3+4\ell_2\ell_3)} \end{array}\right\} \tag{45}$$

4. Unit settlement of support 4:

$$\left.\begin{array}{l} g_1(x_1)=\ell_2\kappa_4\left(-\dfrac{x_1^3}{\ell_1}+\ell_1x_1\right) \\ g_2(x_2)=-\ell_2\kappa_4\left(\dfrac{x_2^3}{\ell_2}-3x_2^2+2\ell_2x_2\right)-2\kappa_4(\ell_1+\ell_2)\left(-\dfrac{x_2^3}{\ell_2}+\ell_2x_2\right) \\ g_3(x_3)=\dfrac{x_3}{\ell_3}-2\kappa_4\left(\dfrac{x_3^3}{\ell_3}-3x_3^2+2\ell_3x_3\right) \\ \text{where:}\quad \kappa_4=\dfrac{1}{\ell_3(4\ell_1\ell_2+3\ell_2^2+4\ell_1\ell_3+4\ell_2\ell_3)} \end{array}\right\} \tag{46}$$

5. Unit rotation of support 2:

$$\left.\begin{array}{l} g_1(x_1)=\dfrac{1}{2\ell_1}\left(-\dfrac{x_1^3}{\ell_1}+\ell_1x_1\right) \\ g_2(x_2)=\left(-\dfrac{1}{6\ell_2}+\dfrac{1}{2\ell_1\ell_2(3\ell_2+4\ell_3)\kappa_1}\right)\left(\dfrac{x_2^3}{\ell_2}-3x_2^2+2\ell_2x_2\right)+\dfrac{1}{3\ell_2+4\ell_3}\left(-\dfrac{x_2^3}{\ell_2}+\ell_2x_2\right) \\ g_3(x_3)=\dfrac{1}{3\ell_2+4\ell_3}\left(\dfrac{x_3^3}{\ell_3}-3x_3^2+2\ell_3x_3\right) \end{array}\right\} \tag{47}$$

6. Unit rotation of support 3:

$$\left.\begin{aligned} g_1(x_1) &= \frac{1}{4\ell_1+3\ell_2}\left(-\frac{x_1^3}{\ell_1}+\ell_1 x_1\right) \\ g_2(x_2) &= \frac{1}{4\ell_1+3\ell_2}\left(\frac{x_2^3}{\ell_2}-3x_2^2+2\ell_2 x_2\right)+\left(\frac{1}{2\ell_2\ell_3(4\ell_1+3\ell_2)\kappa_1}-\frac{1}{2\ell_3}\right)\left(-\frac{x_2^3}{\ell_2}+\ell_2 x_2\right) \\ g_3(x_3) &= \frac{1}{2\ell_3}\left(\frac{x_3^3}{\ell_3}-3x_3^2+2\ell_3 x_3\right) \end{aligned}\right\} \quad \textbf{(48)}$$

LATERAL MOVEMENT OF SUPPORTS

From a first view, the problem of the supports lateral motion is almost identical to the one in the paragraph "The vertical motion of supports" of this chapter, which concerns the vertical motion of supports and thus, the influence functions are also identical. But there are some points that require special attention and study.

Let us consider, again, without loss of generality, the simply supported beam, whose supports 1 and 2 move according to the functions $f_1(t)$ and $f_2(t)$, and always under the condition of small differences of the supports movements. Thus, one can write:

$$\upsilon(x,t) = g_1(x)f_1(t) + g_2(x)f_2(t) + \upsilon_0(x,t) \quad \textbf{(49)}$$

Introducing the expression of υ into eq(63c,d) of chapter 3, we have:

$$\begin{aligned} &EI_z(g_1f_1+g_2f_2+\upsilon_0)'''' - EI_z z_M\theta'''' + c_z(g_1f_1+g_2f_2+\upsilon_0)^{\bullet} + m(g_1f_1+g_2f_2+\upsilon_0)^{\bullet\bullet} = 0 \\ &EC_S\theta'''' - EI_z z_M(g_1f_1+g_2f_2+\upsilon_0)'''' + c_\theta\dot{\theta} - GJ_d\theta'' + \Theta_x\ddot{\theta} = 0 \end{aligned}$$

or after differentiations:

$$\left.\begin{aligned} &EI_z\upsilon_0'''' - EI_z z_M\theta'''' + c_z\dot{\upsilon}_0 + m\ddot{\upsilon}_0 = -m(g_1\ddot{f}_1+g_2\ddot{f}_2) - c_z(g_1\dot{f}_1+g_2\dot{f}_2) \\ &EC_S\theta'''' - EI_z z_M\upsilon_0'''' + c_\theta\dot{\theta} - GJ_d\theta'' + \Theta_x\ddot{\theta} = 0 \end{aligned}\right\} \quad \textbf{(50)}$$

We are searching for a solution in the form:

$$\left.\begin{aligned} \upsilon_0(x,t) &= \sum_n V_n(x)T_n(t) \\ \theta(x,t) &= \sum_n \Phi_n(x)T_n(t) \end{aligned}\right\} \quad \textbf{(51)}$$

where $V_n(x)$ and $\Phi_n(x)$ are the corresponding influence functions, while $T_n(t)$ is the unknown time function, which according to chapter 3, is given by:

$$\left.\begin{aligned} T_n(t) = &-\frac{\int_0^\ell g_1(x)V_n(x)dx}{\bar{\omega}_{\theta n}\left(m\int_0^\ell V_n^2dx+\Theta_x\int_0^\ell \Phi_n^2dx\right)}\cdot\int_0^t e^{-\beta_z(t-\tau)}\left[\ddot{f}_1(\tau)+2\beta_z\dot{f}_1(\tau)\right]\cdot\sin\bar{\omega}_{\theta n}(t-\tau)d\tau \\ &-\frac{\int_0^\ell g_2(x)V_n(x)dx}{\bar{\omega}_{\theta n}\cdot\left(m\int_0^\ell V_n^2dx+\Theta_x\int_0^\ell \Phi_n^2dx\right)}\cdot\int_0^t e^{-\beta_z(t-\tau)}\left[\ddot{f}_2(\tau)+2\beta_z\dot{f}_2(\tau)\right]\cdot\sin\bar{\omega}_{\theta n}(t-\tau)d\tau \\ &\text{where: } \beta_z = \frac{c_z}{2m}\ ,\quad \bar{\omega}_{\theta n} = \sqrt{\omega_{\theta n}^2-\beta_z^2} \end{aligned}\right\} \quad \textbf{(52)}$$

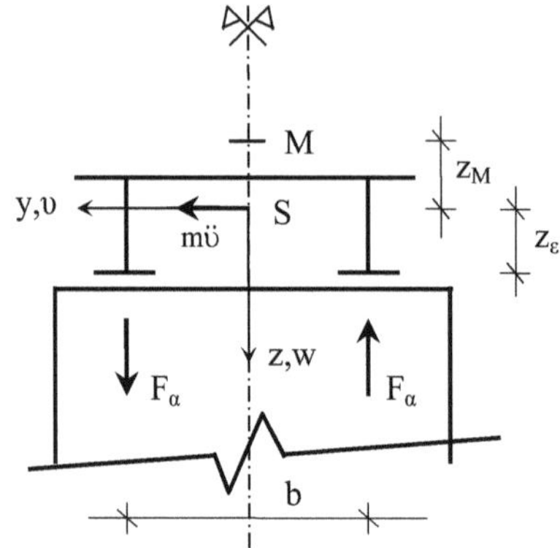

Figure 12: Overturning reaction forces on the pier

The above relations are exact provided that the gravity center of the bridge's cross-section and the supports are on the same level. But usually (Fig. **12**), this is not the case. The gravity center of the bridge's cross-section is at a distance z_ε higher than the supports. Thus, the developing inertia forces $m\ddot{\upsilon}(x,b)$ produce the following reactions:

$$A_1 = \frac{1}{\ell}\cdot\int_0^\ell m x\ddot{\upsilon}dx \quad , \quad A_2 = \frac{1}{\ell}\cdot\int_0^\ell m(\ell - x)\ddot{\upsilon}dx$$

Supposing that the displacement $f_1(t)$ is absolutely bigger than $f_2(t)$, we conclude that $|A_1| > |A_2|$ and therefore, reaction A_1 will produce a bigger overturning moment:

$$M_\alpha = A_1 \cdot z_\varepsilon = \frac{z_\varepsilon}{\ell}\cdot\int_0^\ell m x\ddot{\upsilon}dx$$

The above moment will produce the overturning forces:

$$F_\alpha = \frac{M_\alpha}{b} = \frac{z_\varepsilon}{b\ell}\cdot\int_0^\ell m x\ddot{\upsilon}dx$$

On the other hand, the self-weight produces the reactions:

$$F_B = \frac{mg\ell}{4}$$

The condition expressing the bridge safety against overturning is: $F_B \geq F_\alpha$ or $\frac{mg\ell}{4} \geq \frac{z_\varepsilon m}{b\ell}\cdot\int_0^\ell x\ddot{\upsilon}dx$, and finally:

$$\left.\begin{aligned} &\frac{4z_\varepsilon}{gb\ell^2}\cdot\int_0^\ell x\ddot{\upsilon}dx \leq 1 \\ &\text{where:}\quad \upsilon(x,t) = g_1(x)f_1(t) + g_2(x)f_2(t) + \sum_n V_n(x)T_n(t) \end{aligned}\right\} \qquad \textbf{(53)}$$

For the usual case where $f_1 = f_2 = f$, the above condition is written as:

$$\left.\begin{aligned} &\frac{2z_\varepsilon}{gb\ell}\cdot\int_0^\ell \ddot{\upsilon}dx \leq 1 \\ &\text{where:}\quad \upsilon(x,t) = f(t) + \sum_n V_n(x)T_n(t) \end{aligned}\right\} \qquad \textbf{(54)}$$

Usually, the above overturning forces are not critical. But it has been proved that for bridges located near to the epicenter of an earthquake and for a significant vertical acceleration γ, we must check the above reactions because they may increase from $F_B = \frac{mg\ell}{4}$ to $F_B = \frac{m(g-\gamma)\ell}{4}$.

REFERENCES

[1] S.P. Timoshenko, D.H. Young, W. Weaver Jr. *"Vibration problems in engineering"*, 1974, J. Wiley & Sons, N. York.
[2] A.N. Kounadis *"Dynamics of Continuous Elastic Systems"*, 1983, NTUA Publ., Athens (in Greek).
[3] K. Hirschfeld *"Baustatic"*, 1965, Springer Verlag, Berlin.

CHAPTER 6

Structural Analysis of Bridges

Abstract: This chapter deals with the bridge as structural element. Single span bridges, two-span and three-span bridges, arched bridges, cable-stayed and suspension bridges are analyzed in detail regarding their bending and torsional vibration. The most important relations for studying single cable vibration, harp and fan type cable systems with dense or not arrangement of cables as well as the curved in plane bridges are presented. The determination of modal shapes for some characteristic cases of bridges is also given in detail.

GENERAL

In general, the modeling of a structure, i.e. the formulation of the corresponding linear static system that will attribute more accurately the constructional characteristics but will also approach the functional behavior of the real structure, is a very difficult process.

The simplest example is the modeling of a beam support, where the actual dimensions in the support region (Fig. **1a**) are neglected in the simulation with the well-known support model in Fig. **1b**, a simplification that in some cases may lead to critical states.

Things are even more complicated when the bridge is considered as a three-dimensional structure or when the bridge is curved in plane or even in space.

In the analyses presented in the previous chapters, the method of analysis with the aid of the normal modes has been followed in order to study the structural behavior. The normal modes also called shape functions are obviously depending on the geometrical form of the selected equivalent static system and the correct expression of the support conditions.

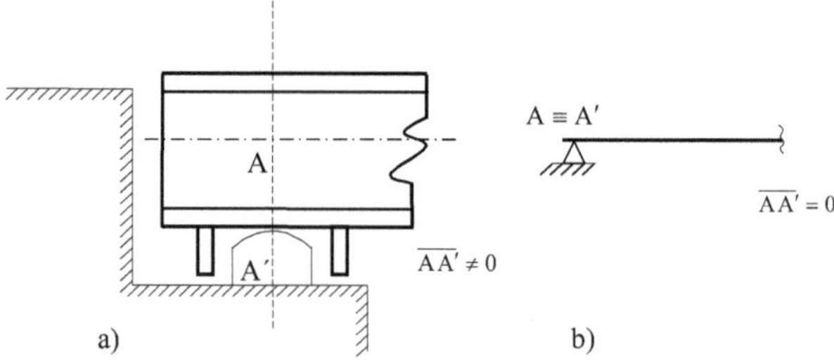

Figure 1: Modeling of support conditions

Let us consider the structural system shown in Fig. **2**, that is arbitrary regarding its shape and the number of supports, with reference to a Cartesian coordinate system $0_i\ x_i\ y_i$ (occasionally one may use also a curvilinear system), with common reference only the system of axes $0_i\ x_i$.

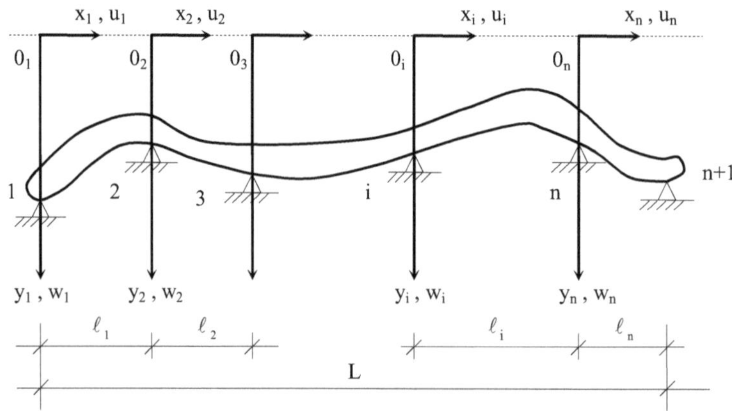

Figure 2: Modeling of a continuous system

George T. Michaltsos and Ioannis G. Raftoyiannis

We also assume that the whole system is not subjected to energy loss at the support points during its movement.

It is possible, using the dynamic equilibrium constitutive equations of the **freely vibrating system** and its support conditions (boundary conditions), to determine the corresponding shape functions. More analytically, the dynamic equilibrium equations of the freely vibrating system shown in Fig. **2** will be:

$$\left.\begin{array}{ll} L_i(x_i, u_i, w_i, t) = 0 & i = 2n \\ R_j = 0 & j = 8n \end{array}\right\} \tag{1}$$

where L_i is an arbitrary linear or nonlinear functional, R_j are the boundary conditions, while the structural line curve expression $y_i = f_i(x_i)$ has been considered known and hence in eq (1) the independents are x_i .

Given that usually the resulting differential equations (1a) are of fourth order, one has to determine 8n integration constants (for the 2 displacements u_i , w_i), which in combination with the boundary conditions eq (1b), lead to a classical eigenvalue problem. Solution of this problem gives the spectrum of eigenfrequencies for the free oscillating system via setting the determinant of the unknown constants for the corresponding linear system equal to zero, as well as the corresponding shape functions under the following form (for the κth shape function):

$$\left.\begin{array}{l} X_\kappa = \left\{\begin{array}{ll} X_{\kappa 1}(x_1) & \text{for span } 1 \\ \ldots\ldots\ldots & \\ X_{\kappa i}(x_i) & \text{for span } i \\ \ldots\ldots\ldots & \\ X_{\kappa n}(x_n) & \text{for span } n \end{array}\right\} \\ \\ \Psi_\kappa = \left\{\begin{array}{ll} \Psi_{\kappa 1}(x_1) & \text{for span } 1 \\ \ldots\ldots\ldots & \\ \Psi_{\kappa i}(x_i) & \text{for span } i \\ \ldots\ldots\ldots & \\ \Psi_{\kappa n}(x_n) & \text{for span } n \end{array}\right\} \end{array}\right\} \tag{2}$$

It is also noted that the above canonical relations are related to each other via the so-called orthogonality conditions in the form:

$$(\omega_\kappa^2 - \omega_\rho^2) \cdot \int_0^L \sum_{i=1}^n m_i (X_{\kappa i} X_{\rho i} + \Psi_{\kappa i} \Psi_{\rho i}) dx = (\omega_\kappa^2 - \omega_\rho^2) \Gamma_{\rho\kappa} = 0 \quad \text{for } \kappa \neq \rho \tag{3}$$

where X_κ , Ψ_κ are the shape functions for the displacements along x and y axes, respectively.

After determining the eigenfrequencies and the corresponding shape functions, we are able henceforth to proceed with studying of the forced vibration due to a given dynamical loading.

The equilibrium equations will then take the form:

$$L_i(x_i, u_i, w_i, t) = (P_{xi}(x_i, t) \quad \text{or} \quad P_{yi}(x_i, t)) \tag{4}$$

where the indexes x and y in the second member correspond to the loading direction P_i. After determining the shape functions, we may seek solutions of eq (4) in the form of separate variables:

$$\left.\begin{array}{l} u = \sum_\rho X_\rho T_\rho(t) \\ w = \sum_\rho \Psi_\rho T_\rho(t) \end{array}\right\} \tag{5}$$

After a suitable process, that was analytically presented in Chapter 3, we arrive at the following equation, which gives the unknown time function T_ρ :

$$\ddot{T}_\rho + 2\beta\dot{T}_\rho + \omega_\rho^2 T_\rho = \frac{\int_0^L \sum_{i=1}^{n} (P_{xi} \cdot X_{\rho i} \text{ or } P_{yi} \cdot \Psi_{\rho i}) dx_i}{\omega_\rho \Gamma_{\rho\rho}} \quad \textbf{(6)}$$

where: $\beta = \dfrac{c_i(x)}{2m_i(x)} = \text{const}$ for the considered system and $c_i(x)$ given in chapter 3.

Solution of eq (6) is given as known by the Duhamel integral. The above process was been already followed analytically in Chapter 3, to study the simply supported beam.

In the present paragraph, we shall determine the frequencies equations and the corresponding shape functions of the most basic structural systems used in bridge engineering, by following the process of chapter 3.

SINGLE SPAN BEAM [1]

Let us consider the single span beam $\overline{12}$ (Fig. **3**), resting on elastic supports with translational constants k_1, k_2 and rotational constants c_1, c_2 respectively. Following the notations and positive signs in Figs. **3b** and **3c**, we formulate the following boundary conditions:

Support 1: $-V_1 + V_{\kappa 1} = 0$

$M_1 + M_{c1} = 0$

Supports 2: $V_2 + V_{\kappa 2} = 0$

$M_2 - M_{c2} = 0$

It is also: $V = -EI_y w'''$

$M = -EI_y w''$

$V_\kappa = kw$

$M_c = cw'$

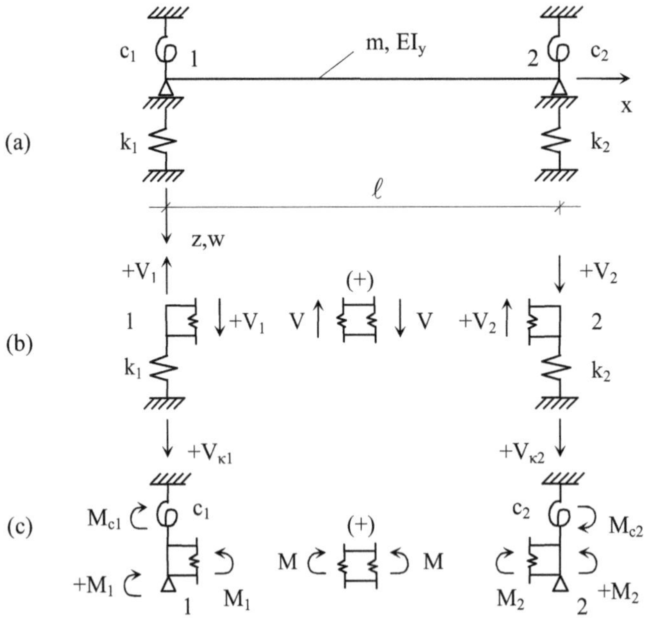

Figure 3: Equilibrium of forces and moments in a single span beam

Seeking a solution in the form $w(x,t) = W(x) \cdot T(t)$, the above boundary conditions can be written in the following form:

$$\left.\begin{aligned} &EI_y W'''(0) + k_1 W(0) = 0 \\ &EI_y W''(0) - c_1 W'(0) = 0 \\ &EI_y W'''(\ell) - k_2 W(\ell) = 0 \\ &EI_y W''(\ell) + c_2 W'(\ell) = 0 \end{aligned}\right\} \qquad (7)$$

Introducing eq(75a) of chapter 3 into the above boundary conditions eq (7), we arrive at the following linear homogeneous system with no second part (i.e., a classic eigenvalue problem):

$$\left.\begin{aligned} &\bar{c}_1 d_1 + \lambda d_2 + \bar{c}_1 d_3 - \lambda d_4 = 0 \\ &(\lambda \sin\lambda\ell - \bar{c}_2 \cos\lambda\ell) d_1 + (\lambda \cos\lambda\ell + \bar{c}_2 \sin\lambda\ell) d_2 + \\ &\qquad + (-\lambda \sinh\lambda\ell - \bar{c}_2 \cosh\lambda\ell) d_3 + (-\lambda \cosh\lambda\ell - \bar{c}_2 \sinh\lambda\ell) d_4 = 0 \\ &-\lambda^3 d_1 + \bar{k}_1 d_2 + \lambda^3 d_3 + \bar{k}_1 d_4 = 0 \\ &(-\lambda^3 \cos\lambda\ell - \bar{k}_2 \sin\lambda\ell) d_1 + (\lambda^3 \sin\lambda\ell - \bar{k}_2 \cos\lambda\ell) d_2 + \\ &\qquad + (\lambda^3 \cosh\lambda\ell - \bar{k}_2 \sinh\lambda\ell) d_3 + (\lambda^3 \sinh\lambda\ell - \bar{k}_2 \cosh\lambda\ell) d_4 = 0 \end{aligned}\right\} \qquad (8)$$

where:

$$\bar{c}_i = \frac{c_i}{EI_y} \quad , \quad \bar{k}_i = \frac{k_i}{EI_y} \quad , \quad \lambda^4 = \frac{m\omega^2}{EI_y} \qquad (9)$$

In order to have other solutions besides the trivial one (zero), the determinant of the system eq (9) d_1, d_2, d_3, d_4 must be set equal to zero. This equation is the frequency equation for bending vibration of a single span beam. It is an equation with unlimited number of roots with respect to the eigenfrequency $\omega = \lambda^2 \cdot \sqrt{\frac{EI_y}{m}}$. From this equation, one can obtain any frequency equation of a single span beam with various boundary conditions. The following are valid:

For pinned support: $\bar{k}_1$ or $\bar{k}_2 \Rightarrow \infty$

For free support: $\bar{k}_1$ or $\bar{k}_2 = 0$

For fixed support: $\bar{c}_1$ or $\bar{c}_2 \Rightarrow \infty$

For free rotation: $\bar{c}_1$ or $\bar{c}_2 = 0$

Simply Supported Beam $\bar{\mathbf{k}}_1 = \bar{\mathbf{k}}_2 = \infty$, $\bar{\mathbf{c}}_1 = \bar{\mathbf{c}}_2 = \mathbf{0}$

Frequency equation:

$$\begin{vmatrix} 0 & 1 & 0 & -1 \\ \sin\lambda_n\ell & \cos\lambda_n\ell & -\sinh\lambda_n\ell & -\cosh\lambda_n\ell \\ 0 & 1 & 0 & -1 \\ -\sin\lambda_n\ell & -\cos\lambda_n\ell & -\sinh\lambda_n\ell & -\cosh\lambda_n\ell \end{vmatrix} = 0 \qquad (10)$$

Shape function:

$$W_n(x) = d_{1n} \cdot \sin\frac{n\pi x}{\ell} \quad , \qquad (n = 1,2,\ldots\ldots) \tag{11}$$

Beam Fixed at Both Ends $\overline{\mathbf{k}}_1 = \overline{\mathbf{k}}_2 = \overline{\mathbf{c}}_1 = \overline{\mathbf{c}}_2 = \infty$

Frequency equation:

$$\begin{vmatrix} 1 & 0 & 1 & 0 \\ -\cos\lambda_n\ell & \sin\lambda_n\ell & -\cosh\lambda_n\ell & -\sinh\lambda_n\ell \\ 0 & 1 & 0 & 1 \\ -\sin\lambda_n\ell & -\cos\lambda_n\ell & -\sinh\lambda_n\ell & -\cosh\lambda_n\ell \end{vmatrix} = 0 \tag{12}$$

Shape function:

$$\left.\begin{aligned} & W_n(x) = d_{2n}\left[\frac{d_{1n}}{d_{2n}}(\sin\lambda_n x - \sinh\lambda_n x) + \cos\lambda_n x - \cosh\lambda_n x\right] \\ & \text{where:} \quad \frac{d_{1n}}{d_{2n}} = \frac{\cos\lambda_n\ell - \cosh\lambda_n\ell}{\sinh\lambda_n\ell - \sin\lambda_n\ell} \end{aligned}\right\} \tag{13}$$

Beam Fixed at End 1 $\overline{\mathbf{k}}_1 = \overline{\mathbf{k}}_2 = \overline{\mathbf{c}}_1 = \infty$, $\overline{\mathbf{c}}_2 = \mathbf{0}$

Frequency equation:

$$\begin{vmatrix} 1 & 0 & 1 & 0 \\ \sin\lambda_n\ell & \cos\lambda_n\ell & -\sinh\lambda_n\ell & -\cosh\lambda_n\ell \\ 0 & 1 & 0 & 1 \\ -\sin\lambda_n\ell & -\cos\lambda_n\ell & -\sinh\lambda_n\ell & -\cosh\lambda_n\ell \end{vmatrix} = 0 \tag{14}$$

Shape function:

$$\left.\begin{aligned} & W_n(x) = d_{2n}\left[\frac{d_{1n}}{d_{2n}}(\sin\lambda_n x - \sinh\lambda_n x) + \cos\lambda_n x - \cosh\lambda_n x\right] \\ & \text{where:} \quad \frac{d_{1n}}{d_{2n}} = \frac{\cos\lambda_n\ell - \cosh\lambda_n\ell}{\sinh\lambda_n\ell - \sin\lambda_n\ell} \end{aligned}\right\} \tag{15}$$

One can observe that eq (13) and eq (15) are identical. Their only difference is on λ_n , which results from different frequency equations.

Cantilever Beam $\overline{\mathbf{k}}_1 = \overline{\mathbf{c}}_1 = \infty$, $\overline{\mathbf{k}}_2 = \overline{\mathbf{c}}_2 = \mathbf{0}$

Frequency equation:

$$\begin{vmatrix} 1 & 0 & 1 & 0 \\ \sin\lambda_n\ell & \cos\lambda_n\ell & -\sinh\lambda_n\ell & -\cosh\lambda_n\ell \\ 0 & 1 & 0 & 1 \\ \cos\lambda_n\ell & \sin\lambda_n\ell & \cosh\lambda_n\ell & \sinh\lambda_n\ell \end{vmatrix} = 0 \tag{16}$$

Shape function:

$$\left.\begin{aligned} &W_n(x) = d_{2n}\left[\frac{d_{1n}}{d_{2n}}(\sin\lambda_n x - \sinh\lambda_n x) - \cos\lambda_n x + \cosh\lambda_n x\right] \\ &\text{where:}\quad \frac{d_{1n}}{d_{2n}} = \frac{\cos\lambda_n \ell + \cosh\lambda_n \ell}{\sinh\lambda_n \ell + \sin\lambda_n \ell} \end{aligned}\right\} \qquad \textbf{(17)}$$

Torsional Vibration

For torsional vibration of the simply supported beam, eqs(53b), (96a,b,c), (98) and (101) of chapter 3 are valid. The equation of motion is:

$$EC_S\Phi'''' - GJ_d\Phi'' - \Theta_x\omega_\theta^2\Phi = 0 \qquad \textbf{(18)}$$

and its solution is:

$$\left.\begin{aligned} &\Phi = d_1\sin\lambda_1 x + d_2\cos\lambda_1 x + d_3\sinh\lambda_2 x + d_4\cosh\lambda_2 x \\ &\text{where:}\quad \lambda_1 = \sqrt{-\frac{GJ_d}{2EC_S} + \sqrt{\left(\frac{GJ_d}{2EC_S}\right)^2 + \frac{\Theta_x\omega_0^2}{EC_S}}} \\ &\qquad\qquad \lambda_2 = \sqrt{\frac{GJ_d}{2EC_S} + \sqrt{\left(\frac{GJ_d}{2EC_S}\right)^2 + \frac{\Theta_x\omega_0^2}{EC_S}}} \end{aligned}\right\} \qquad \textbf{(19a,b,c)}$$

The equation for eigenfrequencies is:

$$\begin{vmatrix} 0 & 1 & 0 & 1 \\ 0 & -\lambda_1^2 & 0 & \lambda_2^2 \\ \sin\lambda_1\ell & \cos\lambda_1\ell & \sinh\lambda_2\ell & \cosh\lambda_2\ell \\ -\lambda_1^2\sin\lambda_1\ell & -\lambda_1^2\cos\lambda_1\ell & \lambda_2^2\sinh\lambda_2\ell & \lambda_2^2\cosh\lambda_2\ell \end{vmatrix} = 0 \qquad \textbf{(20)}$$

Finally, the shape function is:

$$\Phi_n(x) = d_{1n}\left(\sin\lambda_{1n}x - \frac{\sin\lambda_{1n}\ell}{\sinh\lambda_{2n}\ell}\cdot\sinh\lambda_{2n}x\right) \qquad \textbf{(21)}$$

TWO - SPAN BEAM

Bending Vibration [2]

Let us consider the beam shown in Fig. **4** with constant cross-section.

A local coordinate system is attached to each opening as shown in Fig. **4**. The equations of motion are:

$$W_1''''(x_1) - \lambda^4 W_1(x_1) = 0$$
$$W_2''''(x_2) - \lambda^4 W_2(x_2) = 0$$

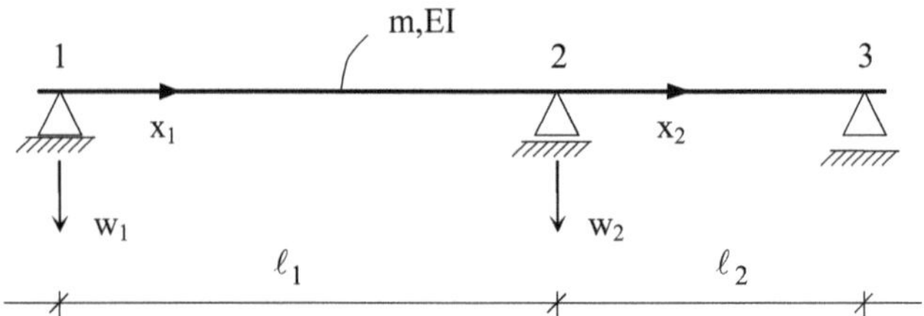

Figure 4: Two-span beam

and the corresponding solutions are:

$$W_1(x_1) = d_1 \sin\lambda x_1 + d_2 \cos\lambda x_1 + d_3 \sinh\lambda x_1 + d_4 \cosh\lambda x_1$$
$$W_2(x_2) = d_5 \sin\lambda x_2 + d_6 \cos\lambda x_2 + d_7 \sinh\lambda x_2 + d_8 \cosh\lambda x_2$$

The boundary conditions are:

$$W_1(0)=0\ ,\quad W_1(\ell_1)=0\ ,\qquad W_2(0)=0\ ,\qquad W_2(\ell_2)=0$$
$$W_1''(0)=0\ ,\quad W_2''(\ell_2)=0\ ,\quad W_1'(\ell_1)=W_2'(0)\ ,\quad W_1''(\ell_1)=W_2''(0)$$

from which the following frequency equation is obtained:

$$\begin{vmatrix} 0 & 1 & 0 & 1 & 0 & 0 & 0 & 0 \\ \sin\lambda\ell_1 & \cos\lambda\ell_1 & \sinh\lambda\ell_1 & \cosh\lambda\ell_1 & 0 & 0 & 0 & 0 \\ 0 & -1 & 0 & 1 & 0 & 0 & 0 & 0 \\ \cos\lambda\ell_1 & -\sin\lambda\ell_1 & \cosh\lambda\ell_1 & \sinh\lambda\ell_1 & -1 & 0 & -1 & 0 \\ -\sin\lambda\ell_1 & -\cos\lambda\ell_1 & \sinh\lambda\ell_1 & \cosh\lambda\ell_1 & 0 & 1 & 0 & -1 \\ 0 & 0 & 0 & 0 & 0 & 1 & 0 & 1 \\ 0 & 0 & 0 & 0 & \sin\lambda\ell_2 & \cos\lambda\ell_2 & \sinh\lambda\ell_2 & \cosh\lambda\ell_2 \\ 0 & 0 & 0 & 0 & -\sin\lambda\ell_2 & -\cos\lambda\ell_2 & \sinh\lambda\ell_2 & \cosh\lambda\ell_2 \end{vmatrix} = 0 \qquad \textbf{(22)}$$

The shape functions are:

$$\left.\begin{aligned} W_n(x_1) &= \frac{1}{\sin\lambda_n\ell_1}\sin\lambda_n x_1 - \frac{1}{\sinh\lambda_n\ell_1}\sinh\lambda_n x_1 \qquad \text{for } 0 \le x_1 \le \ell_1 \\ W_n(x_2) &= -\cot\lambda_n\ell_2 \cdot \sin\lambda_n x_2 + \cos\lambda_n x_2 + \\ &\quad + \coth\lambda_n\ell_2 \cdot \sinh\lambda_n x_2 - \cosh\lambda_n x_2 \qquad \text{for } 0 \le x_2 \le \ell_2 \end{aligned}\right\} \qquad \textbf{(23)}$$

where: $\lambda_n^4 = \dfrac{m\omega_n^2}{EI_y}$

Torsional Vibration

The equations of motion for a two-span beam are:

$$\left.\begin{aligned} EC_S\Phi_1''''(x_1) - GJ_d\Phi_1''(x_1) - \Theta_x\omega_\theta^2\Phi_1(x_1) = 0 \\ EC_S\Phi_2''''(x_2) - GJ_d\Phi_2''(x_2) - \Theta_x\omega_\theta^2\Phi_2(x_2) = 0 \end{aligned}\right\} \qquad \textbf{(24a,b)}$$

The solution of eq (24) is:

$$\left.\begin{aligned}\Phi_1(x_1) &= d_1\sin\lambda_1 x_1 + d_2\cos\lambda_1 x_1 + d_3\sinh\lambda_2 x_1 + d_4\cosh\lambda_2 x_1\\ \Phi_2(x_2) &= d_5\sin\lambda_1 x_2 + d_6\cos\lambda_1 x_2 + d_7\sinh\lambda_2 x_2 + d_8\cosh\lambda_2 x_2\end{aligned}\right\} \quad \textbf{(25)}$$

where λ_1 and λ_2 are given by eq (19 b,c).

The boundary conditions are:

$$\begin{array}{llll}\Phi_1(0)=0 & \Phi_1(\ell_1)=0 & \Phi_2(0)=0 & \Phi_2(\ell_2)=0\\ \Phi_1''(0)=0 & \Phi_2''(\ell_2)=0 & \Phi_1'(\ell_1)=\Phi_2'(0) & \Phi_1''(\ell_1)=\Phi_2''(0)\end{array}$$

from which the following frequency equation is obtained:

$$\begin{vmatrix} 0 & 1 & 0 & 1 & 0 & 0 & 0 & 0\\ \sin\lambda_1\ell_1 & \cos\lambda_1\ell_1 & \sinh\lambda_2\ell_1 & \cosh\lambda_2\ell_1 & 0 & 0 & 0 & 0\\ 0 & -1 & 0 & 1 & 0 & 0 & 0 & 0\\ \lambda_1\cos\lambda_1\ell_1 & -\lambda_1\sin\lambda_1\ell_1 & \lambda_2\cosh\lambda_2\ell_1 & \lambda_2\sinh\lambda_2\ell_1 & -\lambda_1 & 0 & \lambda_2 & 0\\ -\lambda_1^2\sin\lambda_1\ell_1 & -\lambda_1^2\cos\lambda_1\ell_1 & \lambda_2^2\sinh\lambda_2\ell_1 & \lambda_2^2\cosh\lambda_2\ell_1 & 0 & \lambda_1^2 & 0 & -\lambda_2^2\\ 0 & 0 & 0 & 0 & 0 & 1 & 0 & 1\\ 0 & 0 & 0 & 0 & \sin\lambda_1\ell_2 & \cos\lambda_1\ell_2 & \sinh\lambda_2\ell_2 & \cosh\lambda_2\ell_2\\ 0 & 0 & 0 & 0 & -\lambda_1^2\sin\lambda_1\ell_2 & -\lambda_1^2\cos\lambda_1\ell_2 & \lambda_2^2\sinh\lambda_2\ell_2 & \lambda_2^2\cosh\lambda_2\ell_2 \end{vmatrix} = 0 \quad \textbf{(26)}$$

Finally, the corresponding shape functions are:

$$\left.\begin{aligned}\Phi_{1n}(x_1) &= \frac{1}{\sin\lambda_{1n}\ell_1}\cdot\sin\lambda_{1n}x_1 - \frac{1}{\sinh\lambda_{2n}\ell_1}\sinh\lambda_{2n}x_1 \quad , \quad 0\le x_1\le\ell_1\\ \Phi_{2n}(x_2) &= -\cot\lambda_{1n}\ell_2\cdot\sin\lambda_{1n}x_2 + \cos\lambda_{1n}x_2 +\\ &\quad +\coth\lambda_{2n}\ell_2\cdot\sinh\lambda_{2n}x_2 - \cosh\lambda_{2n}x_2 \quad , \quad 0\le x_2\le\ell_2\end{aligned}\right\} \quad \textbf{(27)}$$

THREE - SPAN BEAM

Bending Vibration

Let us consider the beam shown in Fig. **5** with constant cross-section. A local coordinate system is attached to each opening as shown in Fig. **5**. The equations of motion are:

$$\begin{aligned} W_1''''(x_1) - \lambda^4 W_1(x_1) &= 0\\ W_2''''(x_2) - \lambda^4 W_2(x_2) &= 0\\ W_3''''(x_3) - \lambda^4 W_3(x_3) &= 0\end{aligned}$$

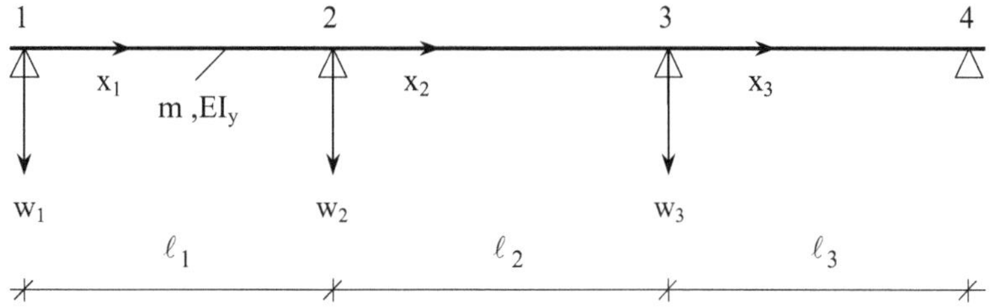

Figure 5: Three-span beam

and the corresponding solutions are:

$$\left.\begin{aligned} W_1(x_1) &= d_1 \sin\lambda x_1 + d_2 \cos\lambda x_1 + d_3 \sinh\lambda x_1 + d_4 \cosh\lambda x_1 \\ W_2(x_2) &= d_5 \sin\lambda x_2 + d_6 \cos\lambda x_2 + d_7 \sinh\lambda x_2 + d_8 \cosh\lambda x_2 \\ W_3(x_3) &= d_9 \sin\lambda x_3 + d_{10} \cos\lambda x_3 + d_{11} \sinh\lambda x_3 + d_{12} \cosh\lambda x_3 \end{aligned}\right\} \tag{28}$$

The boundary conditions are:

$$\left.\begin{array}{lll} W_1(0) = 0 \;, & W_1(\ell_1) = 0 \;, & W_1''(0) = 0 \\ W_1'(\ell_1) = W_2'(0) \;, & W_1''(\ell_1) = W_2''(0) & \\ W_2(0) = 0 \;, & W_2(\ell_2) = 0 & \\ W_2'(\ell_2) = W_3'(0) \;, & W_2''(\ell_2) = W_3''(0) & \\ W_3(0) = 0 \;, & W_3(\ell_3) = 0 \;, & W_3''(\ell_3) = 0 \end{array}\right\} \tag{29}$$

Introducing the expressions (28) into the boundary conditions (29) the following frequency equation is obtained:

$$\begin{vmatrix} 0 & 1 & 0 & 1 & 0 & 0 & 0 & 0 & 0 & 0 & 0 & 0 \\ \sin\lambda\ell_1 & \cos\lambda\ell_1 & \sinh\lambda\ell_1 & \cosh\lambda\ell_1 & 0 & 0 & 0 & 0 & 0 & 0 & 0 & 0 \\ 0 & -1 & 0 & 1 & 0 & 0 & 0 & 0 & 0 & 0 & 0 & 0 \\ \cos\lambda\ell_1 & -\sin\lambda\ell_1 & \cosh\lambda\ell_1 & \sinh\lambda\ell_1 & -1 & 0 & -1 & 0 & 0 & 0 & 0 & 0 \\ -\sin\lambda\ell_1 & -\cos\lambda\ell_1 & \sinh\lambda\ell_1 & \cosh\lambda\ell_1 & 0 & 1 & 0 & -1 & 0 & 0 & 0 & 0 \\ 0 & 0 & 0 & 0 & 0 & 1 & 0 & 1 & 0 & 0 & 0 & 0 \\ 0 & 0 & 0 & 0 & \sin\lambda\ell_2 & \cos\lambda\ell_2 & \sinh\lambda\ell_2 & \cosh\lambda\ell_2 & 0 & 0 & 0 & 0 \\ 0 & 0 & 0 & 0 & \cos\lambda\ell_2 & -\sin\lambda\ell_2 & \cosh\lambda\ell_2 & \sinh\lambda\ell_2 & -1 & 0 & -1 & 0 \\ 0 & 0 & 0 & 0 & -\sin\lambda\ell_2 & -\cos\lambda\ell_2 & \sinh\lambda\ell_2 & \cosh\lambda\ell_2 & 0 & 1 & 0 & -1 \\ 0 & 0 & 0 & 0 & 0 & 0 & 0 & 0 & 0 & 1 & 0 & 1 \\ 0 & 0 & 0 & 0 & 0 & 0 & 0 & 0 & \sin\lambda\ell_3 & \cos\lambda\ell_3 & \sinh\lambda\ell_3 & \cosh\lambda\ell_3 \\ 0 & 0 & 0 & 0 & 0 & 0 & 0 & 0 & -\sin\lambda\ell_3 & -\cos\lambda\ell_3 & \sinh\lambda\ell_3 & \cosh\lambda\ell_3 \end{vmatrix} = 0 \tag{30}$$

The corresponding shape functions are:

$$
\left.
\begin{aligned}
&W_n(x_1) = \frac{1}{\sin\lambda_n \ell_1} \cdot \sin\lambda_n x_1 - \frac{1}{\sinh\lambda_n \ell_1} \cdot \sinh\lambda_n x_1 \qquad \text{for } 0 \le x_1 \le \ell_1 \\
&W_n(x_2) = \left(-\cot\lambda_n \ell_2 + \frac{d}{\sin\lambda_n \ell_2} \right) \sin\lambda_n x_2 + \cos\lambda_n x_2 + \\
&\qquad + \left(\coth\lambda_n \ell_2 - \frac{d}{\sinh\lambda_n \ell_2} \right) \sinh\lambda_n x_2 - \cosh\lambda_n x_2 \quad \text{for } 0 \le x_2 \le \ell_2 \\
&W_n(x_3) = -d \cdot \cot\lambda_n \ell_3 \cdot \sin\lambda_n x_3 + d \cdot \cos\lambda_n x_3 + \\
&\qquad + d \cdot \coth\lambda_n \ell_3 \cdot \sinh\lambda_n x_3 - d \cdot \cosh\lambda_n x_3 \qquad \text{for } \alpha \ 0 \le x_3 \le \ell_3 \\
&\text{where:} \\
&d = \frac{\sin\lambda_n \ell_2 - \sinh\lambda_n \ell_2}{\sin\lambda_n \ell_2 \sinh\lambda_n \ell_2 (\coth\lambda_n \ell_2 + \coth\lambda_n \ell_3 - \cot\lambda_n \ell_2 - \cot\lambda_n \ell_3)}
\end{aligned}
\right\} \qquad (31)
$$

Torsional Vibration

The equations of motion are:

$$
\left.
\begin{aligned}
&EC_S \Phi_1''''(x_1) - GJ_d \Phi_1''(x_1) - \Theta_x \omega_0^2 \Phi_1(x_1) = 0 \\
&EC_S \Phi_2''''(x_2) - GJ_d \Phi_2''(x_2) - \Theta_x \omega_0^2 \Phi_2(x_2) = 0 \\
&EC_S \Phi_3''''(x_3) - GJ_d \Phi_3''(x_3) - \Theta_x \omega_0^2 \Phi_3(x_3) = 0
\end{aligned}
\right\} \qquad (32)
$$

The solutions corresponding to the above equations are:

$$
\left.
\begin{aligned}
&\Phi_1(x_1) = d_1 \sin\lambda_1 x_1 + d_2 \cos\lambda_1 x_1 + d_3 \sinh\lambda_2 x_1 + d_4 \cosh\lambda_2 x_1 \\
&\Phi_2(x_2) = d_5 \sin\lambda_1 x_2 + d_6 \cos\lambda_1 x_2 + d_7 \sinh\lambda_2 x_2 + d_8 \cosh\lambda_2 x_1 \\
&\Phi_3(x_3) = d_9 \sin\lambda_1 x_3 + d_{10} \cos\lambda_1 x_3 + d_{11} \sinh\lambda_2 x_3 + d_{12} \cosh\cdot\lambda_2 x_3
\end{aligned}
\right\} \qquad (33)
$$

The boundary conditions are:

$$
\left.
\begin{aligned}
&\Phi_1(0) = 0 \quad , \quad \Phi_1(\ell_1) = 0 \quad , \quad \Phi_1''(0) = 0 \\
&\Phi_1'(\ell_1) = \Phi_2'(0) \quad , \quad \Phi_1''(\ell_1) = \Phi_2''(0) \\
&\Phi_2(0) = 0 \quad , \quad \Phi_2(\ell_2) = 0 \\
&\Phi_2'(\ell_2) = \Phi_3'(0) \quad , \quad \Phi_2''(\ell_2) = \Phi_3''(0) \\
&\Phi_3(0) = 0 \quad , \quad \Phi_3(\ell_3) = 0 \quad , \quad \Phi_3''(\ell_3) = 0
\end{aligned}
\right\} \qquad (34)
$$

Introducing eq (33) into eq (34) we obtain the following frequency equation:

$$\begin{vmatrix}
0 & 1 & 0 & 1 & 0 & 0 & 0 & 0 & 0 & 0 & 0 & 0 \\
\sin\lambda_1\ell_1 & \cos\lambda_1\ell_1 & \sinh\lambda_2\ell_1 & \cosh\lambda_2\ell_1 & 0 & 0 & 0 & 0 & 0 & 0 & 0 & 0 \\
0 & -\lambda_1^2 & 0 & \lambda_2^2 & 0 & 0 & 0 & 0 & 0 & 0 & 0 & 0 \\
\lambda_1\cos\lambda_1\ell_1 & -\lambda_1\sin\lambda_1\ell_1 & \lambda_2\cosh\lambda_2\ell_1 & \lambda_2\sinh\lambda_2\ell_1 & -1 & 0 & -1 & 0 & 0 & 0 & 0 & 0 \\
-\lambda_1^2\sin\lambda_1\ell_1 & -\lambda_1^2\cos\lambda_1\ell_1 & \lambda_2^2\sinh\lambda_2\ell_1 & \lambda_2^2\cosh\lambda_2\ell_1 & 0 & 1 & 0 & -1 & 0 & 0 & 0 & 0 \\
0 & 0 & 0 & 0 & 0 & 1 & 0 & 1 & 0 & 0 & 0 & 0 \\
0 & 0 & 0 & 0 & \sin\lambda_1\ell_2 & \cos\lambda_1\ell_2 & \sinh\lambda_2\ell_2 & \cosh\lambda_2\ell_2 & 0 & 0 & 0 & 0 \\
0 & 0 & 0 & 0 & \lambda_1\cos\lambda_1\ell_2 & -\lambda_1\sin\lambda_1\ell_2 & \lambda_2\cosh\lambda_2\ell_2 & \lambda_2\sinh\lambda_2\ell_2 & -1 & 0 & -1 & 0 \\
0 & 0 & 0 & 0 & -\lambda_1^2\sin\lambda_1\ell_2 & -\lambda_1^2\cos\lambda_1\ell_2 & \lambda_2^2\sinh\lambda_2\ell_2 & \lambda_2^2\cosh\lambda_2\ell_2 & 0 & 1 & 0 & -1 \\
0 & 0 & 0 & 0 & 0 & 0 & 0 & 0 & 0 & 1 & 0 & 1 \\
0 & 0 & 0 & 0 & 0 & 0 & 0 & 0 & \sin\lambda_1\ell_3 & \cos\lambda_1\ell_3 & \sinh\lambda_2\ell_3 & \cosh\lambda_2\ell_3 \\
0 & 0 & 0 & 0 & 0 & 0 & 0 & 0 & -\lambda_1^2\sin\lambda_1\ell_3 & -\lambda_1^2\cos\lambda_1\ell_3 & \lambda_2^2\sinh\lambda_2\ell_3 & \lambda_2^2\cosh\lambda_2\ell_3
\end{vmatrix} = 0 \qquad (35)$$

The corresponding shape functions are:

$$\left.\begin{aligned}
&\Phi_{1n}(x_1)=\frac{1}{\sin\lambda_{1n}\ell_1}\cdot\sin\lambda_{1n}x_1-\frac{1}{\sinh\lambda_{2n}\ell_1}\cdot\sin\lambda_{2n}x_1 \qquad 0\le x_1\le\ell_1\\
&\Phi_{2n}(x_2)=\left(\cot\lambda_{1n}\ell_2+\frac{d}{\sin\lambda_{1n}\ell_2}\right)\cdot\sin\lambda_{1n}x_2+\cos\lambda_{1n}x_2+\\
&\qquad+\left(\coth\lambda_{2n}\ell_2-\frac{d}{\sinh\lambda_{2n}\ell_2}\right)\cdot\sinh\lambda_{2n}x_2-\cosh\lambda_{2n}x_2 \qquad 0\le x_2\le\ell_2\\
&\Phi_{3n}(x_3)=-d\cdot\cot\lambda_{1n}\ell_3\cdot\sin\lambda_{1n}x_3+d\cdot\cos\lambda_{1n}x_3+\\
&\qquad+d\cdot\coth\lambda_{2n}\ell_3\cdot\sinh\lambda_{2n}x_3-d\cdot\cosh\lambda_{2n}x_3 \qquad 0\le x_3\le\ell_3\\
&\text{where:}\\
&d=\frac{\sin\lambda_{1n}\ell_2-\sinh\lambda_{2n}\ell_2}{\sin\lambda_{1n}\ell_2\cdot\sinh\lambda_{2n}\ell_2(\coth\lambda_{2n}\ell_2+\coth\lambda_{2n}\ell_3-\cot\lambda_{1n}\ell_2-\cot\lambda_{1n}\ell_3)}
\end{aligned}\right\}\qquad \textbf{(36)}$$

THE ARCHED BEAM [3]

Let us consider the arched beam shown in Fig. **6**, resting on points 1 and 2 the curve of which is given by eq (37):

$$z = z(x) \qquad \textbf{(37)}$$

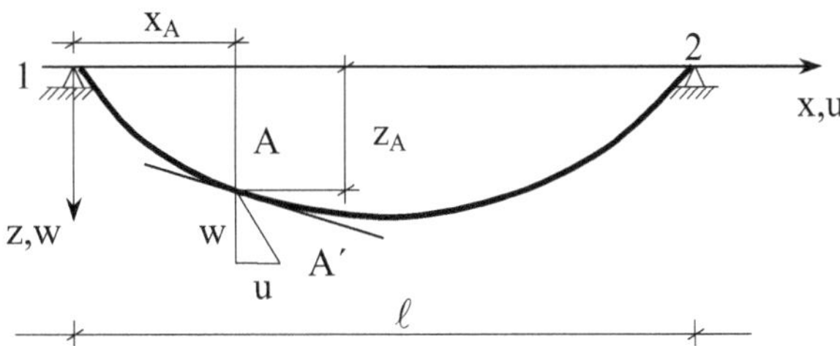

Figure 6: Arched beam

If the infinitesimal arch element AB is displaced into a new position A′B′ (see Fig. **7**), the following relations are valid:

$$\left.\begin{aligned} dx &= ds\cdot\cos\varphi\\ dz &= ds\cdot\sin\varphi \end{aligned}\right\}\qquad \textbf{(38)}$$

Dropping the terms of order higher than 2, such as $\varepsilon\cdot dz\cdot\psi$ and $\varepsilon\cdot dz\cdot dx$ and taking into account that ψ is very small and hence: $\cos\psi\cong 1$, $\sin\psi=\psi$ we will arrive at the following relations:

$$\begin{aligned} dx+du &= (1+\varepsilon)ds\cdot\cos(\varphi+\psi)\cong dx+\varepsilon dx-\psi dz\\ dz+dw &= (1+\varepsilon)ds\cdot\sin(\varphi+\psi)\cong dz+\varepsilon dz+\psi dx \end{aligned}$$

or finally:

$$\left.\begin{aligned} du &= \varepsilon dx - \psi dz \\ dw &= \psi dx + \varepsilon dz \end{aligned}\right\} \qquad \textbf{(39a,b)}$$

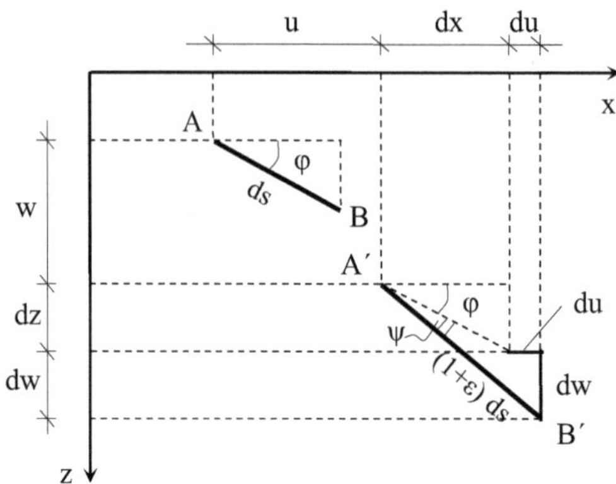

Figure 7: Deformation of an infinitesimal arch element

Assuming that the arch axis is incompressible (i.e., ε=0), we obtain:

$$\left.\begin{aligned} \frac{d\psi}{ds} &= \frac{d^2w}{dx^2}\cdot\frac{dx}{ds} \\ \frac{du}{dx} &= \varepsilon\cdot\left[1+\left(\frac{dz}{dx}\right)^2\right]-\frac{dw}{dx}\cdot\frac{dz}{dx} \end{aligned}\right\} \qquad \textbf{(40a,b)}$$

and hence, from the second of eq (40) we obtain:

$$\int_0^\ell du = u(\ell)-u(0) = \int_0^\ell \varepsilon(1+z'^2)dx - \int_0^\ell z'w'dx$$

After integrating by parts and assuming at least one immovable support (i.e., point 1) we will have: $w(0)=w(\ell)=u(0)=0$ and thus:

$$u(\ell) = \int_0^\ell \varepsilon(1+z'^2)dx + \int_0^\ell z''wdx \qquad \textbf{(41)}$$

For positive signs of forces and moments according to Fig. **8**, it is:

$$\varepsilon = \frac{N}{EF(x)} \;,\quad 1+z'^2 = \frac{1}{\cos^2\varphi} \;,\quad N = \frac{H}{\cos\varphi} \qquad \textbf{(42)}$$

Then, eq (41) gives:

$$\int_0^\ell \varepsilon(1+z'^2)\cdot dx = u(\ell) - \int_0^\ell z''w\cdot dx$$

$$\text{or}\quad \int_0^\ell \frac{N}{EF(x)}\cdot\frac{1}{\cos^2\varphi}dx = \int_0^\ell \frac{H}{EF(x)\cdot\cos^3\varphi}\cdot dx = u(\ell) - \int_0^\ell z''w\cdot dx$$

From Fig. **8**, it follows that: $dH = -p_x dx$ and if (as usual): $p_x = 0$ it will be:

$$H = \text{const.} \tag{43}$$

and thus:

$$\left.\begin{aligned} & H = \frac{1}{L} \cdot \left[u(\ell) - \int_0^\ell z'' w dx \right] \\ & \text{where:} \quad L = \int_0^\ell \frac{dx}{EF(x)\cos^3 \varphi} \end{aligned}\right\} \tag{44}$$

From equilibrium in Fig. **8**, with p_x=0 , we obtain:

$$p_z dx - Hz' + H(z' + dz') = 0 \quad \text{or} \quad p_z = -H \cdot \frac{dz'}{dx} \quad \text{and:} \quad H = -\frac{p_z}{z''} \tag{45}$$

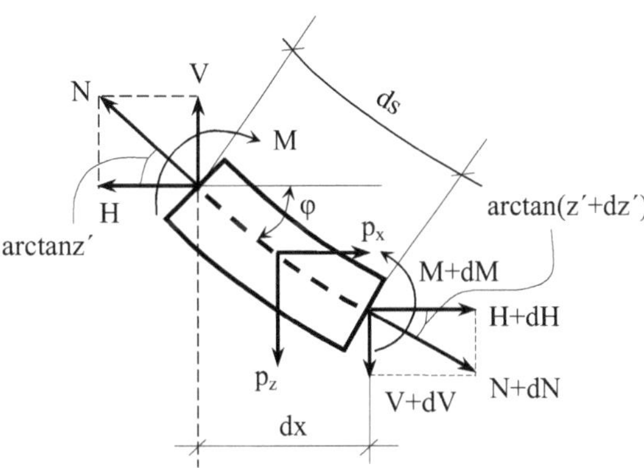

Figure 8: Equilibrium of an infinitesimal arch element

It is also valid:

$$\frac{dV}{dx} = -p_z \quad , \quad -\frac{d^2 M}{dx^2} = H(z - w)'' + p_z \tag{46a,b}$$

Due to eq (40) and (46b), it will be:

$$\frac{d\psi}{ds} = \frac{d^2 w}{dx^2} \cdot \frac{dx}{ds} = -\frac{M}{EI} \quad \text{or} \quad \frac{d^2 w}{dx^2} = -\frac{M}{EI\cos\varphi} \text{ and due to eq (46b):}$$

$$\frac{d^2}{dx^2}\left(EI\cos\varphi \cdot \frac{d^2 w}{dx^2} \right) = H(z - w)'' + p_z \text{ or finally:}$$

$$\frac{d^2}{dx^2}\left(EI\cos\varphi \cdot \frac{d^2 w}{dx^2} \right) + Hw'' = \frac{z''}{L} \cdot \left[u(\ell) - \int_0^\ell z'' w dx \right] + p_z$$

If we consider: $p_z = p - m\ddot{w}$, we will have:

$$\frac{d^2}{dx^2}\left(EI\cos\varphi \cdot \frac{d^2 w}{dx^2} \right) + Hw'' + m\ddot{w} = \frac{z''}{L} \cdot \left[u(\ell) - \int_0^\ell z'' w dx \right] + p_z \tag{47}$$

Finally, from eq (39) with ε = 0 we obtain:

$$u = -\int z'w'dx \tag{48}$$

The above eq (47) is the equation of motion for an arch with shape given by the general expression (37).

The Simply Supported Parabolic Arch [4]

Let us consider the simply supported parabolic arch shown in Fig. **9**, with shape given as follows:

$$z(x) = \frac{4f}{\ell^2}x^2 - \frac{4f}{\ell}x \tag{49}$$

Then, it is:

$$z'' = \frac{8f}{\ell^2} \tag{50}$$

and since $p_z = mg$, it will be:

$$H = -\frac{p_z}{z''} = -\frac{mg\ell^2}{8f} \tag{51}$$

It is also valid:

$$\frac{ds}{dx} = \sqrt{1+z'^2} \cong 1 + \frac{1}{2}z'^2 = 1 + \frac{1}{2}\left(\frac{64f^2}{\ell^4}x^2 - \frac{64f^2}{\ell^3}x + \frac{16f^2}{\ell^2}\right) \quad \text{and}$$

$$s = \int_0^\ell ds = \ell + \left(\frac{32f^2}{\ell^4}\cdot\frac{x^3}{3} - \frac{32f^2}{\ell^3}\cdot\frac{x^2}{2} + \frac{8f^2}{\ell^2}\cdot x\right)_0^\ell = \ell + \frac{8}{3}\cdot\frac{f^2}{\ell}$$

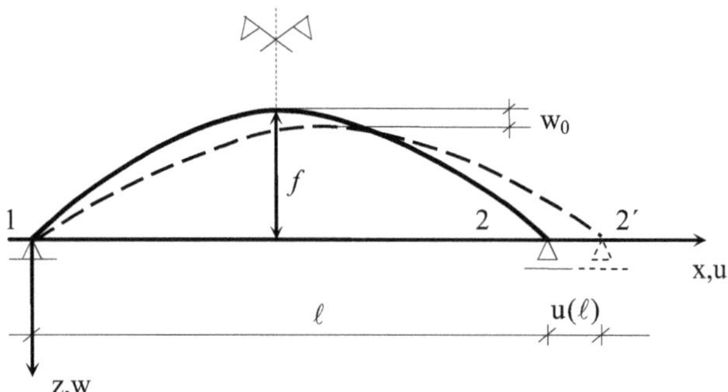

Figure 9: Initial and deformed state of a parabolic arch

Observing that the length s of the arch in both the deformed and the undeformed states remains the same, we will have:

$$s = \ell + \frac{8}{3}\cdot\frac{f^2}{\ell} = \ell + u(\ell) + \frac{8}{3}\cdot\frac{(f+w_0)^2}{\ell}$$

and since: $w_0^2 << f^2$ we finally obtain:

$$u(\ell) = -\frac{16}{3}\cdot\frac{w_0 f}{\ell} \tag{52}$$

It is common practice in the parabolic arches to consider that:

$$I \cdot \cos\varphi = I_c \tag{53}$$

where I_c is the moment of inertia of the cross section at the top point. Then, eq (47) can be written for the case of a freely vibrating parabolic arch as follows:

$$EI_c w'''' - \frac{mg\ell^2}{8f} \cdot w'' + m\ddot{w} = \frac{8f}{\ell^2 L} \cdot \left[-\frac{16}{3} \cdot \frac{w_0 f}{\ell} - \frac{8f}{\ell^2} \cdot \int_0^\ell w dx \right] \tag{54}$$

We seek a solution of the form:

$$w(x,t) = W(x) \cdot e^{i\omega t} \tag{55}$$

Then, eq (54) is written as follows:

$$\left.\begin{aligned} & W'''' - \alpha W'' - k^4 w = -\beta W_0 - \gamma \int_0^\ell W dx \\ & \text{where:} \quad k^4 = \frac{m\omega^2}{EI_c} \ , \quad \alpha = \frac{mg\ell^2}{8f\ EI_c} \ , \quad \beta = \frac{128 f^2}{3\ell^3 LEI_c} \ , \quad \gamma = \frac{64 f^2}{\ell^4 LEI_c} \end{aligned}\right\} \tag{56}$$

Given that $W_0 = W\left(\frac{\ell}{2}\right)$ and $\int_0^\ell W dx$ are constants (not depending on x), the general solution of eq (56) will be:

$$\left.\begin{aligned} & W(x) = c_1 \sin\lambda_1 x + c_2 \cos\lambda_1 x + c_3 \sinh\lambda_2 x + c_4 \cosh\lambda_2 x + \frac{\beta}{k^4} W_0 + \frac{\gamma}{k^4} \int_0^\ell W dx \\ & \text{with:} \quad \lambda_1 = \frac{\alpha}{2} - \sqrt{\frac{\alpha^2}{4} - k^4} \ , \quad \lambda_2 = \frac{\alpha}{2} + \sqrt{\frac{\alpha^2}{4} + k^4} \end{aligned}\right\} \tag{57a,b,c}$$

Integrating eq (57a) we obtain:

$$\int_0^\ell W dx = \frac{k^4}{k^4 - \gamma\ell} \cdot \left\{ -\frac{c_1}{\lambda_1} \cdot (\cos\lambda_1\ell - 1) + \frac{c_2}{\lambda_1} \sin\lambda_1\ell + \frac{c_3}{\lambda_2} \cdot (\cosh\lambda_2\ell - 1) + \frac{c_4}{\lambda_2} \sinh\lambda_2\ell + \frac{\beta\ell}{k_4} \cdot W_0 \right\}$$

and hence:

$$\left.\begin{aligned} & W(x) = \\ & = c_1 \left[\sin\lambda_1 x - \frac{A}{\lambda_1} \cdot (\cos\lambda_1\ell - 1) \right] + c_2 \left[\cos\lambda_1 x + \frac{A}{\lambda_1} \sin\lambda_1\ell \right] + \\ & + c_3 \left[\sinh\lambda_2 x - \frac{A}{\lambda_1} \cdot (\cosh\lambda_2\ell - 1) \right] + c_4 \left[\cosh\lambda_2 x + \frac{A}{\lambda_2} \sinh\lambda_2\ell \right] + \frac{A\gamma}{\beta} W_0 \\ & \text{where:} \quad A = \frac{\gamma}{k^4 - \gamma \cdot \ell} \end{aligned}\right\} \tag{58}$$

Setting finally: $x = \frac{\ell}{2}$ into eq (58), we obtain:

$$W_0 = \frac{\beta}{\beta - A\gamma} \cdot \left\{ c_1 \cdot \left[\sin\frac{\lambda_1 \ell}{2} - \frac{A}{\lambda_1}\left(\cos\lambda_1\ell - 1\right) \right] + c_2 \cdot \left[\cos\frac{\lambda_1 \ell}{2} + \frac{A}{\lambda_1}\sin\lambda_1\ell \right] \right.$$
$$\left. + c_3 \cdot \left[\sinh\frac{\lambda_2 \ell}{2} + \frac{A}{\lambda_2}\cdot\left(\cosh\lambda_2\ell - 1\right) \right] + c_4 \cdot \left[\cosh\frac{\lambda_2 \ell}{2} + \frac{A}{\lambda_2}\sinh\lambda_2\ell \right] \right\}$$

and hence:

$$\left.\begin{aligned}
W(x) = {} & c_1 \cdot \left[\sin\lambda_1 x - \frac{A(B+1)}{\lambda_1}\cdot\left(\cos\lambda_1\ell - 1\right) + B\sin\frac{\lambda_1\ell}{2} \right] \\
& + c_2 \cdot \left[\cos\lambda_1 x + \frac{A(B+1)}{\lambda_1}\cdot\sin\lambda_1\ell + B\cos\frac{\lambda_1\ell}{2} \right] \\
& + c_3 \cdot \left[\sinh\lambda_2 x + \frac{A(B+1)}{\lambda_2}\cdot\left(\cosh\lambda_2\ell - 1\right) + B\sinh\frac{\lambda_2\ell}{2} \right] \\
& + c_4 \cdot \left[\cosh\lambda_2 x + \frac{A(B+1)}{\lambda_2}\cdot\sinh\lambda_2\ell + B\cosh\frac{\lambda_2\ell}{2} \right] \\
\text{where:}\quad & B = \frac{A\gamma}{\beta - A\gamma}
\end{aligned}\right\} \tag{59}$$

Employing the boundary conditions: $W(0) = W(\ell) = W''(0) = W''(\ell) = 0$, we obtain the following frequency equation:

$$\begin{bmatrix}
\left[-\frac{A(B+1)}{\lambda_1}(\cos\lambda_1\ell - \ell) + B\sin\frac{\lambda_1\ell}{2}\right] & \left[1 + \frac{A(B+1)}{\lambda_1}\sin\lambda_1\ell + B\cos\frac{\lambda_1\ell}{2}\right] & \left[\frac{A(B+1)}{\lambda_2}(\cos\lambda_2\ell - 1) + B\sinh\frac{\lambda_2\ell}{2}\right] & \left[1 + \frac{A(B+1)}{\lambda_2}\sinh\lambda_2\ell + B\cosh\frac{\lambda_2\ell}{2}\right] \\
\sin\lambda_1\ell & \left(\cos\lambda_1\ell - 1\right) & \sinh\lambda_2\ell & (\cosh\lambda_2\ell - 1) \\
0 & -\lambda_1^2 & 0 & \lambda_2^2 \\
-\lambda_1^2\sin\lambda_1\ell & -\lambda_1^2\cos\lambda_1\ell & \lambda_2^2\sinh\lambda_2\ell & \lambda_2^2\cosh\lambda_2\ell
\end{bmatrix} = 0 \tag{60}$$

while the shape function is given by eq (59), where:

$$\left.\begin{aligned}
c_2 &= -c_1 \frac{\sin\lambda_1\ell}{\cos\lambda_1\ell - 1} \\
c_3 &= c_1 \left(\frac{\lambda_1}{\lambda_2}\right)^2 \frac{\sin\lambda_1\ell\cdot(\cosh\lambda_2\ell - 1)}{\sinh\lambda_2\ell\cdot(\cos\lambda_1\ell - 1)} \\
c_4 &= -c_1 \left(\frac{\lambda_1}{\lambda_2}\right)^2 \cdot \frac{\sin\lambda_1\ell}{\cos\lambda_1\ell - 1}
\end{aligned}\right\} \tag{61}$$

The Pinned Supported Parabolic Arch

In this case, it will be:

$$u(\ell) = 0 \tag{62}$$

from which it follows that: $\beta=0$ and hence:

$$\left.\begin{aligned}
W(x) = {} & c_1\left[\sin\lambda_1 x - \frac{A}{\lambda_1}\left(\cos\lambda_1\ell - 1\right)\right] + c_2\left[\cos\lambda_1 x + \frac{A}{\lambda_1}\sin\lambda_1\ell\right] \\
& + c_3\left[\sinh\lambda_2 x + \frac{A}{\lambda_2}\left(\cosh\lambda_2\ell - 1\right)\right] + c_4\left[\cosh\lambda_2 x + \frac{A}{\lambda_2}\sinh\lambda_2\ell\right]
\end{aligned}\right\} \tag{63}$$

The boundary conditions are the same as in the previous case. Then, the frequency equation will be:

$$\begin{vmatrix} -\frac{A}{\lambda_1}(\cos\lambda_1\ell-1) & (1+\frac{A}{\lambda_1}\sin\lambda_1\ell) & \frac{A}{\lambda_2}(\cosh\lambda_2\ell-1) & (1+\frac{A}{\lambda_2}\sinh\lambda_2\ell) \\ \sin\lambda_1\ell & (\cos\lambda_1\ell-1) & \sinh\lambda_2\ell & (\cosh\lambda_2\ell-1) \\ 0 & -\lambda_1^2 & 0 & \lambda_2^2 \\ -\lambda_1^2\sin\lambda_1\ell & -\lambda_1^2\cos\lambda_1\ell & \lambda_2^2\sinh\lambda_2\ell & \lambda_2^2\cosh\lambda_2\ell \end{vmatrix} = 0 \tag{64}$$

The corresponding shape function is given by eq (63), where the coefficients c_2 , c_3 , c_4 are taken from eq (61).

Finally, for $u(\ell)=\beta=0$ and boundary conditions as follows:

$$w(0)=w'(0)=w(\ell)=w'(\ell)=0$$

we can determine the eigenfrequencies and shape functions of the fixed-fixed parabolic arch following the same methodology.

The Pinned Supported Circular Arch [5]

The expression for the circular arch shown in Fig. **10**, is given by:

$$z=z_\kappa-\sqrt{R_0^2-(x-x_\kappa)^2} \tag{65}$$

It is obvious, that introducing z´´ and H into eq (47), we will have to solve a linear integral-differential equation with core a very complex function of x.

The problem can be easily solved if we consider a polar coordinate system (R,θ) with sign convention and deformations u(θ), υ(θ) as shown in Fig. **10**. The curvature radius of the deformed arch is given [7] by:

$$\frac{1}{R}=\frac{r^2+2\cdot\left(\frac{dr}{d\theta}\right)^2-r\cdot\frac{d^2r}{d\theta^2}}{\left[r^2+\left(\frac{dr}{d\theta}\right)^2\right]^{\frac{3}{2}}}$$

where: $r=R_0-u$ and hence: $\frac{dr}{d\theta}=-\frac{du}{d\theta}$ or finally:

$$\frac{1}{R}=\frac{R_0^2-2R_0u+u^2+2\left(\frac{du}{d\theta}\right)^2+R_0\frac{d^2u}{d\theta^2}-u\frac{d^2u}{d\theta^2}}{\left[R_0^2-2R_0u+u^2+\left(\frac{du}{d\theta}\right)^2\right]^{\frac{3}{2}}}$$

Neglecting higher order terms we have:

$$\frac{1}{R}=\frac{(R_0-u)^2+R_0\frac{d^2u}{d\theta^2}}{(R_0-u)^3}\cong\frac{R_0-u+\frac{d^2u}{d\theta^2}}{(R_0-u)^2}\cong\frac{R_0-u+\frac{d^2u}{d\theta^2}}{R_0^2}$$

and hence:

$$\frac{1}{R_0} - \frac{1}{R} = \frac{1}{R_0^2}\left(u - \frac{d^2u}{d\theta^2}\right) \tag{66}$$

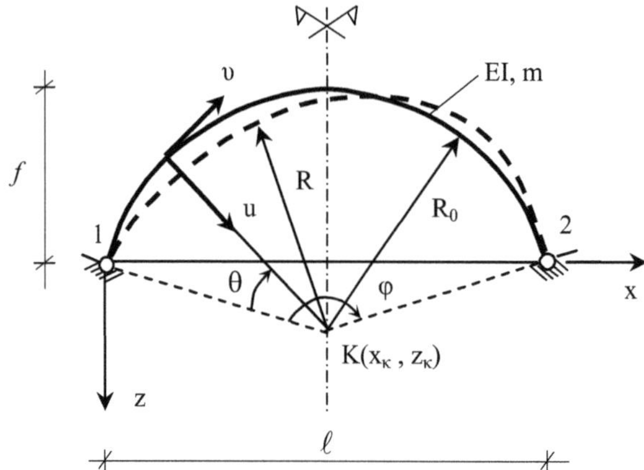

Figure 10: Initial and deformed state of a pinned supported circular arch

The normal strain ε is given by:

$$\varepsilon = -\frac{u}{R_0} + \frac{1}{R_0} \cdot \frac{d\upsilon}{d\theta} \tag{67}$$

Considering the equilibrium of forces and moments on an infinitesimal element ds of the arch (see Fig. **11**) we have.

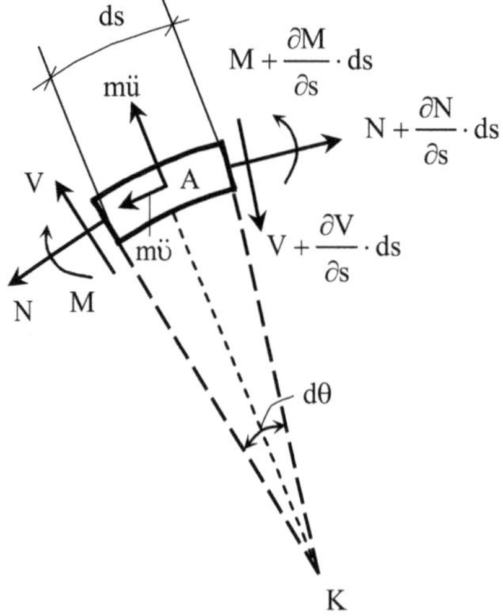

Figure 11: Equilibrium of an infinitesimal arch element

Projection on axis KA:

$$-V + V + \frac{\partial V}{\partial s} \cdot ds + N \cdot \frac{d\theta}{2} + N\frac{d\theta}{2} + \frac{\partial N}{\partial s} ds \frac{d\theta}{2} - m\ddot{u}ds = 0$$

or finally, since $dS = R_0 d\theta$ we will have:

$$\frac{\partial V}{\partial s}+\frac{N}{R_0}-m\ddot{u}=0 \tag{68}$$

Moment equilibrium:

$$V\cdot ds+M-M-\frac{\partial M}{\partial s}\cdot ds+m\ddot{u}ds\frac{ds}{2}=0$$

or finally:

$$V=\frac{\partial M}{\partial s} \tag{69}$$

and hence, eq (69), due to eq (68), becomes:

$$\frac{\partial^2 M}{\partial s^2}+\frac{N}{R_0}-m\ddot{u}=0 \tag{70}$$

The following relations are valid:

$$\left.\begin{array}{ll} & \dfrac{M}{EI}=-\left(\dfrac{1}{R}-\dfrac{1}{R_0}\right)=-\dfrac{1}{R_0^2}\cdot\left(\dfrac{d^2u}{d\theta^2}-u\right) \\ \text{and} & N=EA\varepsilon=\dfrac{EA}{R_0}\cdot\left(-u+\dfrac{d\upsilon}{d\theta}\right)\end{array}\right\} \tag{71}$$

Due to relations (71), eq (70) gives:

$$\frac{EI}{R_0^4}\cdot\frac{\partial^4 u}{\partial\theta^4}-\frac{EI}{R_0^4}\cdot\frac{\partial^2 u}{\partial\theta^2}+\frac{EA}{R_0^4}\cdot\left(u-\frac{\partial\upsilon}{\partial\theta}\right)+m\ddot{u}=0 \tag{72}$$

and considering incompressible axis, i.e., $\varepsilon=0$ or:

$$u=\frac{\partial\upsilon}{\partial\theta} \tag{73}$$

it is: $$\frac{\partial^4 u}{\partial\theta^4}-\frac{\partial^2 u}{\partial\theta^2}+\frac{mR_0^4}{EI}\cdot\frac{\partial^2 u}{\partial t^2}=0 \tag{74}$$

Seeking a solution of the form:

$$u(\theta,t)=U(\theta)\cdot e^{i\omega t} \tag{75}$$

eq (74) can be written as follows:

$$\left.\begin{array}{l} U_\theta''''-U_\theta''-k^4\cdot\ddot{U}_t=0 \\ \text{where:}\quad k^4=\dfrac{mR_0^4\omega^2}{EI}\end{array}\right\} \tag{76}$$

The solution of eq (76) is given by:

$$\left.\begin{aligned} &U(\theta)=c_1\sin\lambda_1\theta+c_2\cos\lambda_1\theta+c_3\sinh\lambda_2\theta+c_4\cosh\lambda_2\theta \\ &\text{with:}\quad \lambda_1=\frac{1}{2}-\sqrt{\frac{1}{4}+k^4}\quad,\quad \lambda_2=\frac{1}{2}+\sqrt{\frac{1}{4}+k^4} \end{aligned}\right\} \tag{77}$$

Employing the boundary conditions: $U(0)=U''(0)=U(\varphi)=U''(\varphi)=0$, we obtain the following frequency equation:

$$\begin{vmatrix} 0 & 1 & 0 & 1 \\ 0 & -\lambda_1^2 & 0 & \lambda_2^2 \\ \sin\lambda_1\varphi & \cos\lambda_1\varphi & \sinh\lambda_2\varphi & \cosh\lambda_2\varphi \\ -\lambda_1^2\cdot\sin\lambda_1\varphi & -\lambda_1^2\cdot\cos\lambda_1\varphi & \lambda_2^2\cdot\sinh\lambda_2\varphi & \lambda_2^2\cdot\cosh\lambda_2\varphi \end{vmatrix}=0 \tag{78}$$

while the shape functions U and V are given by:

$$\left.\begin{aligned} &U(\theta)=c_1\cdot\left(\sin\lambda_1\theta-\frac{\sin\lambda_1\varphi}{\sinh\lambda_2\varphi}\cdot\sinh\lambda_2\theta\right) \\ &V(\theta)=c_1\cdot\left(\lambda_1\cos\lambda_1\theta-\lambda_2\frac{\sin\lambda_1\varphi}{\sinh\lambda_2\varphi}\cdot\cosh\lambda_2\theta\right) \end{aligned}\right\} \tag{79}$$

CABLE - STAYED BRIDGES [8], [9], [10], [11], [12]

Cable-stayed bridges came in use, in a primitive initial form, since the beginning of the 18th century.

Nevertheless, the interest for this type of bridges became particularly big within the last 50 years, mainly as an alternative and cheaper solution (for intermediate openings of 150m up to 600m), of suspension bridges but also as one particularly acceptable aesthetical form.

The main causes that contributed for so long in avoiding this type of bridges were the serious difficulties in dealing with their static and dynamic analysis, the various nonlinearities of materials as well as the structural systems, the required calculation cost (due the essential absence of PCs), the lack of high resistance materials, but also the lack of means and construction techniques.

The presentation and analysis of new high resistance materials that are used nowadays in cable-stayed bridges construction is out of the scope of this book. We will focus mainly on the analysis of the dynamic behavior of these structures. From structural analysis point of view, these structures are characterized and classified into three categories based on the suspension system of cables. Three are the basic types of cables.

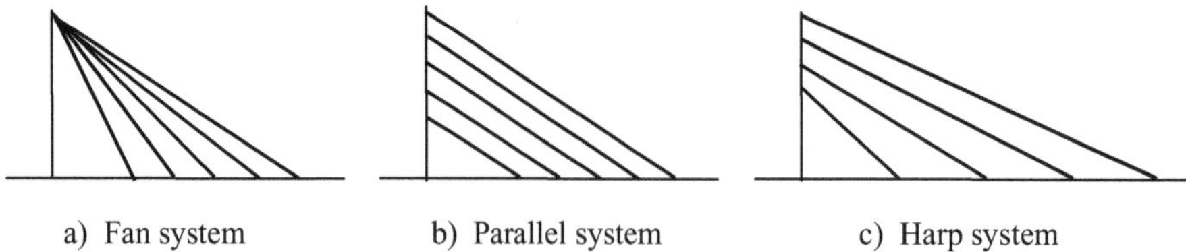

Figure 12: The three basic cable arrangement systems

a) The fan-system, the main characteristic of which is that all cables are intersecting and anchored at a single point on the pylon.

b) The parallel-system, where cables are parallel to each other.

c) The harp-system, where cables are not parallel to each other but are anchored at equal distances on both pylon and deck.

We shall study herein cases a) and c), given that case b) can be considered as a special case of case c).

The Isolated Cable

Let us consider cable i, which is connected to the pylon at height H_i from the bridge deck and to the deck at distance α_i from the pylon.

The deformed state of the pylon-deck-cable system is shown in Fig. **13b**.

The following relations are valid:

$$\left.\begin{aligned} P_{ix} &= P_{iz}\cdot\tan\varphi_i = P_i\cdot\sin\varphi_i \\ P_{iz} &= P_i\cdot\cos\varphi_i \end{aligned}\right\} \quad \textbf{(80)}$$

Assuming that: $u_d \cong 0$, it follows that:

$$f_i - u_d \cong f_i = \frac{(H_0+H_i)^3}{6E_pI_p}\left[2-3\left(\frac{H_0}{H_0+H_i}\right)^2+\left(\frac{H_0}{H_0+H_i}\right)^3\right]\cdot P_i\sin\varphi_i \quad \textbf{(81)}$$

where the indexes "p" refer to the pylon elements, while the indexes "b" refer to the deck elements.

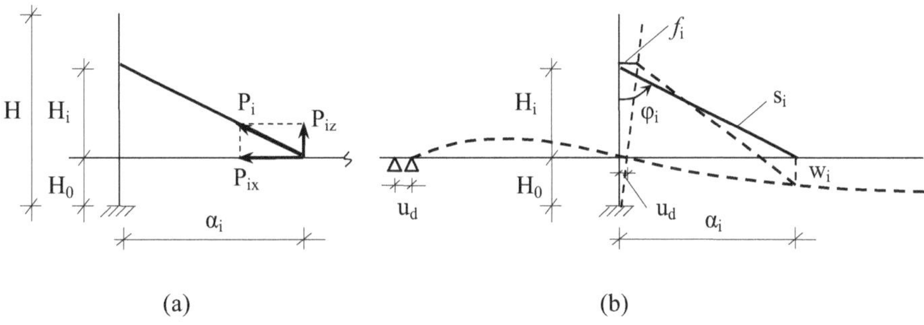

Figure 13: The cable-deck-pylon system

Finally, index s refers to the cable elements. From Fig. **13b** it follows:

$$s_i = f_i\cdot\sin\varphi_i + (s_i+\Delta s_i)\cdot\cos\Delta\varphi_i - w_i\cdot\cos\varphi_i$$

where s_i and Δs_i are the initial length and the deformation of cable i. And because:

$\cos\Delta\varphi_i \cong 1$, we have:

$$f_i\cdot\sin\varphi_i + \Delta_{Si} = w_i\cdot\cos\varphi_i \quad \textbf{(82)}$$

The deformation Δs_i is given by:

$$\Delta s_i = \frac{s_iP_i}{E_sA_i} = \frac{s_iP_{iz}}{E_sA_i}\cdot\frac{1}{\cos\varphi_i} \quad \textbf{(83)}$$

where A_i is the cable-i cross section. Then, eq (82) due to eq (81) and (83), gives:

$$AP_{iz}\cdot\tan\varphi_i\cdot\sin\varphi_i+\frac{s_iP_{iz}}{E_sA_i\cdot\cos\varphi_i}=w_i\cos\varphi_i \quad \text{or finally:}$$

$$\left.\begin{array}{l}\left(A\cdot\dfrac{\sin^2\varphi_i}{\cos\varphi_i}+\dfrac{s_iB_i}{\cos\varphi_i}\right)\cdot P_{iz}=w_i\cos\varphi_i \\ \text{where: } A=\dfrac{(H_0+H_1)^3}{6E_pI_p}\cdot\left[2-3\left(\dfrac{H_0}{H_0+H_i}\right)^2+\left(\dfrac{H_0}{H_0+H_i}\right)^3\right] \\ B=\dfrac{1}{E_sA_i}\end{array}\right\} \qquad \textbf{(84a,b,c)}$$

The Fan System

Let the index j refer to the ρ cables, lying on the left of the pylon and the index i refer to the κ cables, lying on the right of the pylon.

Then, according to Fig. **14**, eq (81) with the symbols in eq (84) gives:

$$f=A\cdot\left\{\sum_{\kappa}P_i\cdot\sin\varphi_i-\sum_{\rho}P_j\cdot\sin\varphi_j\right\} \qquad \textbf{(85)}$$

In this case, eq (82) becomes:

$$\left.\begin{array}{l}\text{left:}\\ A\sin\varphi_j\cdot\left\{\displaystyle\sum_{\rho}P_j\sin\varphi_j-\sum_{\kappa}P_i\sin\varphi_i\right\}+\dfrac{s_j}{E_cA_j}P_j=w_j\cos\varphi_j \\ \text{right:}\\ A\sin\varphi_i\cdot\left\{\displaystyle\sum_{\kappa}P_i\sin\varphi_i-\sum_{\rho}P_j\sin\varphi_j\right\}+\dfrac{s_i}{E_cA_i}P_i=w_i\cos\varphi_i \\ \text{with } j=1 \text{ to } \rho \text{ and } i=(\rho+1) \text{ to } (\rho+\kappa)\end{array}\right\} \qquad \textbf{(86a,b)}$$

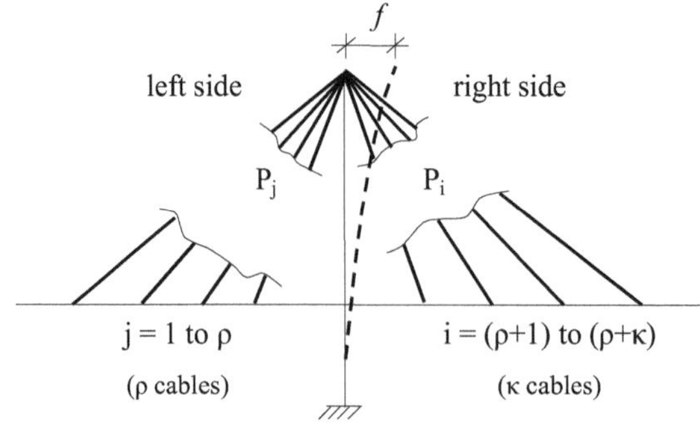

Figure 14: Indexing convention for cables

Thin Arrangement of Cables

With the term thin arrangement, it is meant that the cables are set in big distances to each other so that they can only be considered as concentrated suspension forces. A suitable criterion is: $\delta > \frac{\ell}{10}$ where ℓ is the mid-span length.

Setting: $b_i = \frac{s_i}{E_c A_i} \quad , \quad b_j = \frac{s_j}{E_c A_j}$ **(87)**

equations (86) can be written as:

$$\text{left:} \quad A\frac{\sin^2\varphi_j}{b_j}\cdot\left\{\sum_\rho P_j \sin\varphi_j - \sum_\kappa P_i \sin\varphi_i\right\} + P_j\sin\varphi_j = \frac{w_j}{b_j}\sin\varphi_j\cos\varphi_j$$

$$\text{right:} \quad A\frac{\sin^2\varphi_i}{b_i}\cdot\left\{\sum_\kappa P_i \sin\varphi_i - \sum_\rho P_j \sin\varphi_j\right\} + P_i\sin\varphi_i = \frac{w_i}{b_i}\sin\varphi_i\cos\varphi_i$$

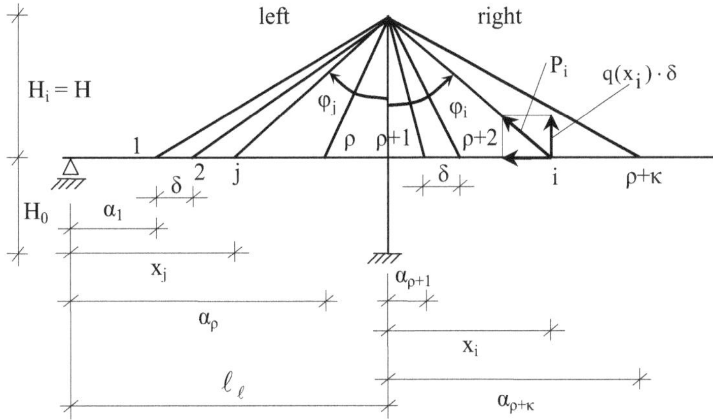

Figure 15: Sign convention for cables

Adding the j equations for the left side and the i equations for the right side, we obtain:

$$\left.\begin{array}{l} A\left(\sum_\rho \frac{\sin^2\varphi_j}{b_j}\right)\cdot\left\{\sum_\rho P_j\sin\varphi_j - \sum_\kappa P_i\sin\varphi_i\right\} + \sum_\rho P_j\sin\varphi_j = \frac{1}{2}\sum_\rho \frac{w_j}{b_j}\sin 2\varphi_j \\ A\left(\sum_\kappa \frac{\sin^2\varphi_i}{b_i}\right)\cdot\left\{\sum_\kappa P_i\sin\varphi_i - \sum_\rho P_j\sin\varphi_j\right\} + \sum_\kappa P_i\sin\varphi_i = \frac{1}{2}\sum_\kappa \frac{w_i}{b_i}\cdot\sin 2\varphi_i \end{array}\right\} \quad \mathbf{(88)}$$

Abstracting by parts eq (88) we obtain:

$$A\left\{\sum_\rho \frac{\sin^2\varphi_j}{b_j} + \sum_\kappa \frac{\sin^2\varphi_i}{b_i}\right\}\left\{\sum_\rho P_j\sin\varphi_j - \sum_\kappa P_i\sin\varphi_i\right\} + \left\{\sum_\rho P_j\sin\varphi_j - \sum_\kappa P_i\sin\varphi_i\right\} = \frac{1}{2}\left\{\sum_\rho \frac{w_j}{b_j}\sin 2\varphi_j - \sum_\kappa \frac{w_i}{b_i}\sin 2\varphi_i\right\}$$

or in a simple form:

$$\left\{\sum_\rho P_j\sin\varphi_j - \sum_\kappa P_i\sin\varphi_i\right\} = \frac{1}{2}\frac{\sum_\rho \frac{w_j}{b_j}\sin 2\varphi_j - \sum_\kappa \frac{w_i}{b_i}\sin 2\varphi_i}{A\cdot\left\{\sum_\rho \frac{\sin^2\varphi_j}{b_j} + \sum_\rho \frac{\sin^2\varphi_i}{b_i}\right\} + 1} \quad \mathbf{(89)}$$

Finally, from eq (86a,b), we can easily determine the forces P as follows:

$$\left.\begin{aligned}
&P_j = \frac{\cos\varphi_j}{b_j} w_j - \frac{\sin\varphi_j}{2b_j} \cdot \frac{\sum\limits_{\rho} \frac{w_j}{b_j}\sin 2\varphi_j - \sum\limits_{\kappa} \frac{w_i}{b_i}\sin 2\varphi_i}{\sum\limits_{\rho} \frac{\sin^2\varphi_j}{b_j} + \sum\limits_{\kappa} \frac{\sin^2\varphi_i}{b_i} + \frac{1}{A}} \\
&P_i = \frac{\cos\varphi_i}{b_i} w_i - \frac{\sin\varphi_i}{2\cdot b_i} \cdot \frac{\sum\limits_{\kappa} \frac{w_i}{b_i}\sin 2\varphi_i - \sum\limits_{\rho} \frac{w_j}{b_j}\sin 2\varphi_j}{\sum\limits_{\rho} \frac{\sin^2\varphi_j}{b_j} + \sum\limits_{\kappa} \frac{\sin^2\varphi_i}{b_i} + \frac{1}{A}} \\
&\text{with } j = 1 \text{ to } \rho \text{ and } i = (\rho+1) \text{ to } (\rho+\kappa)
\end{aligned}\right\} \qquad \textbf{(90a,b)}$$

Dense Arrangement of Cables

Let us consider now that the cables placed in distances $\delta << \frac{\ell}{10}$ between each other, so that their forces can be substituted by a distributed load q(x), from position α_1 to α_ρ and from $\alpha_{\rho+1}$ to $\alpha_{\rho+\kappa}$ while at position x_i it will be for example:

$$q(x_i) = \frac{1}{\delta} \cdot P_i \cdot \cos\varphi_i \qquad \textbf{(91)}$$

It is obvious that the horizontal force applied to the top of the pylon, will be given by the following relation:

$$P_{pylon} = \int\limits_{\alpha_{\rho+1}}^{\alpha_{\rho+\kappa}} q_i(x_i)\tan\varphi_i dx_i - \int\limits_{\alpha_1}^{\alpha_\rho} q_j(x_j)\tan\varphi_j dx_j \qquad \textbf{(92)}$$

Then, eq (85) becomes: $f = A \cdot P_{pylon}$ and due to the following relations:

$$\tan\varphi_i = \frac{x_i}{H} \quad , \quad \tan\varphi_j = \frac{\ell_\ell - x_j}{H} \qquad \textbf{(93)}$$

we can write:

$$f = \frac{A}{H}\left\{ \int\limits_{\alpha_{\rho+1}}^{\alpha_{\rho+\kappa}} q_i(x_i)x_i dx_i - \int\limits_{\alpha_1}^{\alpha_\rho} q_j(x_j)\cdot(\ell_\ell - x_j)dx_j \right\} \qquad \textbf{(94)}$$

Then, eq (84a) gives:

$$\left.\begin{aligned}
&\text{left}: \quad \frac{A}{H}\sin\varphi_j \left\{ \int\limits_{\alpha_1}^{\alpha_\rho} (\ell_\ell - x_j)q_j dx_j - \int\limits_{\alpha_{\rho+1}}^{\alpha_{\rho+\kappa}} x_i q_i dx_i \right\} + \frac{s_j B_j \delta q_j}{\cos\varphi_j} = w_j \cos\varphi_j \\
&\text{right}: \quad \frac{A}{H}\sin\varphi_i \left\{ \int\limits_{\alpha_{\rho+1}}^{\alpha_{\rho+\kappa}} x_i q_i dx_i - \int\limits_{\alpha_1}^{\alpha_\rho} (\ell_\ell - x_j)q_j dx_j \right\} + \frac{s_i B_i \delta q_i}{\cos\varphi_i} = w_i \cos\varphi_i
\end{aligned}\right\} \qquad \textbf{(95a,b)}$$

From the geometry of the structure (Fig. **15**) it follows that:

$$\left.\begin{aligned} s_j &= \frac{\ell_\ell - x_j}{\sin\varphi_j} \quad , \quad s_i = \frac{x_i}{\sin\varphi_i} \\ \sin\varphi_j &= \frac{\ell_\ell - x_j}{\sqrt{H^2 + (\ell_\ell - x_j)^2}} \quad , \quad \sin\varphi_i = \frac{x_i}{\sqrt{H^2 + x_i^2}} \\ \cos\varphi_j &= \frac{H}{\sqrt{H^2 + (\ell_\ell - x_j)^2}} \quad , \quad \cos\varphi_i = \frac{H}{\sqrt{H^2 + x_i^2}} \end{aligned}\right\} \qquad \textbf{(96)}$$

Then, eq (95) become:

$$\left.\begin{aligned} (1 + I_{1\ell})\int_{\alpha_1}^{\alpha_\rho} (\ell_\ell - x_j)\cdot q_j(x_j)dx_j - I_{1\ell}\int_{\alpha_{\rho+1}}^{\alpha_{\rho+\kappa}} x_i q_i(x_i)dx_i &= I_{2\ell} \\ -I_{1R}\int_{\alpha_1}^{\alpha_\rho} (\ell_\ell - x_j)\cdot q_j(x_j)dx_j + (1 + I_{1R})\int_{\alpha_{\rho+1}}^{\alpha_{\rho+\kappa}} x_i q_i(x_i)dx_i &= I_{2R} \end{aligned}\right\} \qquad \textbf{(97a,b)}$$

where:

$$\left.\begin{aligned} I_{1\ell} &= AE_S\int_{\alpha_1}^{\alpha_\rho} \frac{A_j(x_j)\cdot(\ell_\ell - x_j)^2}{[H^2 + (\ell_\ell - x_j)^2]^{3/2}}\cdot dx_j \\ I_{2\ell} &= \int_{\alpha_1}^{\alpha_\rho} F_{1\ell} w_j(x_j)dx_j \\ I_{1R} &= AE_S\int_{\alpha_{\rho+1}}^{\alpha_{\rho+\kappa}} \frac{A_i(x_i)x_i^2}{[H^2 + x_i^2]^{3/2}}\cdot dx_i \quad , \quad I_{2R} = \int_{\alpha_{\rho+1}}^{\alpha_{\rho+\kappa}} F_{1R} w_i(x_i)dx_i \\ F_{1\ell} &= \frac{E_S H^2 A_j(x_j)\cdot(\ell_\ell - x_j)}{[H^2 + (\ell_\ell - x_j)^2]^{3/2}} \quad , \quad F_{1R} = \frac{E_S H^2 A_i(x_i)x_i}{[H^2 + x_i^2]^{3/2}} \\ A_j(x_j) &= \frac{A_j}{\delta} \quad , \quad A_i(x_i) = \frac{A_i}{\delta} \end{aligned}\right\} \qquad \textbf{(98)}$$

Solving the above system of eq (97) we obtain:

$$\left.\begin{aligned} \int_{\alpha_1}^{\alpha_\rho} (\ell_\ell - x_j)q_j(x_j)dx_j &= \frac{I_{2\ell} + I_{1R}I_{2R} + I_{1\ell}I_{2R}}{1 + I_{1\ell} + I_{1R}} \\ \int_{\alpha_{\rho+1}}^{\alpha_{\rho+\kappa}} x_i q_i(x_i)dx_i &= \frac{I_{2R} + I_{1R}I_{2\ell} + I_{1\ell}I_{2R}}{1 + I_{1\ell} + I_{1R}} \\ \text{and} \quad \int_{\alpha_{\rho+1}}^{\alpha_{\rho+\kappa}} x_i q_i(x_i)\cdot dx_i - \int_{\alpha_1}^{\alpha_\rho} (\ell_\ell - x_j)\cdot q_j(x_j)dx_j &= \frac{I_{2R} - I_{2\ell}}{1 + I_{1\ell} + I_{1R}} \end{aligned}\right\} \qquad \textbf{(99)}$$

We can easily, from eq (95), express the stresses of the cables associated with the deformations of the deck as follows:

$$\left.\begin{aligned} q_j(x_j) &= \frac{E_S A A_j(x_j)\cdot(\ell_\ell - x_j)}{[H^2+(\ell_\ell - x_j)^2]^{3/2}}\cdot\frac{I_{2R}-I_{2\ell}}{1+I_{1\ell}+I_{1R}} + \frac{E_S A_j(x_j)H^2}{[H^2+(\ell_\ell - x_j)^2]^{3/2}} w_j(x_j) \\ q_i(x_i) &= \frac{E_S A A_i(x_i)x_i}{[H^2+x_i^2]^{3/2}}\cdot\frac{I_{2\ell}-I_{2R}}{1+I_{1\ell}+I_{1R}} + \frac{E_S A_i(x_i)H^2}{[H^2+x_i^2]^{3/2}} w_i(x_i) \end{aligned}\right\} \quad \textbf{(100a,b)}$$

The Harp System

Cable-stayed bridges with the harp system are the most commonly used in bridge engineering. Such a typical arrangement is shown in Fig. **16**.

It is obvious that the change of cables direction should follow some law that is characteristic for the bridge. Here the most commonly used law will be adopted, as shown in Fig. **16**. According to this law, the cables are anchored at equal distances γ on the pylon, δ_ℓ on the left side of the deck and δ_r on the right side of the deck.

Then, for cable ρ, the following relations are valid:

$$\left.\begin{aligned} \ell_\ell - x_{\rho\ell} &= \delta_{1\ell} + \rho\delta_\ell \\ x_{\rho r} &= \delta_{1r} + \rho\delta_r \\ h_\rho &= h_1 + \rho\gamma \end{aligned}\right\} \quad \textbf{(101a,b,c)}$$

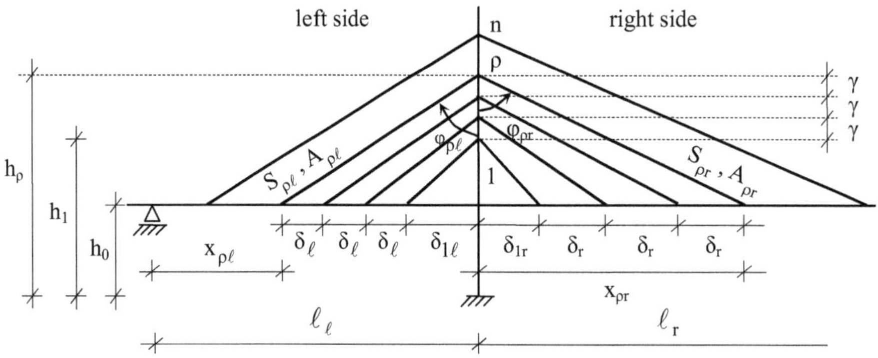

Figure 16: Cable arrangement in the harp system

Note here that for $\dfrac{\gamma}{h_1 - h_0} = \dfrac{\delta_r}{\delta_{1r}} = \dfrac{\delta_\ell}{\delta_{1\ell}}$ the parallel system arrangement is retrieved.

From eq (101a, b) we obtain:

$$\rho = \frac{x_\ell - \delta_{1\ell}}{\delta_\ell} = \frac{x_r - \delta_{1r}}{\delta_r} \quad \textbf{(102)}$$

and eq (101c) gives:

$$h_\rho = h_1 + \frac{x_\ell - \delta_{1\ell}}{\delta_\ell}\cdot\gamma = h_1 + \frac{x_r - \delta_{1r}}{\delta_r}\cdot\gamma \quad \textbf{(103)}$$

Then, we can easily find the following expressions:

$$\left.\begin{aligned}
\tan\varphi_{\rho\ell} &= \frac{\ell_\ell - x_\ell}{h_\rho - h_0} = \frac{\delta_{1\ell}+\rho\delta_\ell}{h_1 - h_0 + \rho\gamma} \\
\sin\varphi_{\rho\ell} &= \frac{\delta_{1\ell}+\rho\delta_\ell}{\sqrt{(\delta_{1\ell}+\rho\delta_\ell)^2+(h_1-h_0+\rho\gamma)^2}} \\
\cos\varphi_{\rho\ell} &= \frac{h_1-h_0+\rho\gamma}{\sqrt{(\delta_{1\ell}+\rho\delta_\ell)^2+(h_1-h_0+\rho\gamma)^2}} \\
\tan\varphi_{\rho r} &= \frac{x_r}{h_\rho - h_0} = \frac{\delta_{1r}+\rho\delta_r}{h_1 - h_0 + \rho\gamma} \\
\sin\varphi_{\rho r} &= \frac{\delta_{1r}+\rho\delta_r}{\sqrt{(\delta_{1r}+\rho\delta_r)^2+(h_1-h_0+\rho\gamma)^2}} \\
\cos\varphi_{\rho r} &= \frac{h_1-h_0+\rho\gamma}{\sqrt{(\delta_{1r}+\rho\delta_r)^2+(h_1-h_0+\rho\gamma)^2}}
\end{aligned}\right\} \qquad \textbf{(104)}$$

Pylon Stressing

In this section we will determine some simple expressions for the deformed state of the pylon that is subjected to the loading shown in Fig. **17**.

Let us consider the pylon – cantilever shown in Fig. **17**, with length h_n moment of inertia J_p , and loaded by the forces P_ρ applied at the points ρ, at distances h_ρ from the fixed base. The reactions at point 0 are:

$$V_0 = \sum_1^n P_\rho \quad , \quad M_0 = \sum_1^n h_\rho P_\rho \qquad \textbf{(105)}$$

The displacements for each section of the pylon are given by the following relations:

$$\left.\begin{aligned}
&\text{section } (\rho-1)\text{: } EI_p f''_{\rho-1} = -M(y) = -\left\{M_0 + yV_0 - \sum_{i=1}^{\rho-2}(y-h_i)P_i\right\} \qquad \text{for } y \ge h_{\rho-2} \\
&\text{section } \rho\text{: } EI_p f''_\rho = -M(y) = -\left\{M_0 + yV_0 - \sum_{i=1}^{\rho-1}(y-h_i)P_i\right\} \qquad \text{for } y \ge h_{\rho-1}
\end{aligned}\right\} \qquad \textbf{(106a,b)}$$

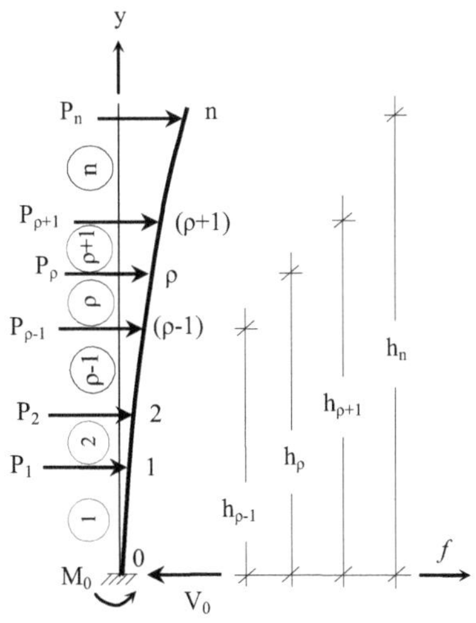

Figure 17: Pylon subjected to horizontal forces from cables

The following conditions must be satisfied:

$$\left.\begin{array}{l} f'_{\rho-1}(h_{\rho-1}) = f'_{\rho}(h_{\rho-1}) \\ f_{\rho-1}(h_{\rho-1}) = f_{\rho}(h_{\rho-1}) \end{array}\right\} \qquad \textbf{(107a,b)}$$

Integrating eq (106) we obtain:

$$\left.\begin{array}{l} EI_{\rho}\cdot f'_{\rho-1} = yM_0 - \dfrac{y^2}{2}V_0 + \dfrac{y^2}{2}\sum_{i=1}^{\rho-2}P_i - y\sum_{i=1}^{\rho-2}h_iP_i + c_{\rho-1} \\ \\ EI_{\rho}f'_{\rho} = yM_0 - \dfrac{y^2}{2}V_0 + \dfrac{y^2}{2}\sum_{i=1}^{\rho-1}P_i - y\sum_{i=1}^{\rho-1}h_iP_i + c_{\rho} \end{array}\right\} \qquad \textbf{(108a,b)}$$

Then, condition (107a) gives:

$$\frac{h_{\rho-1}^2}{2}\sum_{i=1}^{\rho-2}P_i - h_{\rho-1}\sum_{i=1}^{\rho-2}h_iP_i + c_{\rho-1} = \frac{h_{\rho-1}^2}{2}\sum_{i=1}^{\rho-1}P_i - h_{\rho-1}\sum_{i=1}^{\rho-1}h_iP_i + c_{\rho}$$

or finally: $c_{\rho} = c_{\rho-1} + \dfrac{h_{\rho-1}^2}{2}P_{\rho-1}$ and following a similar process we compute the following: $c_{\rho} = c_0 + \dfrac{1}{2}\sum_{i=1}^{\rho-1}h_i^2P_i$.

Also, since for y=0 it is $f'_1(0) = 0$ we will have: c_0=0 and finally:

$$c_{\rho} = \frac{1}{2}\sum_{i=1}^{\rho-1}h_i^2P_i \qquad \textbf{(109)}$$

equations (108) after integration become:

$$\left.\begin{array}{l} EI_{\rho}f_{\rho-1} = \dfrac{y^2}{2}M_0 - \dfrac{y^3}{6}V_0 + \dfrac{y^3}{6}\sum_{i=1}^{\rho-2}P_i - \dfrac{y^2}{2}\sum_{i=1}^{\rho-2}h_iP_i + yc_{\rho-1} + \kappa_{\rho-1} \\ \\ EI_{\rho}f_{\rho} = \dfrac{y^2}{2}M_0 - \dfrac{y^3}{6}V_0 + \dfrac{y^3}{6}\sum_{i=1}^{\rho-1}P_i - \dfrac{y^2}{2}\sum_{i=1}^{\rho-1}h_iP_i + yc_{\rho} + \kappa_{\rho} \end{array}\right\} \qquad \textbf{(110a,b)}$$

Then, condition (107b) gives:

$$\frac{h_{\rho-1}^3}{6}\sum_{i=1}^{\rho-2}P_i - \frac{h_{\rho-1}^2}{2}\sum_{i=1}^{\rho-2}h_iP_i + \frac{h_{\rho-1}}{2}\sum_{i=1}^{\rho-2}h_i^2P_i + \kappa_{\rho-1} = \frac{h_{\rho-1}^3}{6}\sum_{i=1}^{\rho-1}P_i - \frac{h_{\rho-1}^2}{2}\sum_{i=1}^{\rho-1}h_iP_i + \frac{h_{\rho-1}}{2}\sum_{i=1}^{\rho-1}h_i^2P_i + \kappa_{\rho}$$

or finally: $\kappa_{\rho} = \kappa_{\rho-1} - \dfrac{h_{\rho-1}^3}{6}P_{\rho-1}$, and following a similar process we compute: $\kappa_{\rho} = \kappa_0 - \dfrac{1}{6}\sum_{i=1}^{\rho-1}h_i^3P_i$.

Since for y=0 it is $f(0)$=0 we will have κ_0=0 and hence:

$$\kappa_{\rho} = -\frac{1}{6}\sum_{i=1}^{\rho-1}h_i^3P_i \qquad \textbf{(111)}$$

Thus, we arrive at the following expression for pylon deformation $f_\rho(y)$:

$$EI_p f_\rho(y) = \frac{y^2}{2}\sum_{i=1}^{n} h_i P_i - \frac{y^3}{6}\sum_{i=1}^{n} P_i + \frac{y^3}{6}\sum_{i=1}^{\rho-1} P_i - \frac{y^2}{2}\sum_{i=1}^{\rho-1} h_i P_i + \frac{y}{2}\sum_{i=1}^{\rho-1} h_i^2 P_i - \frac{1}{6}\sum_{i=1}^{\rho-1} h_i^3 P_i$$

or finally:

$$EI_p f(y) = -\frac{y^3}{6}\sum_{i=\rho}^{n} P_i + \frac{y^2}{2}\sum_{i=\rho}^{n} h_i P_i + \frac{y}{2}\sum_{i=1}^{\rho-1} h_i^2 P_i - \frac{1}{6}\sum_{i=1}^{\rho-1} h_i^3 P_i \qquad \text{for} \quad y \geq h_{\rho-1} \tag{112}$$

Thin Arrangement of Cables

With the notation in Fig. **16**, eq (82) becomes:

$$f_\rho \cdot \sin\varphi_\rho + \frac{s_\rho}{E_s A_\rho} \cdot P_\rho = w_\rho \cos\varphi_\rho$$

$$\text{Setting:} \quad \alpha = \frac{1}{EI_p} \quad , \quad b_\rho = \frac{s_\rho}{E_c A_\rho} \tag{113}$$

We obtain the following equations:

left side:

$$\left.\begin{aligned} &\alpha \sin\varphi_{\rho\ell} \cdot \left\{ -\frac{h_\rho^3}{6b_{\rho\ell}}\sum_{i=\rho}^{n}(P_{i\ell} - P_{ir}) + \frac{h_\rho^2}{2b_{\rho\ell}}\sum_{i=\rho}^{n} h_i(P_{i\ell} - P_{ir}) + \right. \\ &\left. + \frac{h_\rho}{2b_{\rho\ell}}\sum_{i=1}^{\rho-1} h_i^2(P_{i\ell} - P_{ir}) - \frac{1}{6b_{\rho\ell}}\sum_{i=1}^{\rho-1} h_i^3(P_{i\ell} - P_{ir}) \right\} + P_{\rho\ell} = \frac{w_{\rho\ell}}{b_{\rho\ell}}\cos\varphi_{\rho\ell} \end{aligned}\right\} \tag{114a}$$

right side:

$$\left.\begin{aligned} &\alpha \sin\varphi_{\rho r} \cdot \left\{ -\frac{h_\rho^3}{6b_{\rho r}}\sum_{i=\rho}^{n}(P_{ir} - P_{i\ell}) + \frac{h_\rho^2}{2b_{\rho r}}\sum_{i=\rho}^{n} h_i(P_{ir} - P_{i\ell}) + \right. \\ &\left. + \frac{h_\rho}{2b_{\rho r}}\sum_{i=\rho}^{\rho-1} h_i^2(P_{ir} - P_{i\ell}) - \frac{1}{6b_{\rho r}}\sum_{i=1}^{\rho-1} h_i^3(P_{ir} - P_{i\ell}) \right\} + P_{\rho r} = \frac{w_{\rho r}}{b_{\rho r}}\cos\varphi_{\rho r} \end{aligned}\right\} \tag{114b}$$

Setting:

$$\left.\begin{aligned} &\sum_{i=\rho}^{n}(P_{ir} - P_{i\ell}) = A_0 \\ &\sum_{i=\rho}^{n} h_i(P_{ir} - P_{i\ell}) = A_1 \\ &\sum_{i=1}^{\rho-1} h_i^2(P_{ir} - P_{i\ell}) = A_2 \\ &\sum_{i=1}^{\rho-1} h_i^3(P_{ir} - P_{i\ell}) = A_3 \end{aligned}\right\} \tag{115a,b,c,d}$$

Then, equations (114) can be also written as follows:

left:

$$\alpha\sin\varphi_{\rho\ell}\cdot\left\{\frac{h_\rho^3}{6b_{\rho\ell}}A_0-\frac{h_\rho^2}{2b_{\rho\ell}}A_1-\frac{h_\rho}{2b_{\rho\ell}}A_2+\frac{1}{6b_{\rho\ell}}A_3\right\}+P_{\rho\ell}=\frac{w_{\rho\ell}}{b_{\rho\ell}}\cos\varphi_{\rho\ell} \tag{116a}$$

right:

$$\alpha\sin\varphi_{\rho r}\cdot\left\{-\frac{h_\rho^3}{6b_{\rho r}}A_0+\frac{h_\rho^2}{2b_{\rho r}}A_1+\frac{h_\rho}{2b_{\rho r}}A_2-\frac{1}{6b_{\rho r}}A_3\right\}+P_{\rho r}=\frac{w_{\rho r}}{b_{\rho r}}\cos\varphi_{\rho r} \tag{116b}$$

Applying eq (116a) from ρ to n and adding the results, then applying eq (116b) from ρ to η and adding the results as well, we obtain two equations which abstracted to each other give:

$$\left.\begin{aligned}
&\left\{-\alpha\left[\sum_{i=\rho}^{n}\sin\varphi_{ir}\cdot\frac{h_i^3}{6b_{ir}}+\sum_{i=\rho}^{n}\sin\varphi_{i\ell}\cdot\frac{h_i^2}{2b_{i\ell}}\right]+1\right\}A_0\\
&+\alpha\left[\sum_{i=\rho}^{n}\sin\varphi_{ir}\cdot\frac{h_i^2}{2b_{ir}}+\sum_{i=\rho}^{n}\sin\varphi_{i\ell}\cdot\frac{h_i^2}{2b_{i\ell}}\right]A_1\\
&+\alpha\left[\sum_{i=\rho}^{n}\sin\varphi_{ir}\cdot\frac{h_i}{2b_{ir}}+\sum_{i=\rho}^{n}\sin\varphi_{i\ell}\cdot\frac{h_i}{2b_{i\ell}}\right]\cdot A_2\\
&-\alpha\left[\sum_{i=\rho}^{n}\sin\varphi_{ir}\cdot\frac{1}{6b_{ir}}+\sum_{i=\rho}^{n}\sin\varphi_{i\ell}\cdot\frac{1}{6b_{i\ell}}\right]A_3=\\
&=\sum_{i=\rho}^{n}\frac{w_{ir}}{b_{ir}}\cdot\cos\varphi_{ir}-\sum_{i=\rho}^{n}\frac{w_{i\ell}}{b_{i\ell}}\cdot\cos\varphi_{i\ell}
\end{aligned}\right\} \tag{117a}$$

Multiplying eq (116a) by h_ρ , computing the result from ρ to n and adding the resulting expressions, and doing the same for eq (116b) and abstracting the results, we finally obtain:

$$\left.\begin{aligned}
&-\alpha\left[\sum_{i=\rho}^{n}\sin\varphi_{ir}\cdot\frac{h_i^4}{6b_{ir}}+\sum_{i=\rho}^{n}\sin\varphi_{i\ell}\cdot\frac{h_i^4}{6b_{i\ell}}\right]A_0\\
&+\left\{\alpha\left[\sum_{i=\rho}^{n}\sin\varphi_{ir}\cdot\frac{h_i^3}{2b_{ir}}+\sum_{i=\rho}^{n}\sin\varphi_{i\ell}\cdot\frac{h_i^3}{2b_{i\ell}}\right]+1\right\}A_1\\
&+\alpha\left[\sum_{i=\rho}^{n}\sin\varphi_{ir}\cdot\frac{h_i^2}{2b_{ir}}+\sum_{i=\rho}^{n}\sin\varphi_{i\ell}\cdot\frac{h_i^2}{2b_{i\ell}}\right]A_2\\
&-\alpha\left[\sum_{i=\rho}^{n}\sin\varphi_{ir}\cdot\frac{h_i}{6b_{ir}}+\sum_{i=\rho}^{n}\sin\varphi_{i\ell}\cdot\frac{h_i}{2b_{i\ell}}\right]A_3=\\
&=\sum_{i=\rho}^{n}\frac{h_i w_{ir}}{b_{ir}}\cdot\sin\varphi_{ir}-\sum_{i=\rho}^{n}\frac{h_i w_{i\ell}}{b_{i\ell}}\cdot\sin\varphi_{i\ell}
\end{aligned}\right\} \tag{117b}$$

Multiplying eq (116a) by h_i^2 , computing the result from 1 to ρ-1 and adding the resulting expressions, and doing the same for eq (116b) and abstracting the results, we obtain:

$$\left.\begin{aligned}
&-\alpha\left[\sum_{i=1}^{\rho-1}\sin\varphi_{ir}\cdot\frac{h_i^5}{6b_{ir}}+\sum_{i=1}^{\rho-1}\sin\varphi_{i\ell}\cdot\frac{h_i^5}{6b_{i\ell}}\right]A_0\\
&+\alpha\left[\sum_{i=1}^{\rho-1}\sin\varphi_{ir}\cdot\frac{h_i^4}{2b_{ir}}+\sum_{i=1}^{\rho-1}\sin\varphi_{i\ell}\cdot\frac{h_i^4}{2b_{i\ell}}\right]A_1\\
&+\left\{\alpha\left[\sum_{i=1}^{\rho-1}\sin\varphi_{ir}\cdot\frac{h_i^3}{2b_{ir}}+\sum_{i=1}^{\rho-1}\sin\varphi_{i\ell}\cdot\frac{h_i^3}{2b_{i\ell}}\right]+1\right\}A_2\\
&-\alpha\cdot\left[\sum_{i=1}^{\rho-1}\sin\varphi_{ir}\cdot\frac{h_i^2}{6b_{ir}}+\sum_{i=1}^{\rho-1}\sin\varphi_{i\ell}\cdot\frac{h_i^2}{6b_{i\ell}}\right]A_3=\\
&=\sum_{i=1}^{\rho-1}\frac{h_i^2 w_{ir}}{b_{ir}}\cdot\cos\varphi_{ir}-\sum_{i=1}^{\rho-1}\frac{h_i^2 w_{i\ell}}{b_{i\ell}}\cdot\cos\varphi_{i\ell}
\end{aligned}\right\}\qquad\textbf{(117c)}$$

Following the same procedure, as for eq (117c), but multiplying by h_i^3, we will obtain:

$$\left.\begin{aligned}
&-\alpha\left[\sum_{i=1}^{\rho-1}\sin\varphi_{ir}\cdot\frac{h_i^6}{6b_{ir}}+\sum_{i=1}^{\rho-1}\sin\varphi_{i\ell}\cdot\frac{h_i^6}{6b_{i\ell}}\right]A_0\\
&+\alpha\cdot\left[\sum_{i=1}^{\rho-1}\sin\varphi_{ir}\cdot\frac{h_i^5}{2b_{ir}}+\sum_{i=1}^{\rho-1}\sin\varphi_{i\ell}\cdot\frac{h_i^5}{2b_{i\ell}}\right]A_1\\
&+\alpha\left[\sum_{i=1}^{\rho-1}\sin\varphi_{ir}\cdot\frac{h_i^4}{2b_{ir}}+\sum_{i=1}^{\rho-1}\sin\varphi_{i\ell}\cdot\frac{h_i^4}{2b_{i\ell}}\right]A_2\\
&-\left\{-\alpha\left[\sum_{i=1}^{\rho-1}\sin\varphi_{ir}\cdot\frac{h_i^3}{6b_{ir}}+\sum_{i=1}^{\rho-1}\sin\varphi_{i\ell}\cdot\frac{h_i^3}{6b_{i\ell}}\right]+1\right\}\cdot A_3=\\
&=\sum_{i=1}^{\rho-1}\frac{h_i^3 w_{ir}}{b_{ir}}\cdot\cos\varphi_{ir}-\sum_{i=1}^{\rho-1}\frac{h_i^3 w_{i\ell}}{b_{i\ell}}\cdot\cos\varphi_{i\ell}
\end{aligned}\right\}\qquad\textbf{(117d)}$$

The above system of equations (117a) through (117d), can be written as follows:

$$\left.\begin{aligned}
&\left(-\alpha\frac{Q_3}{6}+1\right)A_0+\alpha\frac{Q_2}{2}A_1+\alpha\frac{Q_1}{2}A_2-\alpha\frac{Q_0}{6}A_3=S_0\\
&-\alpha\frac{Q_4}{6}A_0+\left(\alpha\frac{Q_3}{2}+1\right)A_1+\alpha\frac{Q_2}{2}A_2-\alpha\frac{Q_1}{6}A_3=S_1\\
&-\alpha\frac{R_5}{6}A_0+\alpha\frac{R_4}{2}A_1+\left(\alpha\frac{R_3}{2}+1\right)A_2-\alpha\frac{R_2}{6}A_3=T_2\\
&-\alpha\frac{R_6}{6}A_0+\alpha\frac{R_5}{2}A_1+\alpha\frac{R_4}{2}A_2+\left(-\alpha\frac{R_3}{6}+1\right)A_3=T_3
\end{aligned}\right\}\qquad\textbf{(118)}$$

where:

$$\left.\begin{aligned}
&Q_m=\sum_{i=\rho}^{n}\sin\varphi_{ir}\cdot\frac{h_i^m}{b_{ir}}+\sum_{i=\rho}^{n}\sin\varphi_{i\ell}\cdot\frac{h_i^m}{b_{i\ell}}\\
&R_m=\sum_{i=1}^{\rho-1}\sin\varphi_{ir}\cdot\frac{h_i^m}{b_{ir}}+\sum_{i=1}^{\rho-1}\sin\varphi_{i\ell}\cdot\frac{h_i^m}{b_{i\ell}}\\
&S_m=\sum_{i=\rho}^{n}\sin\varphi_{ir}\cdot\frac{h_i^m}{b_{ir}}w_{ir}-\sum_{i=\rho}^{n}\sin\varphi_{i\ell}\cdot\frac{h_i^m}{b_{i\ell}}w_{i\ell}\\
&T_m=\sum_{i=1}^{\rho-1}\cos\varphi_{ir}\cdot\frac{h_i^m}{b_{ir}}w_{ir}-\sum_{i=1}^{\rho-1}\cos\varphi_{i\ell}\cdot\frac{h_i^m}{b_{i\ell}}w_{i\ell}
\end{aligned}\right\}\qquad\textbf{(119)}$$

From the system of eq (118), we can determine A_0, A_1, A_2, A_3 and from eq (114a,b) the cable stresses as follows:

$$\left.\begin{aligned} P_{\rho\ell} &= \frac{\cos\varphi_{\rho\ell}}{b_{\rho\ell}} w_{\rho\ell} - \alpha\sin\varphi_{\rho\ell}\cdot\left\{\frac{h_\rho^3}{6b_{\rho\ell}}A_0 - \frac{h_\rho^2}{2b_{\rho\ell}}A_1 - \frac{h_\rho}{2b_{\rho\ell}}A_2 + \frac{1}{2b_{\rho\ell}}A_3\right\} \\ P_{\rho r} &= \frac{\cos\varphi_{\rho r}}{b_{\rho r}} w_{\rho r} - \alpha\sin\varphi_{\rho r}\cdot\left\{-\frac{h_\rho^3}{6b_{\rho\ell}}A_0 + \frac{h_\rho^2}{2b_{\rho\ell}}A_1 + \frac{h_\rho}{2b_{\rho\ell}}A_2 - \frac{1}{2b_{\rho\ell}}A_3\right\} \end{aligned}\right\} \quad \textbf{(120)}$$

Dense Arrangement of Cables

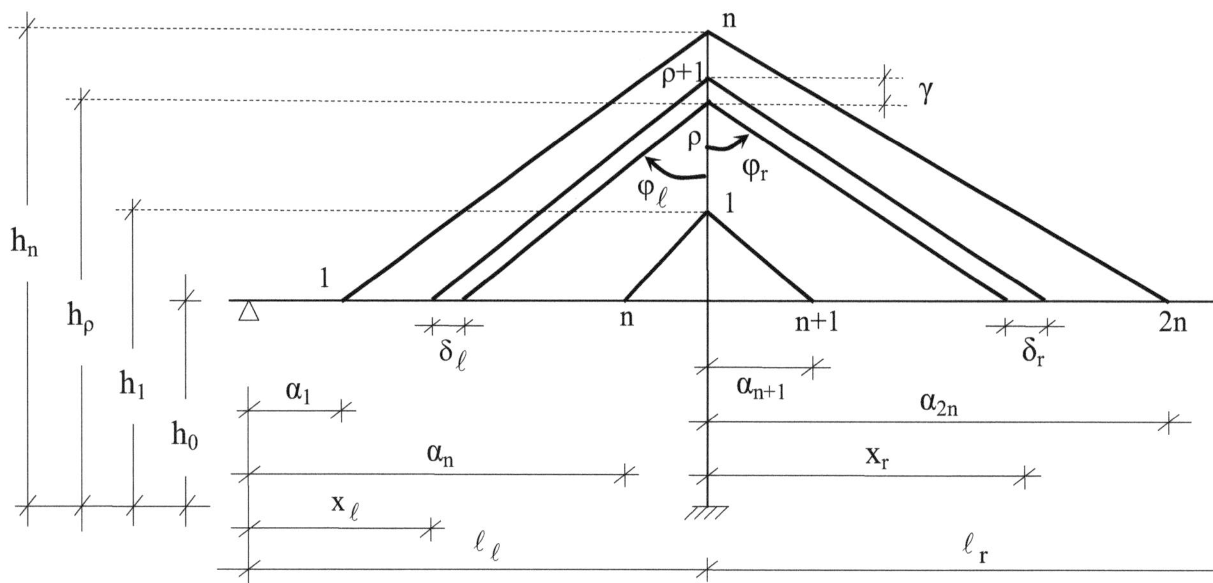

Figure 18: Notation for dense arrangement of cables

Let us consider next that the cables are in a dense arrangement and that the distances δ_ℓ and δ_r between two neighboring cables satisfy the conditions:

$$\left.\begin{aligned} \delta_\ell &<< \alpha_n - \alpha_1 \\ \delta_r &<< \alpha_{2n} - \alpha_{n+1} \end{aligned}\right\} \quad \textbf{(121)}$$

Then, we may consider a distributed load q(x), applied from position α_1 to α_n and from α_{n+1} to α_{2n} , which for instance at position x_ρ will be:

$$q(x_r) = \frac{1}{\delta} P_\rho \cos\varphi_{\rho r} \quad \textbf{(122a)}$$

Following the notation in Fig. **18**, equations (103) and (104) become:

$$\left.\begin{aligned} h_\rho &= h_1 + (x_\ell - \delta_{\ell_1})\cdot\frac{\gamma}{\delta_\ell} = h_1 + [x_\ell - (\ell_\ell - \alpha_n)]\cdot\frac{\gamma}{\delta_\ell} \\ h_\rho &= h_1 + (x_r - \delta_{r_1})\cdot\frac{\gamma}{\delta_r} = h_1 + [x_r - \alpha_{n+1}]\cdot\frac{\gamma}{\delta_r} \end{aligned}\right\} \quad \textbf{(122b)}$$

$$\left.\begin{aligned}
\tan\varphi_{\rho\ell} &= \frac{\ell_\ell - x_\ell}{(h_1-h_0)-\left[x_\ell-(\ell_\ell-\alpha_n)\right]\dfrac{\gamma}{\delta_\ell}}\\
\sin\varphi_{\rho\ell} &= \frac{\ell_\ell - x_\ell}{\sqrt{(\ell_\ell-x_\ell)^2+\left\{(h_1-h_0)-\left[x_\ell-(\ell_\ell-\alpha_n)\right]\dfrac{\gamma}{\delta_\ell}\right\}^2}}\\
\cos\varphi_{\rho\ell} &= \frac{(h_1-h_0)-\left[x_\ell-(\ell_\ell-\alpha_n)\right]\dfrac{\gamma}{\delta_\ell}}{\sqrt{(\ell_\ell-x_\ell)^2+\left\{(h_1-h_0)-\left[x_\ell-(\ell_\ell-\alpha_n)\right]\dfrac{\gamma}{\delta_\ell}\right\}^2}}\\
\tan\varphi_{\rho r} &= \frac{x_r}{(h_1-h_0)-(x_r-\alpha_{n+1})\dfrac{\gamma}{\delta_r}}
\end{aligned}\right\}\qquad \textbf{(123a)}$$

$$\left.\begin{aligned}
\sin\varphi_{\rho r} &= \frac{x_r}{\sqrt{x_r^2+\left\{(h_1-h_0)-(x_\ell-\alpha_{n+1})\dfrac{\gamma}{\delta_r}\right\}^2}}\\
\cos\varphi_{\rho r} &= \frac{(h_1-h_0)-(x_r-\alpha_{n+1})\dfrac{\gamma}{\delta_r}}{\sqrt{x_r^2+\left\{(h_1-h_0)-(x_\ell-\alpha_{n+1})\dfrac{\gamma}{\delta_r}\right\}^2}}
\end{aligned}\right\}\qquad \textbf{(123b)}$$

equations (113) can be written as follows:

$$\left.\begin{aligned}
\alpha = \frac{1}{EI_p}\quad , \quad b_{\rho\ell}(x_\ell) &= \frac{s_{\rho\ell}}{E_cA_{\rho\ell}} = \frac{\ell_\ell - x_\ell}{\sin\varphi_\ell\cdot E_cA_{\rho\ell}}\\
b_{\rho r}(x_r) &= \frac{s_{\rho r}}{E_cA_{\rho r}} = \frac{x_r}{\sin\varphi_r\cdot E_cA_{\rho r}}
\end{aligned}\right\}\qquad \textbf{(124)}$$

Assuming that the cross-section of the cables varies along x, we will have:

$$\left.\begin{aligned}
A_{\rho\ell} &= \delta A_\ell(x_\ell)\\
A_{\rho r} &= \delta A_r(x_r)\\
h_\rho &= h_0+\frac{\ell_\ell - x_\ell}{\tan\varphi_\ell} = h_0+\frac{x_r}{\tan\varphi_r}
\end{aligned}\right\}\qquad \textbf{(125)}$$

Then, equations (119) become:

$$\left.\begin{aligned}
\overline{Q}_m &= E_c\left[\int_{x_r}^{\alpha_{2n}}\left(h_0+\frac{x_r}{\tan\varphi_r}\right)^m\frac{A_r(x_r)\sin^2\varphi_r}{x_r}dx_r\right.\\
&\left.+\int_{x_\ell}^{\alpha_1}\left(h_0+\frac{\ell_\ell-x_\ell}{\tan\varphi_\ell}\right)^m\frac{A_\ell(x_\ell)\sin^2\varphi_\ell}{\ell_\ell-x_\ell}dx_\ell\right]\\
\overline{R}_m &= E_c\left[\int_{\alpha_{n+1}}^{x_r}\left(h_0+\frac{x_r}{\tan\varphi_r}\right)^m\frac{A_r(x_r)\sin^2\varphi_r}{x_r}dx_r\right.\\
&\left.+\int_{\alpha_n}^{x_\ell}\left(h_0+\frac{\ell_\ell-x_\ell}{\tan\varphi_\ell}\right)^m\frac{A_\ell(x_\ell)\sin^2\varphi_\ell}{\ell_\ell-x_\ell}dx_\ell\right]\\
\overline{S}_m &= \frac{E_c}{2}\left[\int_{x_r}^{\alpha_{2n}}\left(h_0+\frac{x_r}{\tan\varphi_r}\right)^m\frac{A_r(x_r)\sin 2\varphi_r}{x_r}w_r(x_r)dx_r\right.\\
&\left.-\int_{x_\ell}^{\alpha_1}\left(h_0+\frac{\ell_\ell-x_\ell}{\tan\varphi_\ell}\right)^m\frac{A_\ell(x_\ell)\sin 2\varphi_\ell}{\ell_\ell-x_\ell}w_\ell(x_\ell)dx_\ell\right]\\
\overline{T}_m &= \frac{E_c}{2}\left[\int_{\alpha_{n+1}}^{x_r}\left(h_0+\frac{x_r}{\tan\varphi_r}\right)^m\frac{A_r(x_r)\sin 2\varphi_r}{x_r}w_r(x_r)dx_r\right.\\
&\left.-\int_{\alpha_n}^{x_\ell}\left(h_0+\frac{\ell_\ell-x_\ell}{\tan\varphi_\ell}\right)^m\frac{A_\ell(x_\ell)\sin 2\varphi_\ell}{\ell_\ell-x_\ell}w_\ell(x_\ell)dx_\ell\right]
\end{aligned}\right\} \quad \textbf{(126)}$$

Setting:

$$\left.\begin{aligned}
\overline{A}_0 &= \int_{x_r}^{\alpha_{2n}}q_r(x_r)\cos\varphi_r dx_r-\int_{x_\ell}^{\alpha_1}q_\ell(x_\ell)\cos\varphi_\ell dx_\ell\\
\overline{A}_1 &= \int_{x_r}^{\alpha_{2n}}\left(h_0+\frac{x_r}{\tan\varphi_r}\right)q_r(x_r)\cos\varphi_r dx_r-\int_{x_\ell}^{\alpha_1}\left(h_0+\frac{\ell_\ell-x_\ell}{\tan\varphi_\ell}\right)q_\ell(x_\ell)\cos\varphi_\ell dx_\ell\\
\overline{A}_2 &= \int_{\alpha_{n+1}}^{x_r}\left(h_0+\frac{x_r}{\tan\varphi_r}\right)^2q_r(x_r)\cos\varphi_r dx_r-\int_{\alpha_n}^{x_\ell}\left(h_0+\frac{\ell_\ell-x_\ell}{\tan\varphi_\ell}\right)^2q_\ell(x_\ell)\cos\varphi_\ell dx_\ell\\
\overline{A}_3 &= \int_{\alpha_{n+1}}^{x_r}\left(h_0+\frac{x_r}{\tan\varphi_r}\right)^3q_r(x_r)\cos\varphi_r dx_r-\int_{\alpha_n}^{x_\ell}\left(h_0+\frac{\ell_\ell-x_\ell}{\tan\varphi_\ell}\right)^3q_\ell(x_\ell)\cos\varphi_\ell dx_\ell
\end{aligned}\right\} \quad \textbf{(127)}$$

we arrive at the following system of equations - similar to eq (118):

$$\left.\begin{aligned}
&\left(-\alpha\frac{\overline{Q}_3}{6}+1\right)\overline{A}_0+\alpha\frac{\overline{Q}_2}{2}\overline{A}_1+\alpha\frac{\overline{Q}_1}{2}\overline{A}_2-\alpha\frac{\overline{Q}_2}{6}\overline{A}_3=\overline{S}_0\\
&-\alpha\frac{\overline{Q}_4}{6}\overline{A}_0+\left(\alpha\frac{\overline{Q}_3}{2}+1\right)\overline{A}_1+\alpha\frac{\overline{Q}_2}{2}\overline{A}_2-\alpha\frac{\overline{Q}_1}{6}\overline{A}_3=\overline{S}_1\\
&-\alpha\frac{\overline{R}_5}{6}\overline{A}_0+\alpha\frac{\overline{R}_4}{2}\overline{A}_1+\left(\alpha\frac{\overline{R}_3}{2}+1\right)\overline{A}_2-\alpha\frac{\overline{R}_2}{6}\overline{A}_3=\overline{T}_2\\
&-\alpha\frac{\overline{R}_6}{6}\overline{A}_0+\alpha\frac{\overline{R}_5}{2}\overline{A}_1+\alpha\frac{\overline{R}_4}{2}\overline{A}_2+\left(-\alpha\frac{\overline{R}_3}{6}+1\right)\overline{A}_3=\overline{T}_3
\end{aligned}\right\} \quad \textbf{(128)}$$

By solving the above system, the coefficients $\bar{A}_0, \bar{A}_1, \bar{A}_2, \bar{A}_3$ are determined, and hence from eq (114a,b) the stresses in cable are also determined as follows:

$$\left.\begin{aligned} q_\ell(x_\ell) &= E_c \sin\varphi_\ell \cos^2\varphi_\ell \frac{A_\ell(x_\ell)}{\ell_\ell - x_\ell} w_\ell(x_\ell) \\ &-\alpha E_c \sin\varphi_\ell \sin 2\varphi_\ell \left\{ \left(h_0 + \frac{\ell_\ell - x_\ell}{\tan\varphi_\ell} \right)^3 \frac{A_\ell(x_\ell)}{12(\ell_\ell - x_\ell)} \bar{A}_0 \right. \\ &- \left(h_0 + \frac{\ell_\ell - x_\ell}{\tan\varphi_\ell} \right)^2 \frac{A_\ell(x_\ell)}{4(\ell_\ell - x_\ell)} \bar{A}_1 \\ &\left. - \left(h_0 + \frac{\ell_\ell - x_\ell}{\tan\varphi_\ell} \right) \frac{A_\ell(x_\ell)}{4(\ell_\ell - x_\ell)} \bar{A}_2 + \frac{A_\ell(x_\ell)}{12(\ell_\ell - x_\ell)} \bar{A}_3 \right\} \end{aligned}\right\} \quad \textbf{(129a)}$$

$$\left.\begin{aligned} q_r(x_r) &= E_c \sin\varphi_r \cos^2\varphi_r \frac{A_r(x_r)}{x_r} w_r(x_r) \\ &-\alpha E_c \sin\varphi_r \sin 2\varphi_r \left\{ - \left(h_0 + \frac{x_r}{\tan\varphi_r} \right)^3 \frac{A_r(x_r)}{12 x_r} \bar{A}_0 \right. \\ &\left. + \left(h_0 + \frac{x_r}{\tan\varphi_r} \right)^2 \frac{A_r(x_r)}{4 x_r} \bar{A}_1 + \left(h_0 + \frac{x_r}{\tan\varphi_r} \right) \frac{A_r(x_r)}{4 x_r} \bar{A}_2 - \frac{A_r(x_r)}{12 x_r} \bar{A}_3 \right\} \end{aligned}\right\} \quad \textbf{(129b)}$$

The Free Vibration of Cable-Stayed Bridges [9], [10]

In the previous sections we have studied the two basic systems of cable-stayed bridges: the fan-system and the harp-system. We have concluded to the determination of the stresses of the cables. These stresses are depending in all cases from the deformation w(x) of the bridge deck and are given in general by:

$$P_i = \gamma_{i1} w(\alpha_1) + \gamma_{i2} w(\alpha_2) + \ldots.. + \gamma_{i\rho} w(\alpha_\rho) \quad \textbf{(130)}$$

for the case of thin arrangement with ρ cables at positions $\alpha_1, \alpha_2, \ldots, \alpha_\rho$ (with $i \leq \rho$) and :

$$q(x) = A(x)w(x) + B(x)\int_{\alpha_1}^{\alpha_n} \Gamma(x)w(x)dx \quad \textbf{(131)}$$

for the case of dense arrangement in α_1 to α_n .

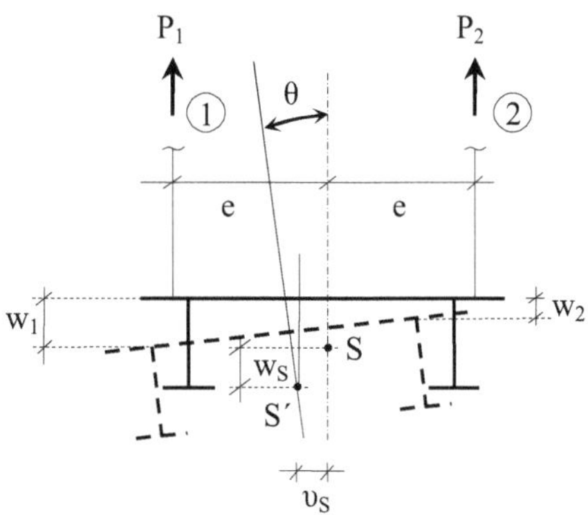

Figure 19: Deformed and undeformed bridge deck suspended form cables

The coefficients γ_i and the functions A,B,Γ are different for each span and are evaluated separately.

It is obvious that the dynamic behavior of cable-stayed bridges is a highly nonlinear problem and cannot be dealed with the classical procedure followed up to now. For this type problems a special methodology will be applied which is presented next.

Thin Arrangement of Cables

The equations for free vibration of a bridge, neglecting the shear deformation and rotational inertia terms for the cross-section as well as the interaction between bending and axial vibration are the following:

$$\left.\begin{aligned}
&EI_y w''''(x,t)+c_y\dot{w}(x,t)+m\ddot{w}(x,t)=\\
&-\sum_{i=1}^{\rho}P_{1i}\cos\varphi_i\delta(x-\alpha_i)-\sum_{i=1}^{\rho}P_{2i}\cos\varphi_i\delta(x-\alpha_i)\\
&EI_z\upsilon''''(x,t)-EI_z z_M\theta''''(x,t)+c_z\dot{\upsilon}(x,t)+m\ddot{\upsilon}(x,t)=0\\
&EC_S\theta''''(x,t)-EI_z z_M\upsilon''''(x,t)-GJ_d\theta''(x,t)+\\
&+c_\theta\dot{\theta}(x,t)+\Theta_x\ddot{\theta}(x,t)=\sum_{i=1}^{\rho}(P_{2i}-P_{1i})e\cos\varphi_i\delta(x-\alpha_i)
\end{aligned}\right\} \qquad \textbf{(132a,b,c)}$$

where P_{1i} and P_{2i} are the cable forces at positions 1 and 2, respectively, while α_i are the positions of cable attachments. Equation (132a), which corresponds to the vertical vibration of the bridge deck, is independent from the other two equations of motion, which are complex and correspond to the transverse and torsional vibration of the bridge deck.

In order to solve this equation, we seek a solution of separate variables type in the form:

$$w(x,t)=W(x)T(t) \qquad \textbf{(133)}$$

Then, as known, eq (132a), due to eq (130), results the following:

$$\left.\begin{aligned}
&W''''(x)+\frac{1}{EI_y}\sum_{i=1}^{\rho}\left[\gamma_{i1}W(\alpha_1)+\gamma_{i2}W(\alpha_2)+\;....\right.\\
&\qquad\qquad ...+\left.\gamma_{i\rho}W(\alpha_\rho)\right]\cos\varphi_i\cdot\delta(x-\alpha_i)-\lambda W(x)=0\\
&\ddot{T}+\frac{c_y}{m}\dot{T}+\omega_\kappa^2T=0 \quad , \quad \text{where: } \lambda=\frac{m\omega_\kappa^4}{EI_y}
\end{aligned}\right\} \qquad \textbf{(134)}$$

In order to apply the Galerkin method, we set:

$$W(x)=c_1\Psi_1(x)+c_2\Psi_2(x)+........+c_n\Psi_n(x) \qquad \textbf{(135)}$$

where $\Psi_j(x)$ are the shape functions of the beam - continuous or not, which have the same characteristics with the bridge-deck without the cables.

Introducing eq (135) into eq (134a), multiplying the outcome by Ψ_1 , Ψ_2 , ...Ψ_n and integrating the resulting relations from 0 to L, we obtain the following linear and homogeneous system of equations with unknown coefficients c_1, c_2,......,c_n .

$$\left.\begin{aligned}&\int_0^L\left(c_1\Psi_1''''(x)+....+c_n\Psi_n''''(x)\right)dx+\frac{1}{EI_y}\Big\{\left[\gamma_{11}\left(c_1\Psi_1(\alpha_1)+...+c_n\Psi_n(\alpha_1)\right)+...\right.\\&...+\gamma_{1\rho}\left(c_1\Psi_1(\alpha_\rho)+...+c_n\Psi_n(\alpha_\rho)\right)\Big]\cos\varphi_1\Psi_\sigma(\alpha_1)+\Big[\gamma_{21}\left(c_1\Psi_1(\alpha_1)+...+c_n\Psi_n(\alpha_1)\right)+...\\&...+\gamma_{2\rho}\left(c_1\Psi_1(\alpha_\rho)+...+c_n\Psi_n(\alpha_\rho)\right)\Big]\cos\varphi_2\Psi_\sigma(\alpha_2)+\\&+...+\\&+\Big[\gamma_{\rho1}\left(c_1\Psi_1(\alpha_1)+...+c_n\Psi_n(\alpha_1)\right)+...+\gamma_{\rho\rho}\left(c_1\Psi_1(\alpha_\rho)+...+c_n\Psi_n(\alpha_\rho)\right)\Big]\cos\varphi_\rho\Psi_\sigma(\alpha_\rho)\Big\}-\\&-\lambda\int_0^L\left(c_1\Psi_1(x)+....+c_n\Psi_n(x)\right)\Psi_\sigma(x)dx=0\quad,\ \text{with}\ \ \sigma=1\ \text{to}\ n\end{aligned}\right\}\qquad\textbf{(136)}$$

The above linear homogeneous system of n-equations with unknown coefficients $c_1,\dots,c_n$, can be also written as follows:

$$\left.\begin{aligned}c_1(A_{\kappa1}-\lambda\cdot B_{\kappa1})+c_2(A_{\kappa2}-\lambda\cdot B_{\kappa2})+......+c_n(A_{\kappa n}-\lambda\cdot B_{\kappa n})=0\\ \text{with}\ \ \kappa=1,2,.....,n\end{aligned}\right\}\qquad\textbf{(137)}$$

In order that the system of eq (137) has a non-trivial solution, the determinant of the unknown coefficients must be zero, i.e.:

$$\left.\begin{aligned}&\left|\Gamma_{\kappa h}\right|=0\\ \text{with}\ \ &\Gamma_{\kappa h}=A_{\kappa h}-\lambda B_{\kappa h}\ \ ,\ \ \kappa,h=1,2,......,n\end{aligned}\right\}\qquad\textbf{(138)}$$

Finally, from the (n-1) remaining equations of the system (137), we can compute the shape functions $W_n(x)$ for the corresponding values of ω_κ:

$$\left.\begin{aligned}&W_n(x)=c_1\sum_{h=2}^{n}\left(\Psi_1+\frac{c_h}{c_1}\Psi_n\right)\\ \text{with}\ \ &\frac{c_h}{c_1}=\frac{\begin{vmatrix}\Gamma_{12}&.....&\Gamma_{11}&.....&\Gamma_{1n}\\.....&.....&.....&.....&....\\\Gamma_{(m-1)2}&.....&\Gamma_{(m-1)1}&.....&\Gamma_{(m-1)n}\end{vmatrix}}{\left|\Gamma_{\kappa h}\right|}\end{aligned}\right\}\qquad\textbf{(139a,b)}$$

Since $z_M\neq0$, equations (132b,c) are not independent and the resulting free torsional vibration will have eigenfrequency ω_σ and hence, we may consider seeking a solution in the following form:

$$\left.\begin{aligned}\upsilon(x,t)=V(x)P(t)\\\theta(x,t)=\Phi(x)P(t)\end{aligned}\right\}\qquad\textbf{(140)}$$

Introducing the above expressions into eq (132b,c), we will obtain the following equations of motion:

$$\left.\begin{array}{l} EI_z V''''(x) - EI_z z_M \Phi''''(x) - m\omega_\sigma^2 V(x) = 0 \\ EC_s \Phi''''(x) - EI_z z_M V''''(x) - GJ_d \Phi'' - \\ \qquad - \Theta_x \omega_\sigma^2 \Phi(x) = \sum_{i=1}^{\rho} (P_{2i} - P_{1i}) e \cos\varphi_i \cdot \delta(x - \alpha_i) \\ \ddot{P} + \frac{c_z}{\Theta_x} \dot{P} + \omega_\sigma^2 P = 0 \end{array}\right\} \quad \textbf{(141a,b,c)}$$

From Fig. **19** it is easily concluded that:

$$\left.\begin{array}{l} w_1 = w + e\theta \\ w_2 = w - e\theta \end{array}\right\} \quad \textbf{(142)}$$

and hence, it is:

$$P_{2i} - P_{1i} = 2e\left(\gamma_{i1}\theta(\alpha_1) + \gamma_{i2}\theta(\alpha_2) + ... + \gamma_{i\rho}\theta(\alpha_\rho)\right) \quad \textbf{(143)}$$

In order to solve the system of eq (141a,b) with the Galerkin method we set:

$$\left.\begin{array}{l} V(x) = d_1 Z_1(x) + ... + d_n Z_n(x) \\ \Phi(x) = f_1 \Omega_1(x) + ... + f_n \Omega_n(x) \end{array}\right\} \quad \textbf{(144)}$$

where d, f are constants to be determined, while $Z_n(x)$ and $\Omega_n(x)$ are the shape functions of the corresponding beam – continuous or not, which has the same characteristics with the bridge deck without the cables.

Introducing eq (144) into eq (141a,b) and (143), multiplying the expression resulting from eq (141a) by $Z_1, Z_2,\ldots, Z_n$, and the expression resulting from eq (141b) by $\Omega_1, \Omega_2,\ldots, \Omega_n$, and integrating the outcomes from 0 to L taking into account the orthogonality conditions, we obtain a system of (2n) equations with unknowns the coefficients $d_1, d_2,\ldots, d_n$ and $f_1, f_2,\ldots, f_n$, which can be written in a concise form as follows:

$$\left.\begin{array}{l} (A_{i\ell} - \omega_\sigma^2 B_{i\ell}) d_1 + ... + (A_{in} - \omega_\sigma^2 B_{in}) d_n - \Gamma_{i\ell} f_1 - ..... - \Gamma_{in} f_n = 0 \\ -A_{\ell_1} d_1 - ... - A_{\ell_n} d_n + (\Gamma_{\ell_1} - \omega_\sigma^2 \Delta_{\ell_1}) f_1 + ... + (\Gamma_{\ell_n} - \omega_\sigma^2 \Delta_{\ell_n}) f_n = 0 \\ \text{where:} \quad i, j, \ell, g = 1 \ \text{to} \ n \end{array}\right\} \quad \textbf{(145)}$$

In order that the system of eq (145) as a non-trivial solution, the determinant of the unknown coefficients must be zero, i.e. ή:

$$\left|E_{op}\right| = 0 \ , \quad \text{where:} \quad o, p = 1, 2, ..., 2n \quad \textbf{(146)}$$

From eq (146), we can determine the combined bending-torsional eigenfrequencies ω_σ, while from the first (2n-1) equations of the system (145), we can determine the corresponding shape functions.

Dense Arrangement of Cables

In this case, the equations of free vibration for the bridge are:

$$\left.\begin{aligned}
&EI_y w''''(x,t) + c_y \dot{w}(x,t) + m\ddot{w}(x,t) = \\
&= -A(x)\left[w_1(x) + w_2(x)\right] - B(x)\int_{\alpha_1}^{\alpha_n} \Gamma(x)\left[w_1(x) + w_2(x)\right] dx \\
&EI_z \upsilon''''(x,t) - EI_z z_M \theta''''(x,t) + c_z \dot{\upsilon}(x,t) + m\ddot{\upsilon}(x,t) = 0 \\
&EC_S \theta''''(x,t) - EI_z z_M \upsilon''''(x,t) - GJ_d \theta''(x,t) + c_0 \dot{\theta}(x,t) + \\
&+\Theta_x \ddot{\theta}(x,t) = 2e^2 A(x)\theta(x) + 2e^2 B(x)\int_{\alpha_1}^{\alpha_n} \Gamma(x)\theta(x)dx
\end{aligned}\right\} \quad \textbf{(147)}$$

The above system of equations is solved following the same procedure presented in the previous paragraph with a suitable modification of the resulting coefficients.

SUSPENSION BRIDGES [7], [14]

Suspension bridges are used since ancient times in a basic form to link small spans. They can be considered as the oldest bridge structures developed. Their use was temporarily stopped in 1940, after the collapse of the newly erected Tacoma-bridge. Since then, both theoretical and experimental investigations revealed the causes of the collapse and offered the know-how that allowed the construction of suspension bridges with even bigger spans.

Suspension bridges are considered as the most economic solutions for spans bigger than 450 m, while for spans bigger than 700 m, it is the only possible solution. In the present, spans in the range of 1.300 to 1.500 m are considered as common solutions.

A characteristic property of suspension bridges, are the relatively big deformations due to live loads, and vibrations due to to dynamic loads and the wind loads. This property makes these structures very sensitive compared to the classical bridge structures.

A typical arrangement for a three-span suspension bridge, where the spans at both ends are usually equal to each other, is shown in Fig. **20**. The main structural parts of such a bridge are also shown in the same figure.

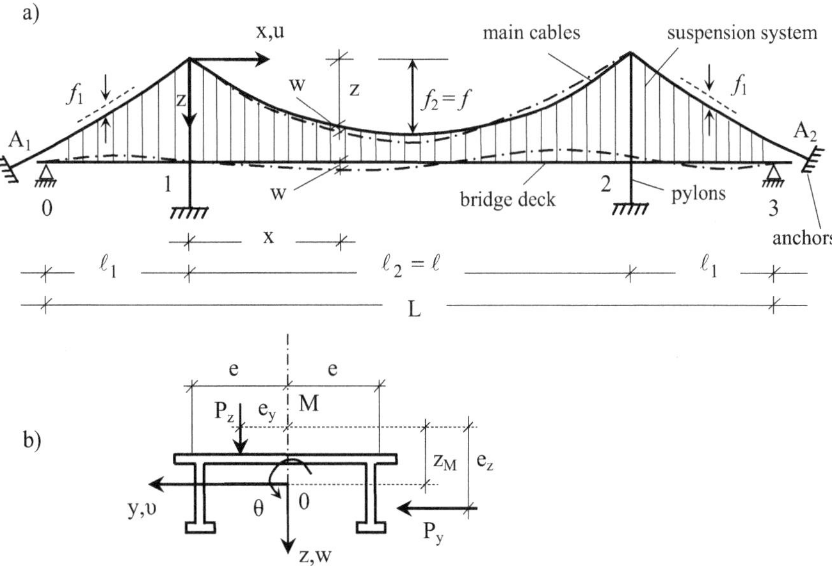

Figure 20: Suspension bridge: a) side view, b) cross-section

For this type of bridges, the following assumptions are valid besides the usual ones valid for thin-walled sections:

1. The self-weight of the deck, the main cables and the suspension system as well as the traffic loads during the erection phase, i.e. the total of permanent loads, is solely undertaken by the suspension system and not by the deck beams.
2. The suspension wires (suspenders) are considered as non-deformable (without axial deformation).
3. The pylons are considered as stable and non-deformable, i.e. their dimensions do not change, and they do not rotate or displace.
4. The anchoring system of the main cables on the pylons is formed in such way that allows free sliding along the direction of the traffic loads, while any other displacement is fixed.
5. The suspension wires are placed in a dense arrangement, compared to the length of the bridge, so that they can be considered as a distributed load.
6. The bridge is modeled as a continuous elastic system with infinite degrees-of-freedom.

Basic Relations

Due to the 1st assumption and keeping in mind that the projection of forces on x-axis that are acting on the infinitesimal element ds of the main cable in Fig. **21** gives:

$$H_g = \text{const.} \tag{148}$$

will result to the following equilibrium equation in the vertical direction:

$$g(x)dx - H_g z' + H_g\left(z' + dz'\right) = 0 \quad \text{or}$$

$$g(x) = -H_g \frac{dz'}{dx} = -H_g z'' \tag{149}$$

or more concisely $z'' = -\dfrac{g}{H_g} = \text{const.}$ **(150)**

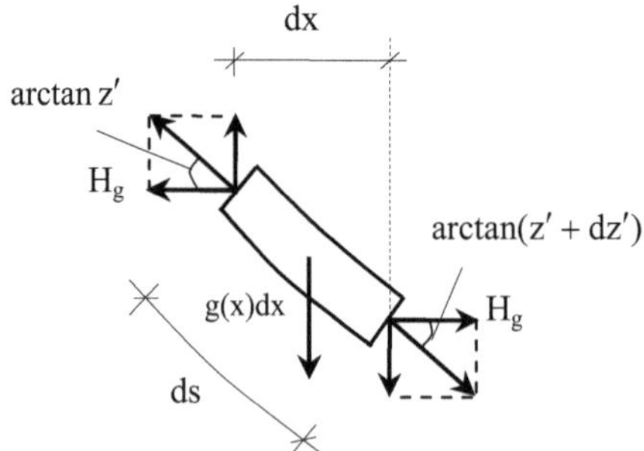

Figure 21: Equilibrium of vertical forces in an infinitesimal cable element ds

Integrating the above relation, we obtain the following expression for the form of the cables (parabolic type):

$$z(x) = \frac{g}{2H_g} \cdot x(\ell - x) \tag{151}$$

and since $z\left(\dfrac{\ell}{2}\right) = f$ we will get:

$$H_g = \frac{g\ell^2}{8f} \tag{152}$$

(where H_g is the tensile force of the cables due to permanent loads and H_p is the tensile force of the cables due to live loads).

Let us consider that the whole structure has taken its final form under the permanent loads, thus the initial deformations will be zero (assumptions 1,2,3).

When the bridge is loaded with the live loads, it will deform by υ,w,θ, and the suspension cables will reach their final stress level $H = H_g + H_p$. The suspension cables are also deformed, but since the anchoring positions A1 and A2 are fixed, the horizontal projection of the change in length of the suspension cables must be zero. That is (Fig. **22**):

$$\int_0^L \Delta dx = 0 \tag{153}$$

where L is the total length of the bridge, as shown in Fig. **20**.

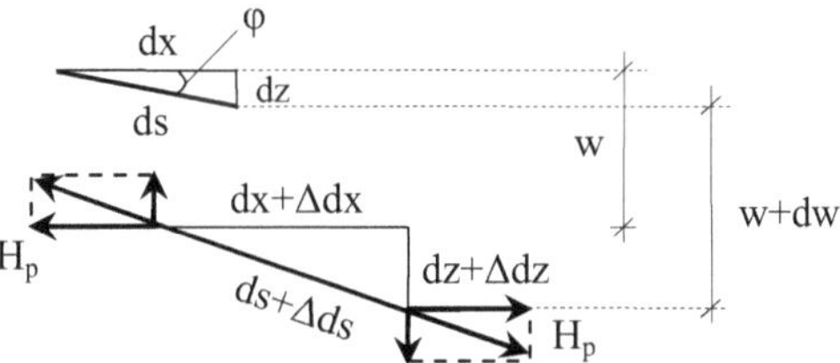

Figure 22: Deformed and undeformed cable element ds

The change in length of an infinitesimal part ds of the cable due to the traffic loads p and a possible temperature change ΔT is:

$$\Delta ds = \frac{H_p ds}{E_c F_c \cos\varphi} + \alpha_T \Delta T ds \tag{154}$$

where E_c and F_c are the elasticity modulus and cross-sectional area of the cables, respectively, while α_T is the thermal coefficient of the cables. From Fig. **22** it follows that:

$$\left.\begin{aligned} ds^2 &= dx^2 + dz^2 \\ (ds + \Delta ds)^2 &= (dx + \Delta dx)^2 + (dz + \Delta dz)^2 \end{aligned}\right\} \tag{155}$$

If we expand the second relation using the first one and neglect the higher order terms, we will obtain:

$$\Delta dx = \Delta ds \frac{ds}{dx} - \Delta dz \frac{dz}{dx} \tag{156}$$

and since: $\frac{dx}{ds} = \cos\varphi$, $\Delta dz = dw$ it will be:

$$\Delta dx = \frac{H_p}{E_c F_c \cos^3\varphi} dx + \frac{\alpha_T \Delta T}{\cos^2\varphi} dx - \frac{dw}{dx}\frac{dz}{dx} dx$$

which due to eq (153) gives:

$$\int_0^L \frac{H_p}{E_c F_c \cos^3\varphi} dx + \int_0^L \frac{\alpha_T \Delta T}{\cos^2\varphi} dx - \int_0^L \frac{dw}{dx}\frac{dz}{dx} dx = 0 \qquad \textbf{(157)}$$

After integrating the third term by parts, and taking into account that: $w(0) = w(L) = 0$ and that due to eq (151) it is g''=const. , we will obtain:

$$\left.\begin{aligned} & H_p \frac{L_c}{E_c F_c} + \alpha_T \Delta T \cdot L_T + z'' \int_0^L w(x) dx = 0 \\ & \text{where:} \quad L_c = \int_0^L \frac{dx}{\cos^3\varphi} \quad , \quad L_T = \int_0^L \frac{dx}{\cos^2\varphi} \end{aligned}\right\} \qquad \textbf{(158)}$$

For the general case shown in Fig. **23** it will be:

$$\left.\begin{aligned} & L_C \cong \ell \left(1 + 8\frac{f^2}{\ell^2} + \frac{3}{2}\tan^2\gamma_0\right) + \frac{S_1}{\cos^2\gamma_1} + \frac{S_2}{\cos^2\gamma_2} \\ & L_T \cong \ell \left(1 + \frac{16}{3}\frac{f^2}{\ell^2} + \tan^2\gamma_0\right) + \frac{S_1}{\cos\gamma_1} + \frac{S_2}{\cos\gamma_2} \end{aligned}\right\} \qquad \textbf{(159)}$$

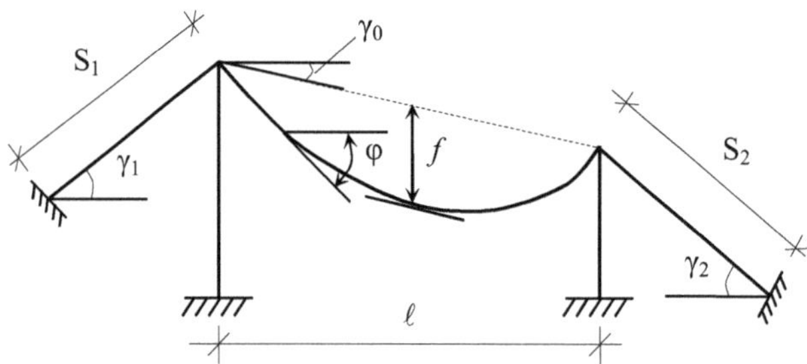

Figure 23: Geometry of a cable-stayed bridge (general case)

When loaded also with the traffic loads, the bridge is deformed as shown in Fig. **24b**, while the cables have tensile forces:

$$\left.\begin{aligned} & H_1 = \frac{H_g}{2} + H_{p_1} \\ & H_2 = \frac{H_g}{2} + H_{p_2} \end{aligned}\right\} \qquad \textbf{(160)}$$

where H_{p_1} and H_{p_2} are the additional forces of cables 1 and 2, respectively, due to live loads.

From Fig. **24b**, and since the angles θ_1, θ_2, φ_1, φ_2 are very small and practically it is:

$\theta_1 = \theta_2$, $\varphi_1 = \varphi_2$ we shall have:

$$\left.\begin{aligned} & \upsilon_1 = \upsilon_2 = \upsilon_S \\ & w_1 = w_S + e\theta \\ & w_2 = w_S - e\theta \end{aligned}\right\} \qquad \textbf{(161a,b,c)}$$

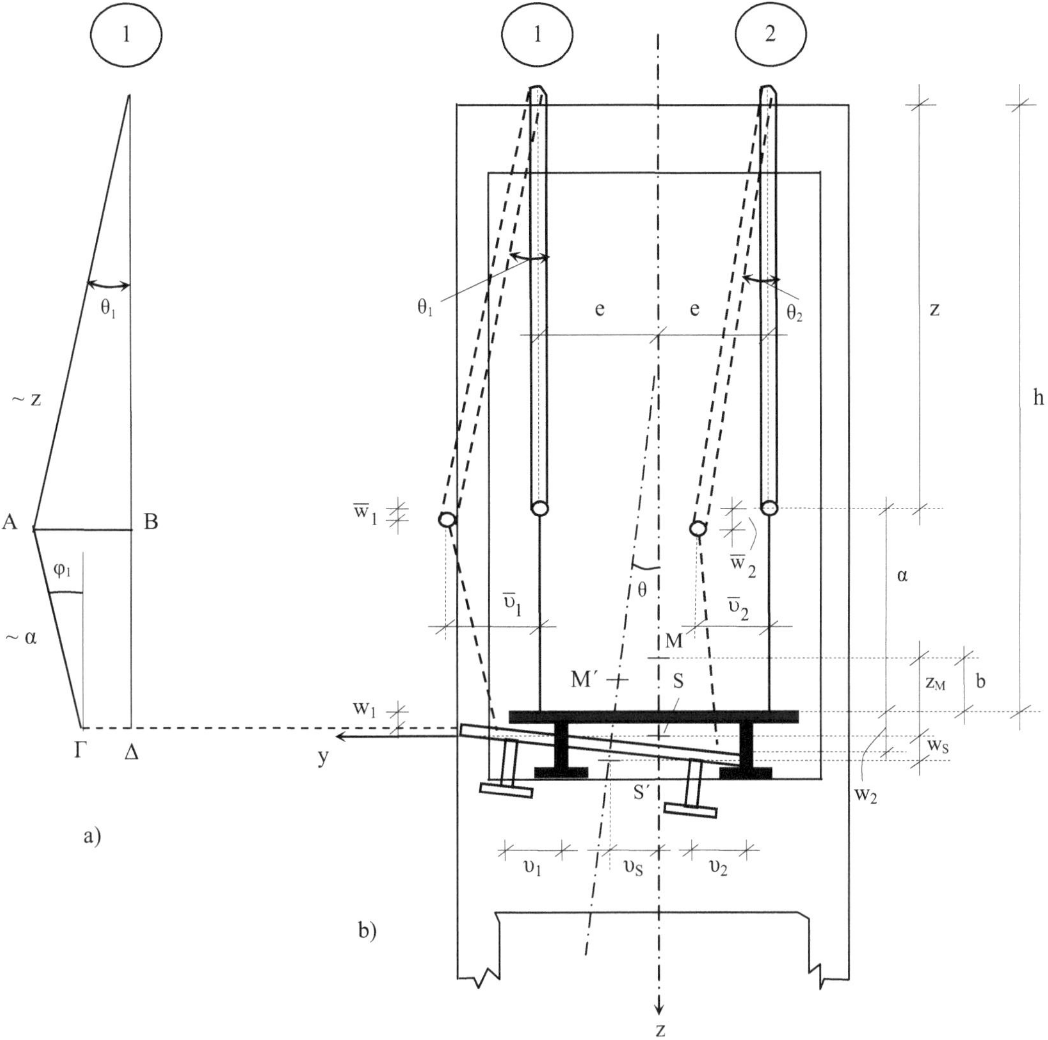

Figure 24: Deformed and undeformed state of the cable-deck system

Isolating cable 1 in the deformed state (Fig. **24a**) we will have: $\overline{1A} \cong z$ and $\overline{A\Gamma} \cong \alpha$.

Since the curve $\overline{1A\Gamma}$ is practically a straight line, it will be: $\dfrac{AB}{\Gamma\Delta} = \dfrac{z}{h}$.

$$\left.\begin{aligned}
&\bar{\upsilon}_1 = \overline{AB} = \upsilon_S \frac{z}{h} && \text{or} && \\
&\bar{\upsilon}_1 = \bar{\upsilon}_2 = \upsilon_S \frac{z}{\alpha + z} && \text{and also:} && \bar{w}_1 = w_1 \\
& && && \bar{w}_2 = w_2
\end{aligned}\right\} \qquad \textbf{(162)}$$

Finally, it is: $\sin\theta_1 = \sin\theta_2 \cong \dfrac{\upsilon_S(x)}{z(x)}$ **(163)**

where the bar denotes the deformation of the cables, while without bar are the deformations of the deck.

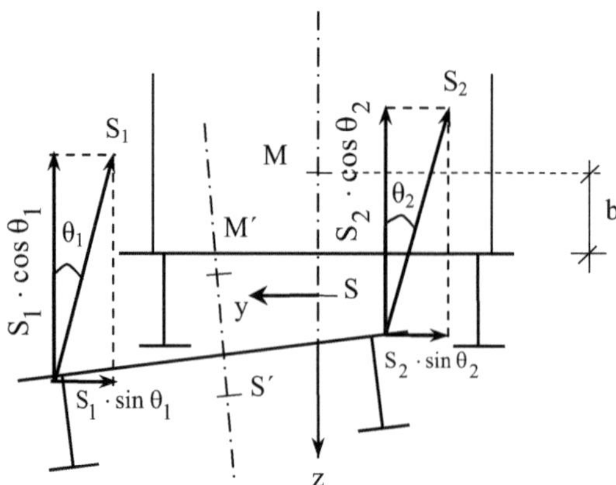

Figure 25: Cable forces acting on the deck

Since the small deformation theory is adopted up to this point, it is: $\theta_1 = \theta_2 \cong 0$ and hence, we shall take into account only the vertical components of the forces S_1 and S_2. Then, we will have:

$$\left.\begin{aligned} S_1 \cos\theta_1 &\cong S_1 = -H_1(z+w_1)'' \\ S_2 \cos\theta_2 &\cong S_2 = -H_2(z+w_2)'' \end{aligned}\right\} \qquad \textbf{(164a,b)}$$

Governing Equations

For the case of a freely vibrating bridge, the forces and moments acting on the deck besides the inertia forces are the forces due to the suspension system. Thus, we shall have:

a) Vertical components:

$$\begin{aligned} \Sigma P_z &= g-(S_1+S_2) = g+H_1(z+w_1)''+H_2(z+w_2)''-m\ddot{w}_S = \\ &= g+z''(H_1+H_2)+H_1w_1''+H_2w_2''-m\ddot{w}_S \end{aligned}$$

which due to eq (150) and (161b,c) can be written as:

$$\Sigma P_z = g+H_gz''+z''(H_{P_1}+H_{P_2})+H_1(w_S+e\theta)''+H_2(w_S-e\theta)''-m\ddot{w}_S$$

and since the effect of the cross-section rotation on the vertical loads is negligible [3,14] and since also $w_S << z$, we finally obtain:

$$\Sigma P_z = z''(H_{P_1}+H_{P_2})+H_gw_S''-m\ddot{w}_S \qquad \textbf{(165)}$$

b) Horizontal components:

$$\Sigma P_y = 0 \qquad \textbf{(166)}$$

c) Torsional components:

$$\begin{aligned} \Sigma m_T &= -S_1e+S_2e-\Theta_x\ddot{\theta} = eH_1(z+w_1)''-eH_2(z+w_2)''-\Theta_x\ddot{\theta} = \\ &= ez''(H_1-H_2)+eH_1w_1''-eH_2w_2''-\Theta_x\ddot{\theta} = \\ &= ez''(H_1-H_2)+eH_1(w_S+e\theta)''-eH_2(w_S-e\theta)''-\Theta_x\ddot{\theta} \end{aligned}$$

The above relation, due to eq (160) and since the effect of the vertical vibration on the combined bending-torsional vibration is negligible, becomes:

$$\Sigma m_T = ez''(H_{p_1} - H_{p_2}) + e^2\theta''(H_1 + H_2) - \Theta_x\ddot{\theta} \qquad \textbf{(167)}$$

Thus, the equations of motion for a freely vibrating cable-stayed bridge can be written as follows:

$$\left.\begin{aligned} & EI_y w_S'''' + c_y\dot{w}_S - H_g w_S'' + m\ddot{w}_S = z''(H_{p_1} + H_{p_2}) \\ & EI_y \upsilon_S'''' - EI_z z_M\theta'''' + c_z\dot{\upsilon}_S + m\ddot{\upsilon}_S = 0 \\ & EC_S\theta'''' - EI_z z_M\upsilon_S'''' + c_\theta\dot{\theta} - GJ_d\theta'' + \Theta_x\ddot{\theta} = \\ & = ez''(H_{p_1} - H_{p_2}) + e^2\theta''(H_1 + H_2) \end{aligned}\right\} \qquad \textbf{(168a,b)}$$

The above equations, along with the following ones:

$$\left.\begin{aligned} & H_{p_1}\frac{L_c}{E_cF_c} + \alpha\Delta TL_T + z''\int_0^L w_1(x)dx = 0 \\ & H_{p_2}\frac{L_c}{E_cF_c} + \alpha\Delta TL_T + z''\int_0^L w_2(x)dx = 0 \end{aligned}\right\} \qquad \textbf{(169a,b)}$$

constitute a integral-differential system of equations, from which the unknowns $w_S, \upsilon_S, \theta, H_{p_1}$ and H_{p_2} can be determined.

It is obvious that a direct closed-form solution of this system is possible only for special cases, as we shall see below. In this section, the eigenfrequencies and eigenshapes (shape functions) will be obtained and studied using special methods.

Pure Vertical Bending Free Vibration

In this case we have: $\upsilon_S = \theta = 0$, $H_{p_1} = H_{p_2}$ and:

$$H_{p_1} + H_{p_2} = -\frac{2z''}{\dfrac{L_c}{E_cF_c}}\cdot\int_0^L w_S(x)dx \qquad \textbf{(170)}$$

It is also:

$$\left.\begin{aligned} & H_g = \frac{g\ell^2}{8f} \quad \text{and} \\ & z'' = -\frac{8f}{\ell^2} \end{aligned}\right\} \qquad \textbf{(171a,b)}$$

and hence, H_g and z'' will be treated as known functions. Then, eq (168a) can be written as:

$$EI_y w_S'''' + c_y\dot{w}_S - H_g w_S'' + m\ddot{w}_S = -\frac{2z''^2}{\dfrac{L_c}{E_cF_c}}\cdot\int_0^L w_S dx \qquad \textbf{(172)}$$

We seek a solution in the form of separate variables as follows:

$$w_S(x,t) = W(x)T(t) \tag{173}$$

Introducing expression (173) into eq (172) and following the usual procedure we obtain:

$$\left.\begin{aligned} & EI_y W'''' - H_g W'' - m\omega_\kappa^2 W = -\frac{2z''^2}{\dfrac{L_c}{E_c F_c}} \cdot \int_0^L W dx \\ & \ddot{T} + \frac{c_y}{m}\dot{T} + \omega_\kappa^2 T = 0 \end{aligned}\right\} \tag{174a,b}$$

The second of eq (174) is the already known equation for the time function in the case of free vibration with damping. The general solution of the differential eq (174a) is:

$$\left.\begin{aligned} & W(x) = c_1 \sin\lambda_1 x + c_2 \cos\lambda_1 x + c_3 \sinh\lambda_2 x + \\ & \qquad + c_4 \cosh\lambda_2 x + \frac{2z''^2}{m\omega_\kappa^2 \left(\dfrac{L_c}{E_c F_c}\right)} \cdot \int_0^L W dx \\ & \text{where:} \quad \lambda_1^2 = -\frac{H_g}{2EI_y} + \sqrt{\left(\frac{H_g}{2EI_y}\right)^2 + \frac{m\omega_\kappa^2}{EI_y}} \\ & \qquad \lambda_2^2 = \frac{H_g}{2EI_y} + \sqrt{\left(\frac{H_g}{2EI_y}\right)^2 + \frac{m\omega_\kappa^2}{EI_y}} \end{aligned}\right\} \tag{175a,b,c}$$

Equation (175a) is a Fredholm integral equation with unknown the function W(x) and trivial core, that is a special case of the Hammerstein [15] equation. After integration, we obtain:

$$\int_0^L W dx = \frac{m\omega_\kappa^2 \left(\dfrac{L_c}{E_c F_c}\right)}{m\omega_\kappa^2 \left(\dfrac{L_c}{E_c F_c}\right) - 2z''^2 L} \int_0^L (c_1 \sin\lambda_1 x + c_2 \cos\lambda_1 x + c_3 \sinh\lambda_2 x + c_4 \cosh\lambda_2 x)\, dx$$

and finally:

$$\left.\begin{aligned} & W(x) = c_1\left(\sin\lambda_1 x - \frac{\kappa\cos\lambda_1 L - \kappa}{\lambda_1}\right) + c_2\left(\cos\lambda_1 x + \frac{\kappa\sin\lambda_1 L}{\lambda_1}\right) + \\ & \quad + c_3\left(\sinh\lambda_2 x + \frac{\kappa\cosh\lambda_2 L - \kappa}{\lambda_2}\right) + c_4\left(\cosh\lambda_2 x + \frac{\kappa\sinh\lambda_2 L}{\lambda_2}\right) \\ & \text{where:} \quad \kappa = \frac{2z''^2}{m\omega_\kappa^2 \left(\dfrac{L_c}{E_c F_c}\right) - 2z''^2 L} \end{aligned}\right\} \tag{176a,b}$$

The boundary conditions for the case shown in Fig. **26** will be: $w_S(0) = w_S(\ell) = w''_S(0) = w''_S(\ell) = 0$ and introducing the solution (176a), we obtain the following system:

$$\left.\begin{array}{l} -c_1\dfrac{\kappa\cos\lambda_1 L-\kappa}{\lambda_1}+c_2\left(1+\dfrac{\kappa\sin\lambda_1 L}{\lambda_1}\right)+c_3\dfrac{\kappa\cosh\lambda_2 L-\kappa}{\lambda_2}+c_4\left(1+\dfrac{\kappa\sinh\lambda_2 L}{\lambda_2}\right)=0 \\ c_1\left(\sin\lambda_1\ell-\dfrac{\kappa\cos\lambda_1 L-\kappa}{\lambda_1}\right)+c_2\left(\cos\lambda_1\ell+\dfrac{\kappa\sin\lambda_1 L}{\lambda_1}\right)+ \\ +c_3\left(\sinh\lambda_2\ell+\dfrac{\kappa\cosh\lambda_2 L-\kappa}{\lambda_2}\right)+c_4\left(\cosh\lambda_2\ell+\dfrac{\kappa\sinh\lambda_2 L}{\lambda_2}\right)=0 \\ 0-c_2\lambda_1^2+0+c_4\lambda_2^2=0 \\ -c_1\lambda_1^2\sin\lambda_1\ell-c_2\lambda_1^2\cos\lambda_1\ell+c_3\lambda_2^2\sinh\lambda_2\ell+c_4\lambda_2^2\cosh\lambda_2\ell=0 \end{array}\right\} \quad (177)$$

This is a linear homogeneous system – without second part – and unknowns the coefficients c_1, c_2, c_3, c_4.

In order for this system to have a solution besides the trivial one, the determinant of the unknown coefficients must be zero. This condition results the eigenfrequencies of the bridge and is after some manipulation:

$$\begin{vmatrix} \dfrac{\kappa}{\lambda_1}(1-\cos\lambda_1 L) & \left(1+\dfrac{\kappa}{\lambda_1}\sin\lambda_1 L\right) & \dfrac{\kappa}{\lambda_2}(\cosh\lambda_2 L-1) & \left(1+\dfrac{\kappa}{\lambda_2}\sinh\lambda_2 L\right) \\ \sin\lambda_1\ell & (\cos\lambda_1\ell-1) & \sinh\lambda_2\ell & (\cosh\lambda_2\ell-1) \\ 0 & -\lambda_1^2 & 0 & \lambda_2^2 \\ -\lambda_1^2\sin\lambda_1\ell & -\lambda_1^2\cos\lambda_1\ell & \lambda_2^2\sinh\lambda_2\ell & \lambda_2^2\cosh\lambda_2\ell \end{vmatrix}=0 \quad (178)$$

Finally, the shape function W(x) is given by eq (176a) if we set:

$$\left.\begin{array}{l} c_2=-\dfrac{\sin\lambda_1\ell}{\cos\lambda_1\ell-1}c_1 \\ c_3=\dfrac{\lambda_1^2\sin\lambda_1\ell\cdot(\cosh\lambda_2\ell-1)}{\lambda_2^2\sinh\lambda_2\ell\cdot(\cos\lambda_1\ell-1)}c_1 \\ c_4=-\dfrac{\lambda_1^2\sin\lambda_1\ell}{\lambda_2^2(\cos\lambda_1\ell-1)}c_1 \end{array}\right\} \quad (179)$$

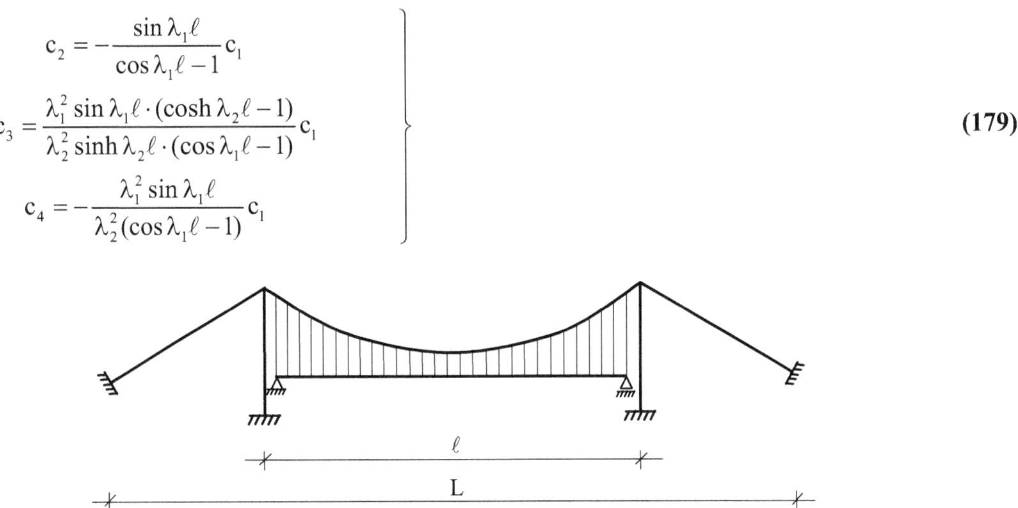

Figure 26: Symmetric suspension bridge model

Pure Transverse Bending-Torsional Free Vibration

In this case, we will have w_S=0 and hence:

$$\left.\begin{aligned} &H_1 + H_2 = H_g + H_{p_1} + H_{p_2} = H_g - \frac{2z''}{\frac{L_c}{E_cF_c}}\int_0^L w_S dx = H_g \\ &H_1 - H_2 = H_{p_1} - H_{p_2} = -\frac{z''}{\frac{L_c}{E_cF_c}}\int_0^L [(w_S - e\theta) - (w_S + e\theta)]dx = \frac{2ez''}{\frac{L_c}{E_cF_c}}\int_0^L \theta dx \end{aligned}\right\} \quad \textbf{(180a,b)}$$

where H_g is given by eq (152).

Then, equations (168b,c) can be written as follows:

$$\begin{aligned} &EI_z\upsilon_S'''' - EI_z z_M\theta'''' + c_z\dot{\upsilon}_S + m\ddot{\upsilon}_S = 0 \\ &EC_S\theta'''' - EI_z z_M\upsilon_S'''' + c_\theta\dot{\theta} - GJ_d\theta'' + \Theta_x\ddot{\theta} = \frac{2e^2z''^2}{\frac{L_c}{E_cF_c}}\int_0^L \theta dx + e^2H_g\theta'' \end{aligned}$$

or finally:

$$\left.\begin{aligned} &EI_z\upsilon_S'''' - EI_z z_M\theta'''' + c_z\dot{\upsilon}_S + m\ddot{\upsilon}_S = 0 \\ &EC_S\theta'''' - EI_z z_M\upsilon_S'''' + c_\theta\dot{\theta} - (GJ_d + e^2H_g)\,\theta'' + \Theta_x\ddot{\theta} = B\cdot\int_0^L \theta dx \\ &\text{with:}\quad B = \frac{2e^2z''^2}{\frac{L_c}{E_cF_c}} \end{aligned}\right\} \quad \textbf{(181a,b,c)}$$

We seek a solution in the form of separate variables as follows:

$$\left.\begin{aligned} &\upsilon(x,t) = V(x)T(t) \\ &\theta(x,t) = \Phi(x)T(x) \end{aligned}\right\} \quad \textbf{(182a,b)}$$

Introducing eq(182) into eqs(181a,b), we obtain:

$$\begin{aligned} &(EI_zV'''' - EI_z z_M\Phi'''')T = -m\left(\frac{c_z}{m}\dot{T} + \ddot{T}\right)V \\ &\left(EC_S\Phi'''' - EI_z z_MV'''' - (GJ_d + e^2H_g)\Phi'' - B\int_0^L \Phi dx\right)T = -\Theta_x\left(\frac{c_\theta}{\Theta_x}\dot{T} + \ddot{T}\right)\Phi \end{aligned}$$

and due to eq(81) of chapter 3, we get:

$$-\frac{\ddot{T} + \frac{c_z}{m}\dot{T}}{T} = \frac{EI_zV'''' - EI_z z_M\Phi''''}{mV} = \frac{EC_S\Phi'''' - EI_z z_MV'''' - (GJ_d + e^2H_g)\Phi'' - B\int_0^L \Phi dx}{\Theta_x\Phi} = \omega_\sigma^2$$

from which we obtain the following system of differential equations:

$$\left.\begin{aligned}&\ddot{T}+\frac{c_z}{m}\dot{T}+\omega_\sigma^2 T=0\\&EI_zV''''-EI_zz_M\Phi''''-m\omega_\sigma^2V=0\\&EC_S\Phi''''-EI_zz_MV''''-(GJ_d+e^2H_g)\Phi''-\Theta_x\omega_\sigma^2\Phi=B\int_0^L\Phi dx\end{aligned}\right\}\qquad\textbf{(183a,b,c)}$$

The above system of equations is very similar to the system (82a,b,c) of chapter 3 with only difference the multiplier of function Φ'' which instead of GJ_d it becomes now $(GJ_d+e^2H_g)$ and also the existence of a second part $B\int_0^L\Phi dx$ in the third equation. Hence, the system can be solved in a similar manner as the system of eqs(82a,b,c) of chapter 3 if we set:

$$\left.\begin{aligned}&\alpha=-E^2I_zC_M\\&\beta=EI_z(GJ_d+e^2H_g)\\&\gamma=EC_Sm\omega_\sigma^2+EI_z\Theta_x\omega_\sigma^2\\&\delta=-(GJ_d+e^2H_g)m\omega_\sigma^2\\&\varepsilon=-\Theta_xm\omega_\sigma^2\end{aligned}\right\}\qquad\textbf{(184)}$$

and hence:

$$\left.\begin{aligned}&\Phi(x)=\sum_i c_ie^{\rho_ix}+\sum_\kappa c_\kappa e^{\alpha_\kappa x}\left(\sin\beta_\kappa x+\cos\beta_\kappa x\right)+\frac{B}{\varepsilon}\int_0^L\Phi dx\\&V=\frac{1}{mz_M\omega_\sigma^2}\cdot\left[EC_M\Phi''''-(GJ_\alpha+e^2H_g)\Phi''-\Theta_x\omega_\sigma^2\Phi-B\int_0^L\Phi dx\right]\end{aligned}\right\}\qquad\textbf{(185a,b)}$$

where $\pm\rho_i$ and $(\alpha_\kappa\pm j\beta_\kappa)$ are the real and the conjugate complex roots of $\alpha\rho^8+\beta\rho^6+\gamma\rho^4+\delta\rho^2+\varepsilon=0$, while $j=\sqrt{-1}$ is the imaginary unit and $c_q(q=i+\kappa=8)$ are the integration constants.

Finally, integrating eq(185a), we obtain $\int_0^L\Phi dx$ as follows:

$$\int_0^L\Phi dx=\frac{\varepsilon}{\varepsilon-BL}\cdot\left\{\sum_i\frac{c_i}{\rho_i}(e^{\rho_iL}-1)+\sum_\kappa c_\kappa\frac{(\beta_\kappa-\alpha_\kappa)+(\alpha_\kappa+\beta_\kappa)e^{\alpha_\kappa L}\sin\beta_\kappa L+(\alpha_\kappa-\beta_\kappa)e^{\alpha_\kappa L}\cos\beta_\kappa L}{\alpha_\kappa^2+\beta_\kappa^2}\right\}\qquad\textbf{(186)}$$

Introducing next the expressions of $V(x)$ and $\Phi(x)$ into the boundary conditions, we obtain a linear system with eight unknowns c_q ($q=i+\kappa=8$). This eigenvalue problem can be solved to obtain the eigenfrequencies ω_σ, by setting zero the determinant of the unknown coefficients c_q of the system. Finally, from this system we can determine the 7 remaining coefficients with respect to the 8th, i.e., the c_1. Then, we obtain the expressions for the shape functions V_n and Φ_n, also known as mode shapes.

The Special Case $z_M = 0$

This special case applies for doubly symmetric cross-sections, and is very common in practice. Equations (181a,b) are independent and take the form:

$$\left.\begin{array}{l} EI_z\upsilon'''' + c_z\dot{\upsilon}_S + m\ddot{\upsilon}_S = 0 \\ EC_S\theta'''' + c_\theta\dot{\theta} - (GJ_d + e^2H_g)\theta'' + \Theta_x\ddot{\theta} = B\int_0^L \theta dx \\ \text{with :} \quad B = \dfrac{\dfrac{2e^2z''}{L_C}}{E_CF_C} \end{array}\right\} \qquad \textbf{(187a,b,c)}$$

Hence, it is obvious that the bridge will vibrate independently in bending along y-axis and torsionally rotating about x-axis. The first of eq (187) has the form of eq (63b) and hence its solution is obtained in a similar manner. For instance, in the case of the bridge shown in Fig. **26** we will have:

$$\left.\begin{array}{ll} & \omega_{y_n}^2 = \dfrac{n^4\pi^4 EI_z}{m\cdot\ell^4} \quad , \quad n = 1,2,3,.... \\ \text{and} & V_n(x) = c_{1n}\sin\dfrac{n\pi x}{\ell} \end{array}\right\} \qquad \textbf{(188a,b)}$$

The first of eq (188) gives the eigenfrequencies for horizontal free vibration, while the second gives the corresponding shape functions. In order to solve eq (187b), we introduce the expression (182b) and we get:

$$EC_S\Phi''''T + c_\theta\Phi\dot{T} - (GJ_d + e^2H_g)\Phi''T + \Theta_x\Phi\ddot{T} = B\int_0^L \Phi dx T \qquad \text{or}$$

$$-\frac{\ddot{T} + \dfrac{c_\theta}{\Theta_x}\dot{T}}{T} = \frac{EC_S\Phi'''' - (GJ_d + e^2H_g)\Phi'' - B\int_0^L \Phi dx}{\Theta_x\Phi} = \omega_\theta^2$$

from which we obtain the following differential equations:

$$\left.\begin{array}{l} \ddot{T} + \dfrac{c_\theta}{\Theta_x}\dot{T} + \omega_\theta^2 T = 0 \\ EC_S\Phi'''' - (GJ_d + e^2H_g)\Phi'' - \Theta_x\omega_\theta^2\Phi = B\int_0^L \Phi dx \end{array}\right\} \qquad \textbf{(189a,b)}$$

The general solution of eq (189b) is given by:

$$\left.\begin{array}{l} \Phi(x) = c_1\sin\lambda_1 x + c_2\cos\lambda_1 x + c_3\sinh\lambda_2 x + c_4\cosh\lambda_2 x - \dfrac{B}{\Theta_x\omega_\theta^2}\int_0^L \Phi dx \\ \text{where:} \quad \lambda_1 = \sqrt{-\dfrac{GJ_d + e^2H_g}{2EC_S} + \sqrt{\left(\dfrac{GJ_d + e^2H_g}{2EC_S}\right)^2 + \dfrac{\Theta_x\omega_\theta^2}{EC_S}}} \\ \qquad\qquad \lambda_2 = \sqrt{\dfrac{GJ_d + e^2H_g}{2EC_S} + \sqrt{\left(\dfrac{GJ_d + e^2H_g}{2EC_S}\right)^2 + \dfrac{\Theta_x\omega_\theta^2}{EC_S}}} \end{array}\right\} \qquad \textbf{(190a,b,c)}$$

Integrating eq (190a), we finally obtain:

$$\int_0^L \Phi dx = \frac{\Theta\omega_\theta^2}{\Theta\omega_\theta^2 + BL} \cdot \left\{ c_1 \left(-\frac{\cos\lambda_1 L - 1}{\lambda_1} \right) + c_2 \frac{\sin\lambda_1 L}{\lambda_1} + c_3 \frac{\cosh\lambda_2 L - 1}{\lambda_2} + c_4 \frac{\sinh\lambda_2 L}{\lambda_2} \right\}$$

and hence:

$$\left.\begin{aligned} &\Phi(x) = c_1\left(\sin\lambda_1 x + \kappa\frac{\cos\lambda_1 L - 1}{\lambda_1}\right) + c_2\left(\cos\lambda_1 x - \kappa\frac{\sin\lambda_1 L}{\lambda_1}\right) + \\ &\quad + c_3\left(\sinh\lambda_2 x - \kappa\frac{\cosh\lambda_2 L - 1}{\lambda_2}\right) + c_4\left(\cosh\lambda_2 x - \kappa\frac{\sinh\lambda_2 L}{\lambda_2}\right) \\ &\text{where:}\quad \kappa = \frac{B}{\Theta\omega_\theta^2 + BL} \end{aligned}\right\} \qquad \textbf{(191a,b)}$$

Introducing eq (191a) into the boundary conditions of the bridge in Fig. **26**, we obtain:

$$\left.\begin{aligned} &c_1\frac{\kappa\cos\lambda_1 L - \kappa}{\lambda_1} + c_2\left(1 - \frac{\kappa\sin\lambda_1 L}{\lambda_1}\right) - c_3\frac{\kappa\cos\lambda_2 L - \kappa}{\lambda_2} + c_4\left(1 - \frac{\kappa\sinh\lambda_2 L}{\lambda_2}\right) = 0 \\ &c_1\left(\sin\lambda_1\ell + \frac{\kappa\cos\lambda_1 L - \kappa}{\lambda_1}\right) + c_2\left(\cos\lambda_1\ell - \frac{\kappa\sin\lambda_1 L}{\lambda_1}\right) \\ &+c_3\left(\sinh\lambda_2\ell - \frac{\kappa\cosh\lambda_2 L - \kappa}{\lambda_2}\right) + c_4\left(\cosh\lambda_2\ell - \frac{\kappa\sinh\lambda_2 L}{\lambda_2}\right) = 0 \\ &-c_2\lambda_1^2 + c_4\lambda_2^2 = 0 \\ &-c_1\lambda_1^2\sin\lambda_1\ell - c_2\lambda_1^2\cos\lambda_1\ell + c_3\lambda_2^2\sinh\lambda_2\ell + c_4\lambda_2^2\cosh\lambda_2\ell = 0 \end{aligned}\right\} \qquad \textbf{(192)}$$

Setting the determinant of the unknown coefficients equal to zero, we obtain the equation for the eigenfrequencies of the bridge:

$$\begin{vmatrix} \frac{\kappa}{\lambda_1}(\cos\lambda_1 L - 1) & \left(1 - \frac{\kappa}{\lambda_1}\sin\lambda_1 L\right) & -\frac{\kappa}{\lambda_2}(\cosh\lambda_2 L - 1) & \left(1 - \frac{\kappa}{\lambda_2}\sinh\lambda_2 L\right) \\ \sin\lambda_1\ell & (\cos\lambda_1\ell - 1) & \sinh\lambda_2\ell & (\cosh\lambda_2\ell - 1) \\ 0 & -\lambda_1^2 & 0 & \lambda_2^2 \\ -\lambda_1^2\sin\lambda_1\ell & -\lambda_1^2\cos\lambda_1\ell & \lambda_2^2\sinh\lambda_2\ell & \lambda_2^2\cosh\lambda_2\ell \end{vmatrix} = 0 \qquad \textbf{(193)}$$

Finally, the shape function $\Phi(x)$ can be determined from eq (191a) by setting:

$$\left.\begin{aligned} c_2 &= -\frac{\sin\lambda_1\ell}{\cos\lambda_1\ell - 1}c_1 \\ c_3 &= \frac{\lambda_1^2\sin\lambda_1\ell(\cosh\lambda_2\ell - 1)}{\lambda_2^2\sinh\lambda_2\ell(\cos\lambda_1\ell - 1)}c_1 \\ c_4 &= -\frac{\lambda_1^2\sin\lambda_1\ell}{\lambda_2^2(\cos\lambda_1\ell - 1)}c_1 \end{aligned}\right\} \qquad \textbf{(194)}$$

CURVED IN-PLANE BRIDGES [16], [17], [18]

In some cases, when the terrain is highly curved in the site of the bridge, it is mandatory to design and erect a curved in-plane bridge (Fig. **27**). The curvature of such a bridge is usually very small, in order to assume that: "the distances y_M , z_M between the shear center and the gravity center are very small compared to the curvature radius R of the bridge" (for $R \geq 10b$). This leads to the assumption that the torsional moment m_x is applied on the gravitational axis, while in fact it is acting on the shear center axis.

The technical theory of beams with thin-walled section can be extended to curved beams, when the cross-section of the beam does not distort, i.e. its shape does not change when deformed.

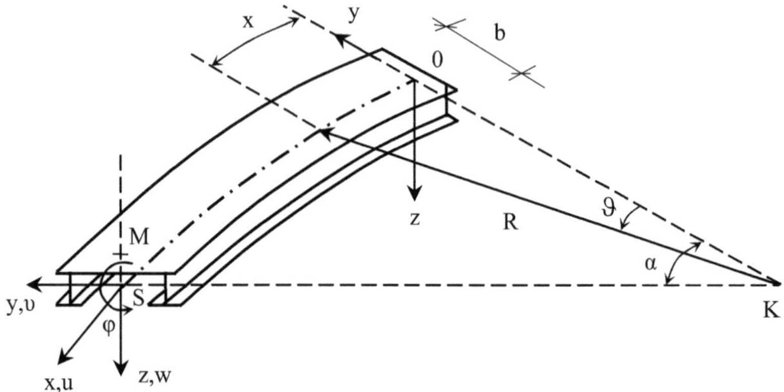

Figure 27: Curved in-plane deck-beam

Let us consider the deck-beam shown in Fig. **27**, which gravitational axis 0S is circular with radius R and lies on the plane Sxy. We also assume that the beam refers to the system Sxyz, where axis 0x is also circular with radius R passing through the gravitational center S of each cross-section, while axes Sy, and Sz form in reference with axis Sx a clockwise system. Finally, axis Sy lies on the plane of the curved beam and hence, it passes through the center K of the circular part of axis 0sx. We finally assume that the length of the curved beam is defined by the angle α. Isolating an infinitesimal element dx of the beam and formulate the equilibrium equations for forces and moments in axes α-α, K-β and 0z, as shown in Fig. **28a,b**, we will obtain:

$$\left.\begin{aligned}
&\frac{dQ_x}{dx}+\frac{Q_y}{R}+q_x=0\\
&-\frac{Q_x}{R}+\frac{dQ_y}{dx}+q_y=0\\
&\frac{dQ_z}{dx}+q_z=0\\
&\frac{dM_x}{dx}+\frac{M_y}{R}+m_x=0\\
&\frac{dM_y}{dx}-\frac{M_x}{R}-Q_z+m_y=0\\
&\frac{dM_z}{dx}+Q_y+m_z=0
\end{aligned}\right\} \qquad (195)$$

Eliminating Q_x, Q_y, Q_z for the most common case where $m_y=m_z=0$ we will get:

$$\left.\begin{aligned} & M_x' + \frac{M_y}{R} + m_x = 0 \\ & M_y'' - \frac{M_x'}{R} + q_z = 0 \\ & M_z''' + \frac{M_z'}{R^2} - \frac{q_x}{R} - q_y' = 0 \end{aligned}\right\} \qquad \textbf{(196)}$$

Denoting by k_y and k_z the curvatures of the beam in the planes Sxz and Syx, by k_x the torsion curvature, while by ε_x the axial strain along axis Sx of the element dx we will have:

$$\left.\begin{aligned} & \varepsilon_x = u' + \frac{\upsilon}{R} \\ & k_x = \varphi' \\ & k_y = -w'' \\ & k_z = +\upsilon'' + \frac{\upsilon}{R^2} + \frac{1}{R} \end{aligned}\right\} \qquad \textbf{(197)}$$

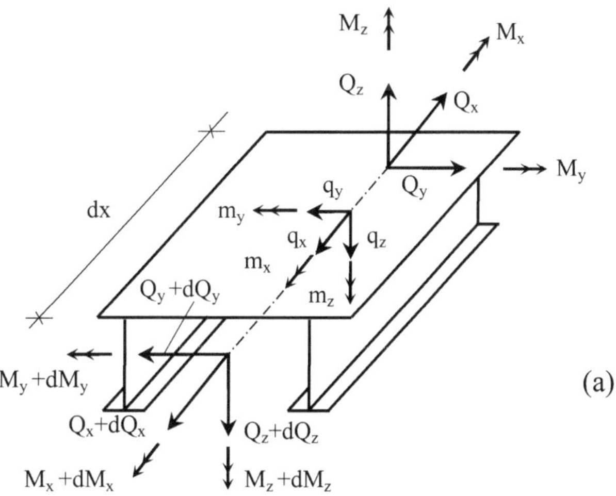

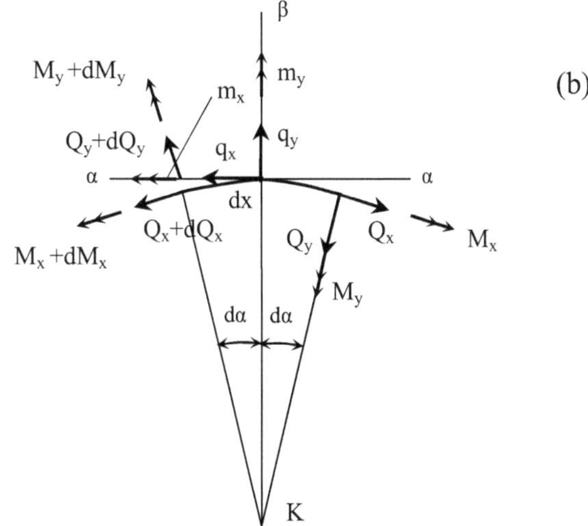

Figure 28: Equilibrium in a curved element dx

The corresponding expressions for curvatures, with reference to the new positions S′x′y′z′ of the axes Sxyz after deformation, are:

$$k'_x = k_x \cos\widehat{xx'} + k_y \cos\widehat{yx'} + k_z \cos\widehat{zx'}$$

$$k'_y = k_x \cos\widehat{xy'} + k_y \cos\widehat{yy'} + k_z \cos\widehat{zy'}$$

$$k'_z = k_x \cos\widehat{xz'} + k_y \cos\widehat{yz'} + k_z \cos\widehat{zz'}$$

where the directional cosines in the above expressions, for small displacements values u,w,φ, are given as follows:

	x	y	z
x′	1	υ′	w′
y′	- υ′	1	φ
z′	- w′	- φ	1

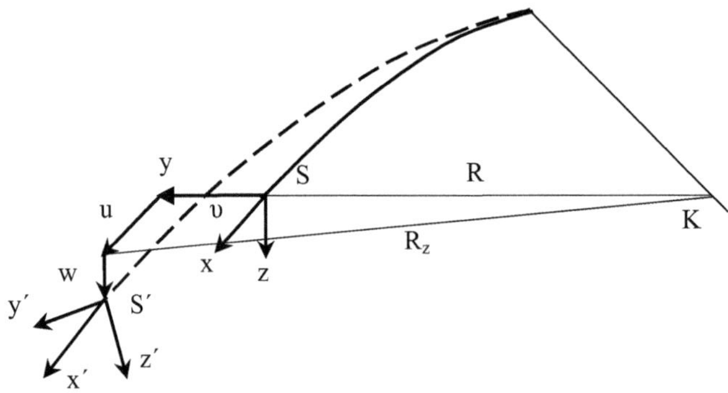

Figure 29: Coordinate systems in the deformed and undeformed state

and hence, the above expressions due to eq (197) become:

$$\left.\begin{aligned} \varepsilon_x &= u' + \frac{\upsilon}{R} \\ k'_x &= \varphi' + \frac{w'}{R} \\ k'_y &= -w'' + \frac{\varphi}{R} \\ k'_z &= +\upsilon'' + \frac{\upsilon}{R^2} + \frac{1}{R} \end{aligned}\right\} \qquad \mathbf{(198)}$$

where higher order terms have been neglected. The following relations are valid:

$$\begin{aligned} Q_x &= EAu' \\ M_y &= EI_y k_y \\ M_z &= EI_z (k_z - \frac{1}{R}) \\ M_x &= -EC_M k''_x + GJ_d k_x \end{aligned}$$

which due to eq (198) become as follows:

$$\left.\begin{aligned} Q_x &= EA\left(u' + \frac{\upsilon}{R}\right) \\ M_y &= -EI_y\left(w'' - \frac{\varphi}{R}\right) \\ M_z &= EI_z\left(\upsilon'' + \frac{\upsilon}{R^2}\right) \\ M_x &= -EC_M\left(\varphi''' + \frac{w'''}{R}\right) + GJ_d\left(\varphi' - \frac{w'}{R}\right) \end{aligned}\right\} \quad (199)$$

Introducing eq (199) into eq (196) and into the relation resulting after the elimination of Q_y between eq (195a) and (195b) and adding the inertia forces and moments acting onto the part, we finally obtain:

$$\left.\begin{aligned} & EA\left(u''' + \frac{\upsilon''}{R} + \frac{u'}{R^2} + \frac{\upsilon}{R^3}\right) - c_x\dot{u}' - m\ddot{u}' + \frac{m\ddot{\upsilon}}{R} = -q_x' + \frac{q_y}{R} \\ & EI_z\left(\upsilon^{(5)} + \frac{2\upsilon'''}{R^2} + \frac{\upsilon'}{R^4}\right) + c_z\dot{\upsilon}' + m\ddot{\upsilon}' + \frac{m\ddot{u}}{R} = q_y' + \frac{q_x}{R} \\ & \left(EI_y - \frac{EC_M}{R^2}\right)w'''' + \frac{GJ_d}{R^2}w'' - \frac{EI_y - GJ_d}{R}\varphi'' - \frac{EC_M}{R}\varphi'''' + c_y\dot{w} + m\ddot{w} = q_z \\ & EC_M\varphi'''' - GJ_d\varphi'' - \frac{EI_y}{R^2}\varphi + \frac{EC_M}{R}w'''' + \frac{EI_y - GJ_d}{R}w'' + c_\theta\dot{\varphi} + \Theta_x\ddot{\varphi} = m_x \end{aligned}\right\} \quad (200)$$

If R is sufficiently big, so that bending does not significantly affect transverse deformations, while the last also does not affect the axial deformation, equations (200) for free vibration can be written after integration of the first two as follows:

$$\left.\begin{aligned} & EA\left(u'' + \frac{u}{R^2}\right) - c_x\dot{u} - m\ddot{u} = 0 \\ & EI_z\left(\upsilon'''' + \frac{2\upsilon''}{R^2} + \frac{\upsilon}{R^4}\right) + c_z\dot{\upsilon} + m\ddot{\upsilon} = 0 \\ & \left(EI_y - \frac{EC_M}{R^2}\right)w'''' + \frac{GJ_d}{R^2}w'' - \frac{EI_y - GJ_d}{R}\varphi'' - \frac{EC_M}{R}\varphi'''' + c_y\dot{w} + m\ddot{w} = 0 \\ & EC_M\varphi'''' - GJ_d\varphi'' - \frac{EI_y}{R^2}\varphi + \frac{EC_M}{R}w'''' + \frac{EI_y - GJ_d}{R}w'' + c_\theta\dot{\varphi} + \Theta_x\ddot{\varphi} = 0 \end{aligned}\right\} \quad (201)$$

Free Axial Vibration

The governing equation for axial vibration is (201a). Seeking a solution in the form of separate variables as follows: $u(x,t) = U(x)T(t)$, and employing the usual procedure, we obtain the following:

$$\left.\begin{aligned} & U''(x) + \lambda^2 U(x) = 0 \\ & \ddot{T}(t) + \frac{c_x}{m}\dot{T}(t) + \omega_x^2 T(t) = 0 \\ & \text{where: } \lambda^2 = \frac{m\omega_x^2}{EA} + \frac{1}{R^2} \end{aligned}\right\} \quad (202)$$

The above eq (202) constitute the typical solution for axial free vibration of a straight bar (see also paragraph "The longitudinal motion" in page 64) and hence, it will be (i.e. for a free bar):

$$\left.\begin{aligned} &U(x) = c_2 \cos\lambda_n x \\ \text{and} \quad &\omega_{xn}^2 = \frac{EA}{m}\left(\frac{n^2\pi^2}{\ell^2} - \frac{1}{R^2}\right) \end{aligned}\right\} \tag{203}$$

In order that the eigenfrequencies ω_{xn} are real (stable system), it must be: $\frac{n^2\pi^2}{\ell^2} - \frac{1}{R^2} > 0$ and since: $\ell = \alpha \cdot R$ (see also Fig. 27), it must be: $\alpha < n\pi$ or for n=1: $\alpha < \pi$, which is valid.

Horizontal Bending Free Vibration

Seeking again a solution in the form of separate variables: $\upsilon(x,t) = V(x) \cdot T(t)$, eq (201b) will result the following:

$$\left.\begin{aligned} &V'''' + \frac{2}{R^2}V'' + \frac{1}{R^4}V - \frac{m\omega_z^2}{EI_z}V = 0 \\ &\ddot{T} + \frac{c_y}{m}\dot{T} + \omega_y^2 T = 0 \end{aligned}\right\} \tag{204}$$

Equation (204a) can be written as: $V'''' + \frac{2}{R^2}V'' + \left(\frac{1}{R^4} - \frac{m\omega_z^2}{EI_z}\right)V = 0$, which has the following solution:

$$\left.\begin{aligned} &V(x) = c_1 \sin\lambda_1 x + c_2 \cos\lambda_1 x + c_3 \sinh\lambda_2 x + c_4 \cosh\lambda_2 x \\ &\text{where:} \quad \lambda_1 = \sqrt{\frac{1}{R^2} + \sqrt{\frac{m\omega_z^2}{EI_z}}} \quad , \quad \lambda_2 = \sqrt{-\frac{1}{R^2} + \sqrt{\frac{m\omega_z^2}{EI_z}}} \end{aligned}\right\} \tag{205}$$

Employing the boundary conditions for the simply supported beam, we obtain the following system:

$$\left.\begin{aligned} &V(0) = c_2 + c_4 = 0 \\ &V''(0) = -\lambda_1^2 c_2 + \lambda_2^2 c_4 = 0 \\ &V(\ell) = c_1 \sin\lambda_1\ell + c_2 \cos\lambda_1\ell + c_3 \sinh\lambda_2\ell + c_4 \cosh\lambda_2\ell = 0 \\ &V''(\ell) = -c_1\lambda_1^2 \sin\lambda_1\ell - c_2\lambda_1^2 \cos\lambda_1\ell + c_3\lambda_2^2 \sinh\lambda_2\ell + c_4\lambda_2^2 \cosh\lambda_2\ell = 0 \end{aligned}\right\} \tag{206}$$

In order that the above system of equations has a non-trivial solution, the determinant of the unknown coefficients must be zero. This condition gives the following equation for eigenfrequencies:

$$(\lambda_1^2 + \lambda_2^2)^2 \sin\lambda_1\ell \sinh\lambda_2\ell = 0$$

which since: $\lambda_1 \neq \lambda_2 \neq 0$ gives: $\lambda_1\ell = n\pi$ and hence:

$$\omega_{zn}^2 = \frac{EI_z}{m}\left(\frac{n^2\pi^2}{\ell^2} - \frac{1}{R^2}\right)^2 \tag{207}$$

In order to have real values of ω_{zn}, we find again the condition in paragraph "Free Axial Vibration" in page 191:

$$\alpha < n\pi \tag{208}$$

Finally, from eq (206a,b,c) we can easily determine the corresponding shape functions:

$$V_n(x) = c_1\left(\sin\lambda_1 x + \frac{\sin\lambda_1\ell}{\sinh\lambda_2\ell}\sinh\lambda_2 x\right) \tag{209}$$

Vertical Bending-Torsional Free Vibration

In order to solve the system of eq (201c,d), we seek again a solution in the form of separate variables as follows:

$$\left.\begin{aligned} w(x,t) &= W(x)T(t) \\ \varphi(x,t) &= \Phi(x)T(t) \end{aligned}\right\} \tag{210}$$

Introducing expressions (210) into eq (201c and d) and following the usual procedure, we finally arrive at the following system:

$$\left.\begin{aligned} &\alpha_1 w^{(4)} + \alpha_2 w^{(2)} + \alpha_3 w + \alpha_4 \varphi^{(4)} + \alpha_5 \varphi^{(2)} = 0 \\ &\beta_1 w^{(4)} + \beta_2 w^{(2)} + \beta_3 \varphi^{(4)} + \beta_4 \varphi^{(2)} + \beta_5 \varphi = 0 \\ &\ddot{T} + \frac{c_y}{m}\dot{T} + \omega_{z\varphi}^2 T = 0 \\ &\text{where:}\quad \alpha_1 = \frac{-EC_M}{R^2} + EI_y \quad,\quad \beta_1 = \frac{EC_M}{R} \\ &\qquad\qquad \alpha_2 = \frac{GJ_d}{R^2} \quad,\quad \beta_2 = -\alpha_5 \end{aligned}\right\} \tag{211}$$

$$\left.\begin{aligned} &\alpha_3 = -m\omega_{z\varphi}^2 \quad,\quad \beta_3 = EC_M \\ &\alpha_4 = -\frac{EC_M}{R} \quad,\quad \beta_4 = -GJ_d \\ &\alpha_5 = -\frac{EI_y - GJ_d}{R} \quad,\quad \beta_5 = -\frac{EI_y}{R^2} - \Theta_x\omega_{z\varphi}^2 \end{aligned}\right\} \tag{211}$$

From the first two of eq (211) we determine:

$$w^{(4)} = -\frac{(\alpha_4\beta_2 - \alpha_2\beta_3)\varphi^{(4)} + (\alpha_5\beta_2 - \alpha_2\beta_4)\varphi^{(2)} - \alpha_2\beta_5\varphi + \alpha_3\beta_2 w}{\alpha_1\beta_2 - \alpha_2\beta_1}$$

$$w^{(2)} = \frac{(\alpha_4\beta_1 - \alpha_1\beta_3)\varphi^{(4)} + (\alpha_5\beta_1 - \alpha_1\beta_4)\varphi^{(2)} - \alpha_1\beta_5\varphi + \alpha_3\beta_1 w}{\alpha_1\beta_2 - \alpha_2\beta_1}$$

which can be also written as follows:

$$\left.\begin{aligned}&(\alpha_1\beta_2-\beta_1\alpha_2)w^{(4)}+\alpha_3\beta_2 w=-(\alpha_4\beta_2-\alpha_2\beta_3)\varphi^{(4)}-\\&\qquad\qquad -(\alpha_5\beta_2-\alpha_2\beta_4)\varphi^{(2)}+\alpha_2\beta_5\varphi\\&(\alpha_1\beta_2-\beta_1\alpha_2)w^{(2)}+\alpha_3\beta_1 w=(\alpha_4\beta_1-\alpha_1\beta_3)\varphi^{(4)}+\\&\qquad\qquad +(\alpha_5\beta_1-\alpha_1\beta_4)\varphi^{(2)}-\alpha_1\beta_5\varphi\end{aligned}\right\}\qquad \textbf{(211a,b)}$$

Differentiating eq (211b) twice, we get:

$$\begin{aligned}&(\alpha_1\beta_2-\beta_1\alpha_2)w^{(4)}+\alpha_3\beta_1 w^{(2)}=(\alpha_4\beta_1-\alpha_1\beta_3)\varphi^{(6)}+\\&\qquad\qquad +(\alpha_5\beta_1-\alpha_1\beta_4)\varphi^{(4)}-\alpha_1\beta_5\varphi^{(2)}\end{aligned}\qquad \textbf{(211c)}$$

The above system of eq (211a,b,c) is linear with respect to $w^{(4)}$, $w^{(2)}$, w, and can be solved only with respect to φ and its derivatives.

Introducing the obtained results for $w^{(4)}$, $w^{(2)}$ into eq (211b), we arrive at an 8th order differential equation with respect to φ, which is:

$$A_1\varphi^{(8)}+A_2\varphi^{(6)}+A_3\varphi^{(4)}+A_4\varphi^{(2)}+A_5\varphi=0\qquad \textbf{(212)}$$

This equation is similar to (eq 85a of chapter 3) and when solved, according to the paragraph "The Lateral-Torsional Free Motion" in page 68, produces the eigenfrequencies and corresponding shape functions.

REFERENCES

[1] S.P. Timoshenko, D.H. Young, W. Weaver Jr. *"Vibration problems in engineering"*, 1974, J. Wiley & Sons, N. York.
[2] G.L. Rogers *"Dynamics of framed Structures"*, 1959, J. Wiley &Sons Inc., London.
[3] A. Hawranek, O. Steinhardt *"Theorie und Berechnung der Stahlbrücken"*, 1958, Springer Verlag, Berlin.
[4] G.T. Michaltsos, D.S. Sophianopoulos, T.G. Konstantakopoulos *"Buckling and postbuckling analysis of a low parabolic arch"*, 1997, A. Techn. Session and Meeting of SSRC, Torondo.
[5] G.T. Michaltsos, J. Ermopoulos *"Free vibrations of a low arch"*, 1986, Techn. Chronicles, 6(2) (in Greek).
[6] G.T. Michaltsos *"Erzwungene Schwingung einer flachen kreisformigen Bogenbrücke"*, 1994, Bauingenieur (4).
[7] G. Burgermeister, H. Steup, H. Kretzschmar *"Stabilitätstheorie"*, 1966, Akademie Verlag, Berlin.
[8] M.S. Troitsky *"Cable-Stayed Bridges"*, 1977, Granada Publ. Limited, London.
[9] G.T. Michaltsos *"A simplified model for the dynamic analysis of cable - stayed bridges"*, 2001, Facta Universitatis, Vol. 3 (11).
[10] G.T. Michaltsos, T.G. Konstantakopoulos *"A simplified dynamic analysis for cable-stayed bridges"*, 2000, 5th Symposium of Applied Mechanics, Nîs.
[11] G.T. Michaltsos, J.C.Ermopoulos, T.G. Konstantakopoulos *"Preliminary design of cable-stayed bridges for vertical static loads"*, 2003, Journal of Structural Engineering and Mechanics, Vol. 15(6).
[12] D. Bruno, A. Grimaldi *"Non linear behaviour of long-span cable-stayed bridges"*, 1985, Mechanica, J. of the Italian Association for Theoretical and Applied Mechanics, 20(4).
[13] P.K. Chatterje, T.K. Datta, C.S. Surana *"Vibration of cable-stayed bridges under moving vehicles"*, 1994, J. Structure Ing. International 2.
[14] G.T. Michaltsos *"Contribution on the static and dynamic solution problem of suspension bridges"*, 1976, PhD Dissertation NTUA, Athens (in Greek).
[15] I.G. Petrovsky *"Lectures on the theory of Integral equations"*, 1971, Mir Publishers, Moscow.
[16] L.T. Stavridis, G.T. Michaltsos *"Eigenfrequency analysis of thin-walled girders curved in plan"*, 1999, Journal of Sound and Vibration, 227.
[17] G.T. Michaltsos *"Thin-walled Metal Structures – Calculation Procedures"*, 2008, Symeon Publ., Athens (in Greek).
[18] B.Z. Vlassov *"Pieces longues en voiles minces"*, 1962, Ed. Eurolles, Paris.

CHAPTER 7

Aeroelasticity

Abstract: This chapter deals with the basic principles of aeroelasticity theory (Theodorsen theory) and how these can be applied to bridge structures with sensitive deck shapes. The most common aerodynamical models with two or three forces as well as buffeting forces are briefly presented. Special effects such as galloping instability, torsional divergence and flutter are also presented. The technique of experimental design employing test results from scaled models is also described.

GENERAL

In 1854, the suspension bridge in Wheeling – West Virginia, designed by Charles Ellet Jr. collapsed due to severe windstorm according to the local press. At that time, this bridge was the biggest suspension bridge ever constructed with main (central) span about 303m, which was small compared to nowadays openings. This catastrophic collapse (both the deck and the suspension system collapsed), shacked off the bridge designers who concluded that some stiffening system is required for increasing the stiffness of the bridge, and which must probably had to be added in all existing suspension bridges.

The famous mathematician and engineer George Airy (known also from the Airy stress function), was inspired by this problem and published an article entitled "*On the use of the suspension bridge with stiffened roadway for railway and other bridges of great span*" [1].

Even though this article brought Airy the Telford prize of the Institution of Civil Engineers in Great Britain in 1867, it did not contribute to the right direction for solution of the problem. Airy concluded his work by suggesting the use of rigid elements, and justifying that these elements can help in a better distribution of severe concentrated loads. Wind forces were not even mentioned in his work.

As the bridge design was in a highly conservative period, while the needs for heavier loads and bigger spans were increasing, the use of stability trusses was spreading to all suspension bridges, reaching a peak with the construction of the Williamsburg bridge on East River in 1903, the deck of which has a 12m high cross section. This bridge looked mostly like a heavy truss structure than a suspension bridge. The weird aesthetical result and the unnatural proportions of the bridge from architectural point of view, have been the main causes for engineers to design and study bridges with decks of gradually smaller heights, so that the design solutions be more acceptable. Experimental investigation as well as improvement of computational methods contributed to this course significantly.

The George Washington bridge set in use in 1931 had no stiffeners, thus becoming the most slender bridge structure. Nevertheless, in 1940, the Tacoma bridge with stiffeners and a not particularly big span and a bridge deck of 2,4 m height, collapsed due to wind loads. This fact affected severely the known computational and design methods in bridge engineering. The main question was what were the causes for this failure since both the Ellets Wheeling bridge without stiffeners as well as the Tacoma with stiffeners were collapsed.

On the contrary, the George Washington bridge, without stiffeners and with an extremely slender deck had no problem at all, and after 34 year of service had proved that the use of stiffeners was wrong (note that in 60s the bridge was added a second deck for traffic loads).

Today it is known that the collapse of the Tacoma bridge was due to the aerodynamic shape of the stiffeners' system itself [2].

With the above data, designers began to suspect that the combination of some characteristics of the bridge with the characteristics of wind flow are more critical than the stiffness of the bridge deck and to study the behavior of bridges under the action of wind.

George T. Michaltsos and Ioannis G. Raftoyiannis

Before proceeding with the investigation and the determination of this dangerous combination, we will present the mechanism that contributes to the development of some specific forces acting on the deck of bridge due to wind flow. In chapter 2, page 26, the mechanism producing oscillations to a structure being into wind flow as well as the developing pressures and subpressures has been described.

It is obvious that these pressures and subpressures due to the turbulence developing on the top and bottom face of the cross-section in Fig. **11** of Chapter 2 are not possible to balance at all times even if the cross-section is doubly symmetric, due to the nature of these forces. Unbalanced pressures and subpressures on the top and the bottom face of the bridge deck will result to the appearance of transverse forces with respect to the horizontal axis of the bridge inducing oscillations to the structure.

Let us consider that the subpressures will prompt the structure to move downwards (for simplicity we consider only vertical displacements for the cross-section). If a-a is the resting level of the structure and the cross-section is located into a wind flow with velocity υ, then a horizontal force D will develop (Fig. **1-A**).

Let z be the downward displacement of the cross-section (Fig. **1-B**). Due to the displacement z, the cross-section will be moving with velocity $\dot{z}$ and consequently, the relative flow of the wind current with respect to the cross-section will be:

$$\bar{\upsilon}_r = \bar{\upsilon} + \dot{\bar{z}} \tag{1}$$

by an angle φ with respect to the horizontal direction.

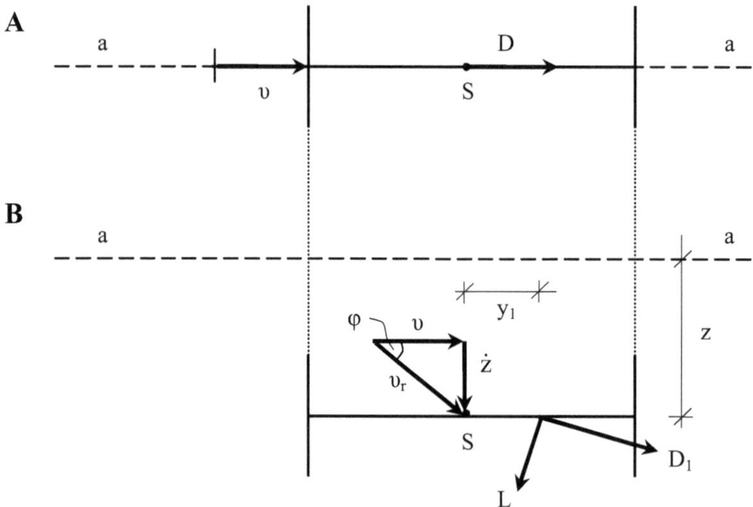

Figure 1: Direction of pressures developing on a structure into a wind flow

The curved movement of the cross-section about the instantaneous pole of rotation corresponds to acceleration $\dot{\upsilon}_r$, the direction of which is perpendicular to the velocity vector υ_r, and hence the corresponding force L causing this movement will be also perpendicular to υ_r and is acting at a distance y_1 from the gravity center S. Finally, the relative movement of the cross-section into the wind flow υ_r will produce a force D_1 on the cross-section, which is parallel to υ_r and hence, perpendicular to L.

Finally, the above aerodynamic loads D and L create a moment known as aerodynamic moment, which increases with the wind flow velocity and the effective angle φ.

For a cross-section that rotates about the elastic center such the one shown in Fig. **1**, the relations giving the aerodynamic loads are the following:

$$\left.\begin{aligned} L &= \frac{1}{2}\rho\upsilon^2 BC_L(\varphi) \\ D &= \frac{1}{2}\rho\upsilon^2 BC_D(\varphi) \\ M &= \frac{1}{2}\rho\upsilon^2 B^2 C_M(\varphi) \end{aligned}\right\} \qquad \textbf{(2a,b,c)}$$

where ρ is the air density, υ is the velocity, B is the width of the bridge deck and $C_L(\varphi)$, $C_D(\varphi)$, $C_M(\varphi)$ are the aerodynamic coefficients that are experimentally determined.

In Fig. **2** one can see the variation of these coefficients with respect to the effective angle φ for some well-known bridges.

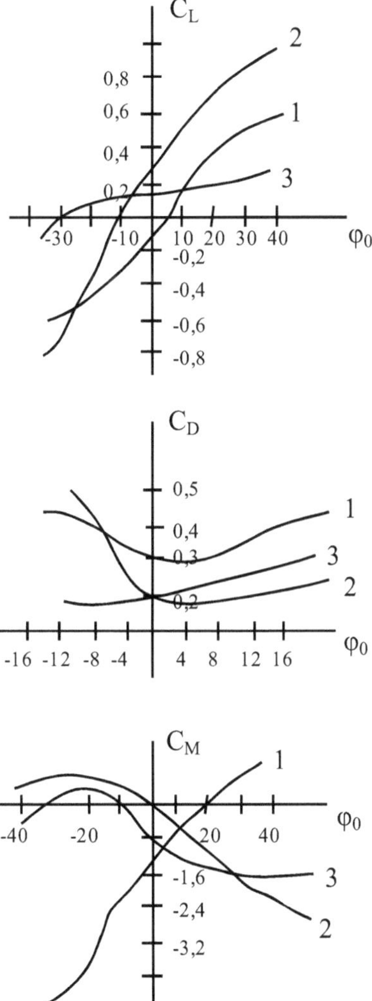

Figure 2: Coefficients C_L, C_D, C_M for some well-known bridges 1. Tacoma Narrows 2. Golden Gate 3. MacKinac

The above described mechanism shows that forces L and D_1 are caused from and depend on the movement z of the cross-section. It is obvious that if besides the displacement z, we consider also the rotation θ about point S for the cross-section, the above forces will depend on both movement z and rotation θ.

The aerodynamic analysis of bridges is based on equations of classic aeroelasticity theory. These equations are used mainly for the analysis and the design of airplanes and only recently they are also employed in other fields, such as in Civil Engineering. At this point, the following observation should be made:

Airplanes are designed with efficient aerodynamic shape and hence the aeroelasticity theory in the aeronautics is based on the application of dynamic flow theory, since the airflow around the airplane body is smooth.

On the contrary, the shape of a bridge is not aerodynamic and decks of bridges fall into the category of aerodynamic bluff - bodies with proportional aerodynamic behavior.

It is not explicit how the potential flow theory can be applied to study the aerodynamics of bridges and hence, numerous analytical models for the determination of aerodynamic forces exist. The most popular analyses are based on the relative theory of Theodorsen.

THE THEORY OF THEODORSEN

This approach is widely used for the design of aircrafts and is proved to be useful for the determination of the aerodynamic loads on a wing moving into a fluid with velocity υ in the time interval from $\tau = 0$ to $\tau = t$. The geometry of the wing is considered as that of a plate. In Fig. **3**, one can see the geometrical properties of the plate. Assuming that the rotation angle φ is small, it is $\overline{\Gamma\Delta} \cong 2b$, while $e \cdot b$ is the eccentricity of the elastic center K (where e is an experimentally determined factor).

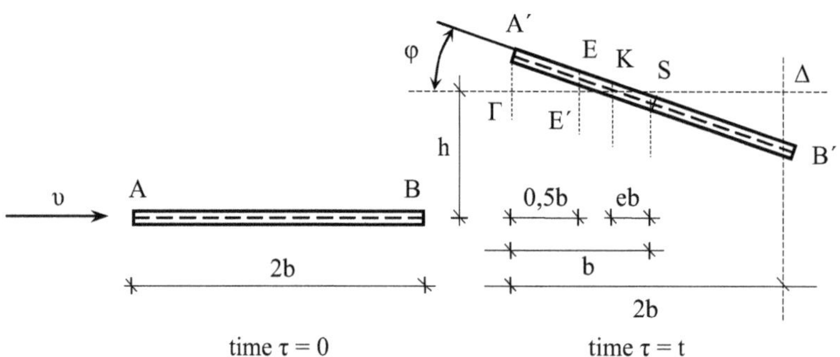

Figure 3: Plate moving into a wind flow

At time $\tau = 0$ the plate starts moving due to the impulse created by the wind flow with speed υ, and its increase around the plate produces a vertical force L_1 (per unit length of the plate), given by:

$$L_1 = 2\pi b\rho\upsilon w(\tau)\Phi(\tau) \qquad \textbf{(3)}$$

where $w(\tau)$ is the vertical velocity of the plate, τ is the nondimensional time given by:

$$\tau = \frac{\upsilon t}{b} \qquad \textbf{(4)}$$

and $\Phi(\tau)$ is a function known as the Wagner function given by:

$$\left.\begin{array}{l} \Phi(z) = 1 - \int\limits_0^\infty [(K_0 - K_1)^2 + \pi^2(I_0 - I_1)^2]^{-1} e^{x\tau} x^2 dx \\ \qquad \text{with} \quad \Phi(\tau) = 0 \quad \text{for } \tau < 0 \end{array}\right\} \qquad \textbf{(5)}$$

The terms K_0, K_1, I_0, I_1 are modified Bessel functions of the first and second order. Numerical values for the Wagner function are given in the diagram of Fig. **4**. From the diagram in Fig. **4**, we observe that half of the total force L_1, develops immediately when the plate starts moving and then increases asymptotically reaching the maximum value at time $\tau = \infty$. The pressure center E, that is the point where the force L_1 is applied, lies at distance 0,5b form the frontal end A′ of the plate (for supersonic velocities υ, this point almost coincides with the elastic center S of the plate).

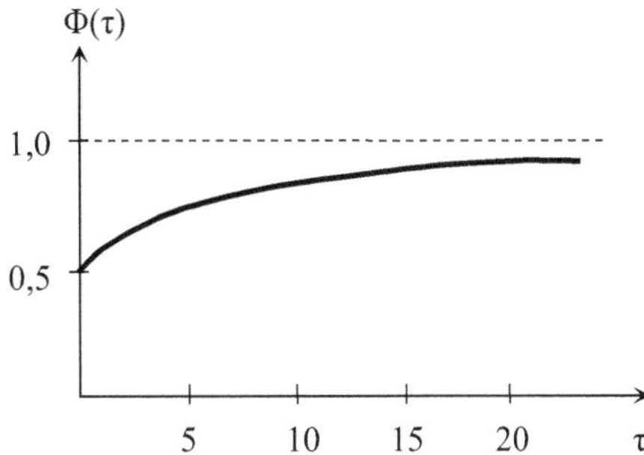

Figure 4: The Wagner function

Let us now consider the case where the wing does not only move vertically due to the relative flow movement with velocity υ, but is moving in a more complex manner determined by the vertical displacement h and the rotation angle φ, also called pitching, with point K being the pole of rotation.

Taking into account all the above movements, the velocity components at the pressure center E will be:

a. The vertical component due to velocity υ:

$$w_V = \upsilon \sin\varphi \cong \upsilon \cdot \varphi \qquad \textbf{(6)}$$

b. The vertical component due to displacement h:

it is: $\dot{h} = \frac{dh}{dt} = \frac{dh}{d\tau} \cdot \frac{d\tau}{dt}$ or due to eq(4):

$$\dot{h} = \frac{dh}{dt} = h' \cdot \frac{\upsilon}{b} \qquad \textbf{(7)}$$

c. The vertical component due to rotation φ:

$$w_\varphi = \frac{d\overline{EE'}}{dt} = \frac{d(\overline{E'K} \cdot \varphi)}{dt} = \frac{d(0,5b - eb)\varphi}{dt} = (0,5-e)b\frac{d\varphi}{dt} = (0,5-e)b\frac{d\varphi}{d\tau} \cdot \frac{d\tau}{dt} = (0,5-e)b\frac{\upsilon}{b}\varphi'$$

or finally:

$$w_\varphi = (0,5-e)\upsilon\varphi' \qquad \textbf{(8)}$$

Hence, the expression for w(τ) can be written as follows:

$$w(\tau) = \upsilon\varphi + \frac{\upsilon}{b}h' + (0,5-e)\upsilon\varphi' \qquad \textbf{(9)}$$

In the time interval $(\tau_0, \tau_0 + d\tau_0)$, the vertical velocity $w(\tau_0)$ increases by $\frac{dw(\tau_0)}{d\tau_0} \cdot d\tau_0$, and if $d\tau_0$ is sufficiently small, the movement can be considered as impulsive and the increase of the vertical force L_1 per unit length will be:

$$dL_1 = 2\pi b\rho\upsilon\Phi(\tau-\tau_0)\cdot\frac{dw(\tau_0)}{d\tau_0}d\tau_0 \quad \text{for} \quad \tau>\tau_0 \tag{10}$$

Thus, the value of L_1 will be:

$$L_1 = 2\pi b\rho\upsilon\int_{-\infty}^{\tau}\Phi(\tau-\tau_0)\frac{dw(\tau_0)}{d\tau_0}\cdot d\tau_0 \tag{11}$$

which, due to eq(9) takes the following form:

$$L_1(\tau) = 2\pi b\rho\upsilon^2\int_{-\infty}^{\tau}\Phi(\tau-\tau_0)\cdot\left[\varphi'(\tau_0)+\frac{1}{b}h''(\tau_0)+(0,5-e)\varphi''(\tau_0)\right]\cdot d\tau_0 \tag{12}$$

that is the vertical force L_1 caused by the increase of flow movement around the plate due to the flow velocity υ and the elastic deformations h and φ.

However, besides the above force due to the air movement, inertia forces exist as well. Following a similar methodology, the vertical force applied on the center of the plate can be determined from:

$$L_2 = \rho\pi\upsilon^2\left(h''-eb\varphi''\right) \tag{13}$$

and the vertical force applied on the pressure center E from:

$$L_3 = \rho\pi b\upsilon^2\varphi' \tag{14}$$

The first term in eq(13) is due to the vertical acceleration, while the second term is due to the vertical acceleration from angular acceleration and consequently, the total vertical force L is:

$$L = L_1 + L_2 + L_3 \tag{15}$$

The corresponding induced total moment is:

$$\left.\begin{aligned} &M = (0,5-e)bL_1 + ebL_2 - (0,5-e)bL_3 + M_\alpha \\ &\text{where:}\quad M_\alpha = -\frac{\rho\pi b^4}{8}\cdot\ddot{\varphi} = -\frac{\rho\pi b^2\upsilon^2}{8}\cdot\varphi'' \end{aligned}\right\} \tag{16a,b}$$

The term M_{α}, is the total moment due to the angular velocity. Introducing the expressions of L_1, L_2, L_3 into eqs(15) and (16a) we obtain:

$$\left.\begin{aligned} L(\tau) &= 2\pi b\rho\upsilon^2\int_{-\infty}^{\tau}\Phi(\tau-\tau_0)\cdot\left[\varphi'(\tau_0)+\frac{1}{b}h''(\tau_0)+(0,5-e)\varphi''(\tau_0)\right]d\tau_0 \\ &\quad+\rho\pi\upsilon^2\left(h''-eb\varphi''\right)+\rho\pi b\upsilon^2\varphi' \\ M(\tau) &= (0,5-e)2\pi b^2\rho\upsilon^2\int_{-\infty}^{\tau}\Phi(\tau-\tau_0)\left[\varphi'(\tau_0)+\frac{1}{b}h''(\tau_0)+(0,5-e)\varphi''(\tau_0)\right]d\tau_0 \\ &\quad+eb\rho\pi\upsilon^2\left(h''-eb\varphi''\right)-(0,5-e)\rho\pi b^2\upsilon^2\varphi'-\frac{\rho\pi b^2\upsilon^2}{8}\varphi'' \end{aligned}\right\} \tag{17a,b}$$

After the determination of the aerodynamic loads L and M, we can formulate the equations for dynamic equilibrium of the wing.

The main parameters involved in these equations are the mass m (per unit length), the moment of inertia I, the static moment $S = e \cdot m$, the vertical k_h and rotational k_φ stiffness of the wing and the vertical c_h and rotational c_φ damping of the system.

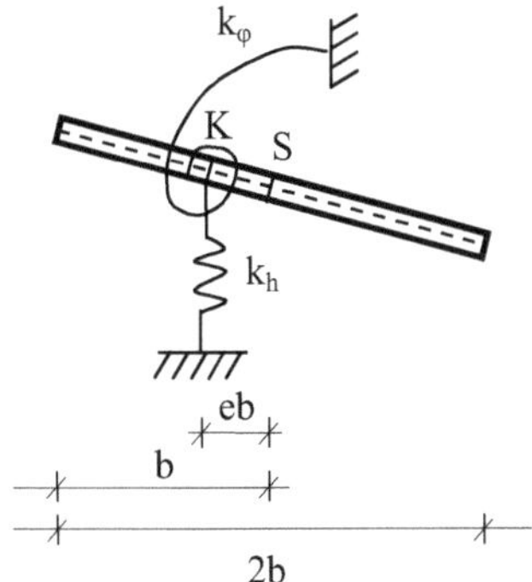

Figure 5: Two degrees-of-freedom wing model

Thus, the equations of motion will be:

$$\left.\begin{aligned} m\ddot{h} + S\ddot{\varphi} + c_h\dot{h} + hk_h &= L \\ S\ddot{h} + I\ddot{\varphi} + c_\varphi\dot{\varphi} + \varphi k_\varphi &= M \end{aligned}\right\} \qquad \textbf{(18a,b)}$$

Introducing the natural frequencies due to the corresponding stiffness: $\omega_h^2 = \frac{k_h}{m}$, $\omega_\varphi^2 = \frac{k_\varphi}{I}$ as well as the damping ratios:

$$\zeta_h = \frac{c_h}{2m\omega_h} \quad , \quad \zeta_\varphi = \frac{c_\varphi}{2I\omega_\varphi} \qquad \textbf{(19)}$$

equations (18) can be written as follows:

$$\left.\begin{aligned} \ddot{h} + e\ddot{\varphi} + 2\zeta_h\omega_h\dot{h} + \omega_h^2 h &= \frac{L}{m} \\ \frac{e}{r_g}\ddot{h} + \ddot{\varphi} + 2\zeta_\varphi\omega_\varphi\dot{\varphi} + \omega_\varphi^2\varphi &= \frac{M}{I} \\ \text{where:} \quad r_g &= \frac{I}{m} \end{aligned}\right\} \qquad \textbf{(20a,b,c)}$$

Assuming that forces L and M vary harmonically, deformations h and φ will also vary harmonically and hence, we can set:

$$h = h_0 \cdot e^{i\omega t} \quad , \quad \varphi = \varphi_0 \cdot e^{i\omega t} \qquad \textbf{(21a,b)}$$

where h_0 and φ_0 are complex constants corresponding to the deformation range and the angular phase between deformations and applied forces. If $\overline{\omega} = \frac{\omega b}{\upsilon}$ is the nondimensional frequency, eqs(21) can be written as:

$$h = h_0 \cdot e^{i\bar{\omega}\tau} \quad , \quad \varphi = \varphi_0 \cdot e^{i\bar{\omega}\tau} \quad , \quad \bar{\omega} = \frac{\omega b}{\upsilon} \tag{22a,b,c}$$

Then, the integral in relations (17) used for the determination of the aerodynamic forces, becomes:

$$\left.\begin{aligned} &\lim_{\varepsilon \to 0} \int_{-\infty}^{\tau} \Phi(\tau - \tau_0)\left[\varphi'(\tau_0) + \frac{1}{b}h''(\tau_0) + (0,5+e)\cdot\varphi''(\tau_0)\right]\cdot e^{i\bar{\omega}\tau_0 + \varepsilon\tau_0}\cdot d\tau_0 = \\ &\qquad = c(\bar{\omega})\cdot\left[\varphi_0 + \frac{i}{b}\bar{\omega}h_0 + (0,5-e)\ i\bar{\omega}\varphi_0\right]\cdot e^{i\bar{\omega}\tau} \end{aligned}\right\} \tag{23}$$

where $c(\bar{\omega})$ is the Theodorsen function, which is related to the Wagner function as follows:

$$c(\bar{\omega}) = F(\bar{\omega}) + i\cdot G(\bar{\omega}) = \frac{k_1(i\bar{\omega})}{k_0(i\bar{\omega}) + k_1(i\bar{\omega})} \tag{24}$$

The terms k_0 and k_1 in the above relation, are the Hunkel functions. The functions F and G are given by Fung [4] in a graphical form (Fig. **6**).

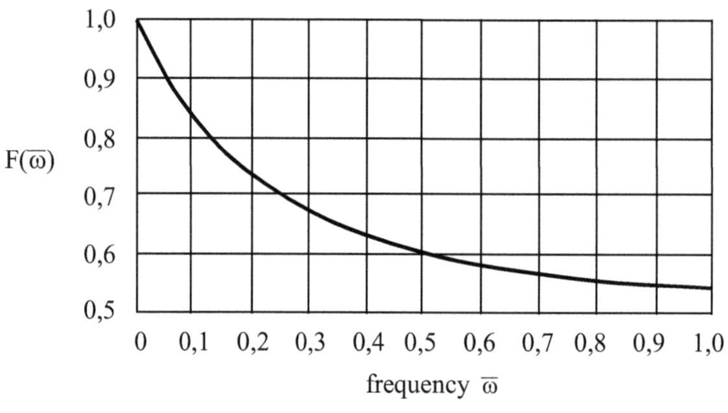

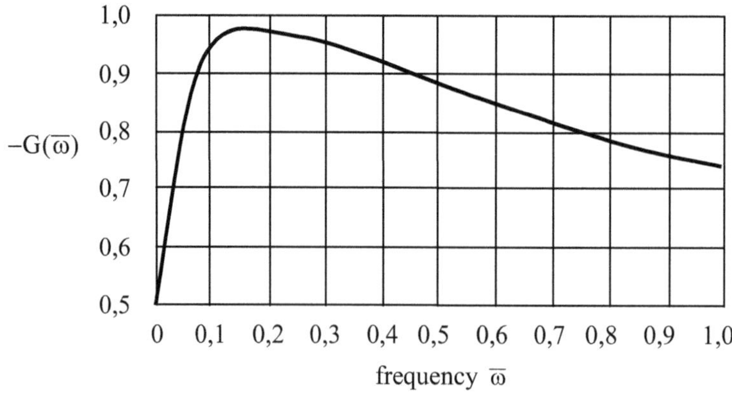

Figure 6: Functions F and G related to the Hunkel functions

According to the above, the expressions for the aerodynamic loads are the following:

$$\left.\begin{aligned} L &= 2\pi b \rho \upsilon^2 \left\{ (F+iG)\cdot\left[\varphi_0 + \frac{i}{b}\overline{\omega} h_0 + (0,5-e) i\overline{\omega}\varphi_0\right] \right. \\ &\quad \left. -\frac{1}{2}\overline{\omega}^2\left(\frac{h_0}{b} - e\varphi_0\right) - \frac{1}{2} i\overline{\omega}\varphi_0 \right\}\cdot e^{i\overline{\omega}\tau} \\ M &= 2\pi b \rho^2 \upsilon^2 \left\{ (0,5+e)(F+iG)\cdot\left[\varphi_0 + \frac{i}{b}\overline{\omega} h_0 + (0,5-e) i\overline{\omega}\varphi_0\right] \right. \\ &\quad \left. -\frac{1}{2}\overline{\omega}^2 e\left(\frac{h_0}{b} - e\varphi_0\right) + (0,5-e) i\overline{\omega}\varphi_0 + \frac{\overline{\omega}^2}{8}\varphi_0 \right\}\cdot e^{i\overline{\omega}\tau} \end{aligned}\right\} \quad \textbf{(25a,b)}$$

These relations, even though widely used in aerodynamics, require some modifications in order to become suitable for studying bridge structures. The deck width is $B = 2\cdot b$ and hence, we set: $\Omega = \frac{\omega B}{\upsilon} = \frac{2b\omega}{\upsilon} = 2\overline{\omega}$. We also introduce the dimensional time t instead of the nondimensional τ. Finally, after some manipulations, eqs(25) can be written as follows:

$$\left.\begin{aligned} L &= \frac{1}{2}\rho\upsilon^2 B\left\{ 2\pi F\frac{\dot{h}}{\upsilon} + \frac{\pi}{2}\left[1 + \frac{4G}{\Omega} + 2(0,5-e)F\right]\cdot\frac{B}{\upsilon}\dot{\varphi} \right. \\ &\quad \left. + \pi\left[2F - (0,5-e)G\Omega + \frac{\Omega^2 e}{4}\right]\varphi - \frac{\pi}{2}\Omega^2\left[1 + \frac{4G}{\Omega}\right]\cdot\frac{h}{B} \right\} \\ M &= \frac{1}{2}\rho\upsilon^2 B^2\left\{ \pi F(0,5+e)\frac{\dot{h}}{\upsilon} \right. \\ &\quad - \frac{\pi}{2}\left[0,5\cdot(0,5+e) - (0,5+e)\frac{2G}{\Omega} + F(e^2 - 0,25)\right]\cdot\frac{B}{\upsilon}\dot{\varphi} \\ &\quad + \frac{\pi}{2}\left[\frac{\Omega^2}{4}(e^2 + 0,125) + 2F(0,5+e) + G\Omega(e^2 - 0,25)\right]\varphi \\ &\quad \left. - \frac{\pi}{2}\left[\frac{\Omega^2 e}{2} + (0,5+e)2G\Omega\right]\cdot\frac{h}{B} \right\} \end{aligned}\right\} \quad \textbf{(26a,b)}$$

AERODYNAMIC LOADS IN BRIDGES

The analytical procedure for the determination of aerodynamic loads in bridges presented in the previous paragraph cannot be performed directly.

Bridge decks do not have an aerodynamic shape and thus, it is necessary to follow an approach based on the aeroelasticity theory and on data from experimental studies.

Extensive theoretical and experimental research has been carried out by R. Scanlan and his associates and the results have been presented in a number of publications.

In 1996, Scanlan and Simiu presented a technically correct methodology for the determination of the aerodynamic loads for a model with two loads and a model with three loads. The theoretical analysis was similar to the one presented above for the determination of eqs(26).

Model with Two Aerodynamic Loads

This model, also referred as 2D-model, has two degrees-of-freedom, namely the vertical displacement h and the rotation φ. Then, the developing aerodynamic forces L and M are functions of the displacements h, φ and their velocities $\dot{h}$, $\dot{\varphi}$ and take the following form:

$$\left.\begin{aligned} L_{\alpha e} &= \frac{1}{2}\rho\upsilon^2 B\cdot\left[\Omega H_1\frac{\dot{h}}{\upsilon}+\Omega H_2\frac{B\dot{\varphi}}{\upsilon}+\Omega^2 H_3\varphi+\Omega^2 H_4\frac{h}{B}\right] \\ M_{\alpha e} &= \frac{1}{2}\rho\upsilon^2 B^2\cdot\left[\Omega A_1\frac{\dot{h}}{\upsilon}+\Omega A_2\frac{B\dot{\varphi}}{\upsilon}+\Omega^2 A_3\varphi+\Omega^2 A_4\frac{h}{B}\right] \end{aligned}\right\} \qquad \textbf{(27a,b)}$$

The terms H_i and A_i ($i = 1 \div 4$) are called flutter derivatives and depend on the shape of the deck. These terms are determined experimentally with the use of wind tunnels. In the special case of a flat plate, eqs(27) and (26) are coincident. Hence, in a preliminary design we can determine A_i and H_i, before proceeding to a more detailed experimental investigation, as follows

$$\left.\begin{aligned} &H_1 = \frac{2\pi F}{\Omega} \quad , \quad H_2 = \frac{2\pi}{\Omega}\cdot\left(\frac{1}{2}+\frac{G}{\Omega}+\frac{F}{2}\right) \ , \\ &H_3 = \frac{2\pi}{\Omega^2}\cdot\left(F-\frac{\Omega G}{2}\right) \quad , \quad H_4 = -2\pi\cdot\left(\frac{1}{2}+\frac{G}{\Omega}\right) \\ &A_1 = \frac{\pi F}{\Omega} \quad , \quad A_2 = -\frac{\pi}{\Omega}\cdot\left(\frac{G}{\Omega}+\frac{F-1}{2}\right) \\ &A_3 = \frac{\pi}{\Omega^2}\cdot\left(F-\frac{\Omega G}{2}+\frac{\Omega^2}{8}\right) \quad , \quad A_4 = -\frac{\pi G}{\Omega} \end{aligned}\right\} \qquad \textbf{(28)}$$

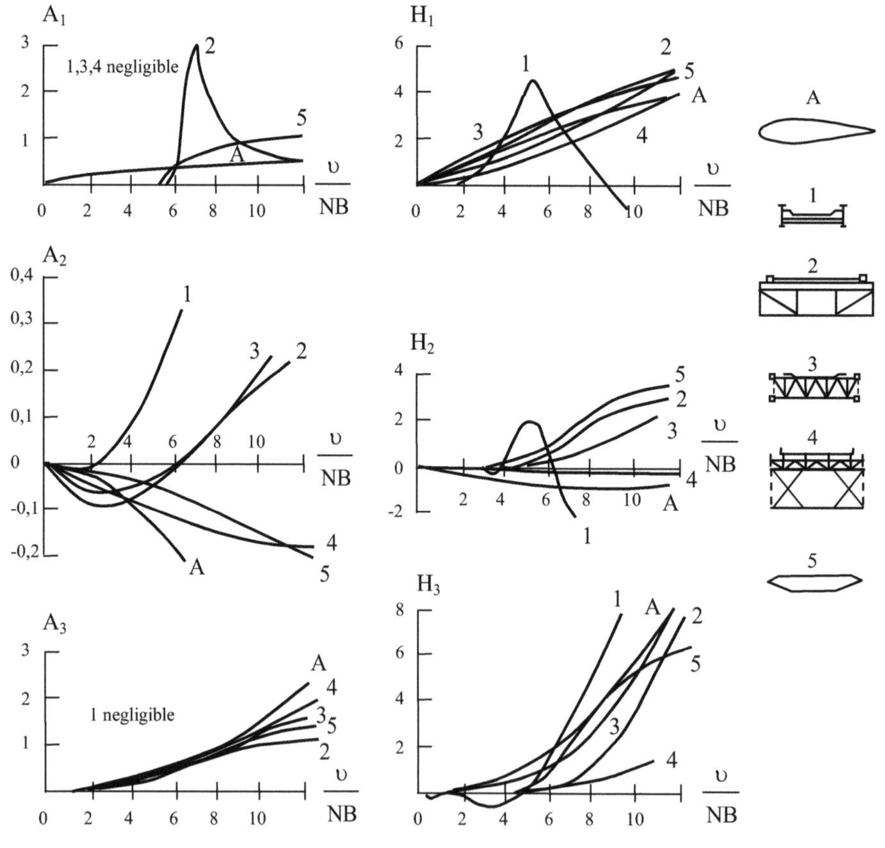

Figure 7: Coefficients A_i and H_i versus υ / NB

where F and G are the real and imaginary parts of the Theodorsen function of eq(24) presented in graphical form in a number of references [4].

The terms H_4 and A_4 in eqs(27) have a negligible influence and are usually omitted. Here, these terms are presented for completeness.

Some typical experimental results for the terms H_i and A_i (i=1,2,3) and various deck section types of bridges are given in the diagrams shown in Fig. 7.

Models with Three Aerodynamic Loads

In the doctoral theses of Singh [11] and Jain [12] presented at the Johns Hopkins University in Baltimore, the previous model is extended and the case of three deformations or three aeroelastic forces is considered.

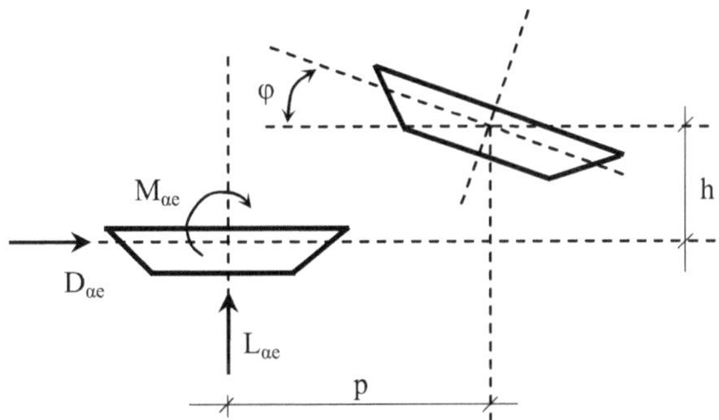

Figure 8: Aerodynamic model with three degrees of freedom

Thus, we have a three-dimensional problem where the displacements depend on the cross-sectional properties, the position and time. In other words, the displacements h,p,φ are functions of x and t,*.i.e.* h(x,t), p(x,t), φ(x,t), respectively. In this case, a third force D with horizontal direction causing subpressure to the rear part of the cross-section must be accounted for, which is also known as back-drag force.

Then, the aerodynamic loads can be expressed through a summation of terms including the above deformations and their derivatives as well as 18 flutter coefficients such as:

$$\left.\begin{aligned}
L_{\alpha e} &= \frac{1}{2}\rho\upsilon^2 B\left(\Omega H_1\frac{\dot{h}}{\upsilon}+\Omega H_2\frac{B\dot{\varphi}}{\upsilon}+\Omega^2 H_3\varphi+\Omega^2 H_4\frac{h}{B}+\Omega H_5\frac{\dot{p}}{\upsilon}+\Omega^2 H_6\frac{p}{B}\right)\\
D_{\alpha e} &= \frac{1}{2}\rho\upsilon^2 B\left(\Omega P_1\frac{\dot{p}}{\upsilon}+\Omega P_2\frac{B\dot{\varphi}}{\upsilon}+\Omega^2 P_3\varphi+\Omega^2 P_4\frac{p}{B}+\Omega P_5\frac{\dot{h}}{\upsilon}+\Omega^2 P_6\frac{h}{B}\right)\\
M_{\alpha e} &= \frac{1}{2}\rho\upsilon^2 B^2\left(\Omega A_1\frac{\dot{h}}{\upsilon}+\Omega A_2\frac{B\dot{\varphi}}{\upsilon}+\Omega^2 A_3\varphi+\Omega^2 A_4\frac{h}{B}+\Omega A_5\frac{\dot{p}}{\upsilon}+\Omega^2 A_6\frac{p}{B}\right)
\end{aligned}\right\} \quad \textbf{(29)}$$

In the above expressions, the last terms H_6, P_6, A_6 have a negligible influence and are usually omitted.

Buffeting Forces

Buffeting forces are usually caused by abrupt changes of the wind velocity (see also page 20, Fig. 6 and eq.19 of Chapter 2). These forces resulting from wind gusts are generally considered to be independent from the motion of the structure and are due to the variation of the wind velocity, with the additional assumption that changes in a storm action are slow.

The wind velocity υ can be expressed by the three components, u, v, w on the Cartesian axes system. Usually, the component v is practically negligible and the buffeting forces can be expressed as follows:

$$\left.\begin{aligned} L_b &= \frac{1}{2}\rho\upsilon^2 B\cdot\left[c_L\left(\frac{2u}{\upsilon}\right)+(c_L'+c_D)\frac{w}{\upsilon}\right]=\frac{1}{2}\rho\upsilon^2 B\mathcal{L}_b(x,\varphi,t)\\ D_b &= \frac{1}{2}\rho\upsilon^2 B\cdot\left[c_D\left(\frac{2u}{\upsilon}\right)+c_D'\frac{w}{\upsilon}\right]=\frac{1}{2}\rho\upsilon^2 B\mathcal{D}_b(x,\varphi,t)\\ M_b &= \frac{1}{2}\rho\upsilon^2 B^2\cdot\left[c_M\left(\frac{2u}{\upsilon}\right)+c_M'\frac{w}{\upsilon}\right]=\frac{1}{2}\rho\upsilon^2 B^2\mathcal{M}_b(x,\varphi,t) \end{aligned}\right\} \quad \textbf{(30a,b,c)}$$

where: $c_L' = \dfrac{dc_L}{d\varphi}$, $c_D' = \dfrac{dc_D}{d\varphi}$, $c_M' = \dfrac{dc_M}{d\varphi}$ are the derivatives of the aerodynamic coefficients in page 195 with respect to angle φ.

In order to have dynamic equilibrium of the bridge deck, we must add the aeroelastic forces and the buffeting forces:

$$L = L_{\alpha e} + L_b \ , \ D = D_{\alpha e} + D_b \ , \ M = M_{\alpha e} + M_b \quad \textbf{(31)}$$

DYNAMIC STABILITY OF BRIDGES

The loads in eqs(29) or (30) may cause various types of dynamic instability to the bridge.

The most common dynamic instability types are: bending instability in the vertical direction (galloping instability), torsional instability, and finally, flutter instability. The last type can be due to the action of aeroelastic forces, or the action of buffeting forces, or even the combined action of both.

Galloping Instability

This instability type is typical for structural elements with orthogonal or circular section. It may appear for instance in a cable of a cable-stayed bridge (see also Ch. 8), where ice has been accumulated and the cable is subjected to the action of strong wind. Structures sensitive to this type of instability are vibrating perpendicularly to the wind direction with big amplitudes, substantially bigger than the characteristic dimensions of the structure.

Cable-stayed bridges, suspension bridges as well as power transmission towers are the most sensitive structures in having such a behavior.

If we consider instantaneously that only the vertical degree-of-freedom exists in bridge while the buffeting forces are zero, then the acting aeroelastic force (setting $h = w_S$) is:

$$L_{\alpha e} = \frac{1}{2}\rho\upsilon^2 B\left(\Omega H_1\frac{\dot{w}_S}{\upsilon}+\Omega^2 H_4\frac{w_S}{B}\right) \quad \textbf{(32)}$$

In this case, the equation of motion is:

$$EI_y w_S'''' + c_y\dot{w}_S + m\ddot{w}_S = \frac{1}{2}\rho\upsilon^2 B\left(\Omega H_1\frac{\dot{w}_S}{\upsilon}+\Omega H_4\frac{w_S}{B}\right) \quad \textbf{(33)}$$

We will seek a solution in the form of separate variables as follows:

$$w_S(x,t) = \sum_n W_n(x)T_n(t) \quad \textbf{(34)}$$

Introducing eq(34) into eq(33) and following the usual procedure, we get:

$$\left.\begin{aligned}&\ddot{T}_n+\left(2\zeta_{y_n}\omega_{y_n}-\frac{\rho\upsilon B\Omega H_1}{2m}\right)\cdot\Gamma_{y_n}\dot{T}_n+\left(\omega_{y_n}^2-\frac{\rho\upsilon^2\Omega^2H_4}{2m}\right)\cdot\Gamma_{y_n}T_n=0\\&\text{where:}\quad \zeta_{y_n}=\frac{c_y}{2m\omega_{y_n}}\quad,\quad \Gamma_{y_n}=\frac{\int_0^\ell W_n dx}{\int_0^\ell W_n^2 dx}\end{aligned}\right\}\qquad\textbf{(35a,b,c)}$$

Equation (35a) gives the time function for free vibration of the bridge and has the same form as eq(73) of chapter 3, with different unknown coefficients. The second term in eq(35a) is the damping term with a coefficient different than the one in eq(73) of chapter 3. In order that the oscillation decreases, the factor of the damping term must be positive, otherwise we will have a progressive increase of the oscillation amplitude. In this case, dynamic instability of the structure will occur when:

$$2\zeta_{y_n}\omega_{y_n}-\frac{\rho\upsilon B\Omega H_1}{2m}<0\ \text{ or for wind velocity values: }\upsilon>\frac{4\zeta_{y_n}\omega_{y_n}m}{\rho B\Omega H_1}\qquad\textbf{(36)}$$

Torsional Divergence

This instability case can occur only when $z_M = 0$ (in doubly symmetric sections), where the system of equations (123c,d) of chapter 3 is not conjugated. Considering that only the rotational degree of freedom of the bridge exists while the buffeting forces are zero, the acting aeroelastic force will be:

$$M_{\alpha e}=\frac{1}{2}\rho\upsilon^2B^2\left(\Omega A_2\frac{B\dot{\varphi}}{\upsilon}+\Omega^2A_3\varphi\right)\qquad\textbf{(37)}$$

Then, the equation of motion will be:

$$EC_S\varphi''''+c_\theta\dot{\varphi}-GJ_d\varphi''+\Theta_x\ddot{\varphi}=\frac{1}{2}\rho\upsilon^2B^2\left(\Omega A_2\frac{B\dot{\varphi}}{\upsilon}+\Omega^2A_3\varphi\right)\qquad\textbf{(38)}$$

We will seek a solution in the following form:

$$\varphi(x,t)=\sum_n\Phi_n(x)T_n(t)\qquad\textbf{(39)}$$

Introducing eq(39) into eq(38) and following the usual procedure, we obtain:

$$\left.\begin{aligned}&\ddot{T}_n+\left(2\zeta_{\theta_n}\omega_{\theta_n}-\frac{\rho\upsilon B^3\Omega A_2}{2\Theta_x}\right)\Gamma_{\theta_n}\dot{T}+\left(\omega_{\theta_n}^2-\frac{\rho\upsilon^2B^2\Omega^2A_3}{2\Theta_x}\right)\Gamma_{\theta_n}T_n=0\\&\text{where:}\quad \zeta_{\theta_n}=\frac{c_\theta}{2m\omega_{\theta_n}}\quad,\quad \Gamma_{\theta_n}=\frac{\int_0^\ell \Phi_n dx}{\int_0^\ell \Phi_n^2 dx}\end{aligned}\right\}\qquad\textbf{(40a,b,c)}$$

From the second term corresponding to damping, we can determine wind velocity values that lead to dynamic instability:

$$\upsilon > \frac{4\zeta_{\theta_n}\omega_{\theta_n}\Theta_x}{\rho B^3 \Omega A_2} \tag{41}$$

Flutter

In general, the dynamic behavior of a bridge is described by the three deformations υ_S, w_S and φ, *i.e.* the displacements along axes y, z and the rotation about axis x. The general equations of motion 51 in chapter 3 show that the motion in z-direction is independent from the combined motion (for $z_M \neq 0$) in y-direction and about x-axis. This is valid since loads q_z, q_y, mx in eqs(51) of chapter 3 do not depend on the deformations w_S, υ_S, φ.

Studying though the aerodynamic instability of a bridge, we observe that the loads $L_{\alpha e}$, $D_{\alpha e}$ and $M_{\alpha e}$ in the previous sections are functions of the deformations w_S, υ_S and φ. Thus, we must consider a system of three combined equations with unknowns the three above deformations. By setting zero the buffeting forces (without loss of generality of the method), the equations of motion can be written as follows:

$$\left.\begin{aligned} &EI_y w'''' + c_y \dot{w}_S + m\ddot{w}_S = L_{\alpha e}(w_S, \upsilon_S, \varphi) \\ &EI_z \upsilon'''' - EI_z z_M \varphi'''' + c_z \dot{\upsilon}_S + m\ddot{\upsilon}_S = D_{\alpha e}(w_S, \upsilon_S, \varphi) \\ &EC_S \varphi'''' - EI_z z_M \upsilon_S'''' + c_\theta \dot{\varphi} - GJ_d \varphi'' + \Theta_x \ddot{\varphi} = M_{\alpha e}(w_S, \upsilon_S, \varphi) \end{aligned}\right\} \tag{42a,b,c}$$

where $L_{\alpha e}$, $D_{\alpha e}$, $M_{\alpha e}$ are given by eq(29) for the 3D-model, or by eq(27) for the 2D-model.

Since the system of eqs(42) is combined, we may assume that the bridge will vibrate in directions y, z and about x with a single time function and hence, we seek solutions of the following form:

$$\left.\begin{aligned} w_S(x,t) &= \sum_n W_n(x) T_n(t) \\ \upsilon_S(x,t) &= \sum_n V_n(x) T_n(t) \\ \varphi(x,t) &= \sum_n \Phi_n(x) T_n(t) \end{aligned}\right\} \tag{43a,b,c}$$

where W_n, V_n, Φ_n are the shape functions of the bridge (a priori known).

Introducing expressions (43) into eqs(42) and taking into account that the shape functions satisfy eqs(73b) of chapter 3 and (82b,c) of chapter 3 for free vibration, we obtain:

$$\left.\begin{aligned} &m\sum_n \omega_{y_n}^2 W_n T_n + 2m\sum_n \zeta_{y_n}\omega_{y_n} W_n \dot{T}_n + m\sum_n W_n \ddot{T}_n = L_{\alpha e} \\ &m\sum_n \omega_{\theta_n}^2 V_n T_n + 2m\sum_n \zeta_{\theta_n}\omega_{\theta_n} V_n \dot{T}_n + m\sum_n V_n \ddot{T}_n = D_{\alpha e} \\ &\Theta_x\sum_n \omega_{\theta_n}^2 \Phi_n T_n + 2\Theta_x\sum_n \zeta_{\theta_n}\omega_{\theta_n} \Phi_n \dot{T}_n + \Theta_x\sum_n \Phi_n \ddot{T}_n = M_{\alpha e} \end{aligned}\right\} \tag{44a,b,c}$$

Multiplying the first of eqs(44) by W_n, integrating the outcome from 0 to ℓ and applying the orthogonality conditions given by equations 115 of chapter 3, we obtain:

$$\ddot{T}_n + 2\zeta_{y_n}\omega_{y_n}\dot{T}_n + \omega_{y_n}^2 T_n = \frac{\int_0^\ell L_{\alpha e} W_n dx}{m\int_0^\ell W_n^2 dx}$$

Introducing eq(29a) into the above equation, we finally get:

$$\left.\begin{aligned}
&\ddot{T}_n + C_{y_n}\dot{T}_n + K_{y_n}T_n = 0\\
&\text{where:}\\
&C_{y_n} = 2\zeta_{y_n}\omega_{y_n} - \frac{\rho\upsilon\Omega BH_1}{2m} - \frac{\rho\upsilon\Omega B^2H_2}{2m}\cdot\frac{\int_0^\ell \sum_i \Phi_i W_n dx}{\int_0^\ell W_n^2 dx} - \frac{\rho\upsilon\Omega BH_5}{2m}\cdot\frac{\int_0^\ell \sum_i V_i W_n dx}{\int_0^\ell W_n^2 dx}\\
&K_{y_n} = \omega_{y_n}^2 - \frac{\rho\upsilon\Omega^2 BH_4}{2m} - \frac{\rho\upsilon\Omega^2 BH_3}{2m}\cdot\frac{\int_0^\ell \sum_i \Phi_i W_n dx}{\int_0^\ell W_n^2 dx} - \frac{\rho\upsilon^2\Omega^2 H_6}{2m}\cdot\frac{\int_0^\ell \sum_i V_i W_n dx}{\int_0^\ell W_n^2 dx}
\end{aligned}\right\} \tag{45}$$

Multiplying the second of eqs(44) by V_n and integrating the outcome from 0 to ℓ, then multiplying the third of eqs(44) by Φ_n and integrating from 0 to ℓ, and adding the resulting equations applying the orthogonality conditions given by equation 120 of chapter 3, we obtain:

$$\ddot{T}_n + 2\zeta_{\theta_n}\omega_{\theta_n}\dot{T}_n + \omega_{\theta_n}^2 T_n = \frac{\int_0^\ell D_{\alpha e}V_n dx + \int_0^\ell M_{\alpha e}\Phi_n dx}{m\int_0^\ell V_n^2 dx + \Theta_x\int_0^\ell \Phi_n^2 dx}$$

Introducing eqs(29b,c) into the above equation, we finally get:

$$\left.\begin{aligned}
&\ddot{T}_n + C_{\theta_n}\dot{T}_n + K_{\theta_n}T_n = 0\\
&\text{where:}\\
&C_{\theta_n} = 2\zeta_{\theta_n}\omega_{\theta_n} - \frac{\rho\upsilon B\Omega P_1}{2G}\cdot\int_0^\ell V_n^2 dx - \frac{\rho\upsilon B^2\Omega P_2}{2G}\cdot\int_0^\ell \sum_i \Phi_i V_n dx\\
&\quad - \frac{\rho\upsilon B\Omega P_5}{2G}\cdot\int_0^\ell \sum_i W_i V_n dx - \frac{\rho\upsilon B^2\Omega A_1}{2G}\cdot\int_0^\ell \sum_i W_i \Phi_n dx\\
&\quad - \frac{\rho\upsilon B^3\Omega A_2}{2G}\cdot\int_0^\ell \Phi_n^2 dx - \frac{\rho\upsilon B^2\Omega A_5}{2G}\cdot\int_0^\ell \sum_i V_i \Phi_n dx\\
&K_{\theta_n} = \omega_{\theta_n}^2 - \frac{\rho\upsilon^2 B\Omega^2 P_3}{2G}\cdot\int_0^\ell \sum_i \Phi_i V_n dx - \frac{\rho\upsilon^2\Omega^2 P_4}{2G}\cdot\int_0^\ell V_n^2 dx\\
&\quad - \frac{\rho\upsilon^2\Omega^2 P_6}{2G}\cdot\int_0^\ell \sum_i W_i V_n dx - \frac{\rho\upsilon^2 B^2\Omega^2 A_3}{2G}\cdot\int_0^\ell \Phi_n^2 dx\\
&\quad - \frac{\rho\upsilon^2 B\Omega^2 A_4}{2G}\cdot\int_0^\ell \sum_i W_i \Phi_n dx - \frac{\rho\upsilon^2 B\Omega^2 A_6}{2G}\cdot\int_0^\ell \sum_i V_i \Phi_n dx\\
&G = m\int_0^\ell V_n^2 dx + \Theta_x\int_0^\ell \Phi_n^2 dx
\end{aligned}\right\} \tag{46}$$

Since the vibration is damped, the coefficients of terms $\dot{T}_n$ in the first of eqs(45) and in the first of eqs(46) must be positive. Hence, we should have:

$$C_{y_n} > 0 \quad , \quad C_{\theta_n} > 0 \qquad \textbf{(47a,b)}$$

From the above conditions and for given values of Ω and υ, we can determine the frequencies ω_{y_n}, ω_{θ_n}. Note that for a bridge with known properties, eqs(47) give the corresponding critical velocities characterized by two values: the upper and lower critical velocity, the first of which is very important in the design.

EXPERIMENTAL INVESTIGATION – MODEL THEORY

As it has been described in previous sections, the various "**flutter coefficients**" are determined through experimental investigation involving wind tunnels (Fig. **9**). Such tests are not only performed in order to determine the above "flutter coefficients", but also the general dynamic behavior of a bridge subjected to aeroelastic or aerodynamic loads. The main purpose of these tests is to determine the aerodynamic characteristics of a bridge, the damping coefficients and the critical velocities, *i.e.* the upper and lower critical velocity. Most important is the upper critical velocity, the determination of which is the main parameter in the design and structural analysis of bridges with long spans.

Figure 9: Wind tunnel testing

Testing of models serves the purpose of design of a specific bridge or more rarely of a theoretical investigation.

Testing of Specific Bridges

In this case, testing involves a combination of partial-scale and full-scale models (see Figs **10** and **11**).

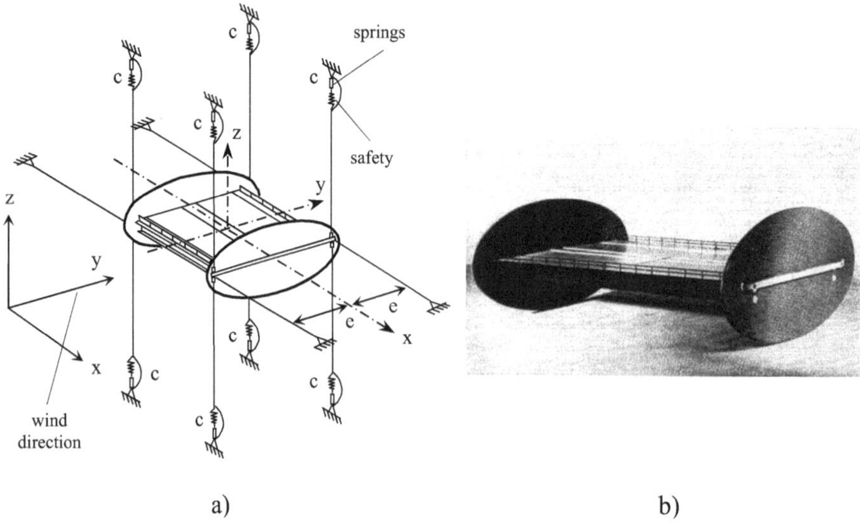

Figure 10: Scaled model of a bridge deck

Thus, it is initially constructed a partial scaled model, usually in 1/100 scale, the first data of which will allow us to determine the elements of the cross-section of the bridge. Next, a full-scale model is build, again in 1/100 scale, based on the previous partial-scale model data already obtained on the structural behavior of the bridge to be designed.

From this full-scale model, we obtain a full aspect of the dynamic behavior of the bridge and we determine the eigenfrequencies. With these new data, we proceed to the construction of a bigger scale model (*i.e.* 1/30), from which we determine the critical velocity for the bridge as well as the resulting deformations.

The results from the partial-scale model are not absolutely reliable, since it is not possible to conceive completely the influence of some factors such as: the stability of pylons, the suspension forces, the shape of the deformation wave, the direction of wind (horizontal or not) and many others. These effects are determined by the suspension springs constants (see Fig. **10a**).

Even in the full-scale model, where it would be possible to eliminate the above deficiencies, the above reported influences do not become fully perceptible, since the achievement of complete resemblance of individual elements of the structure is cumbersome and in general impossible, while it requires time and serious cost expenses.

Figure 11: Scaled bridge model

Despite all these, a measurements correlation between the full and partial-scale models would give us the certainty that these measurements could be considered as reliable.

The determination of measured vibration sizes for the particular construction and in reverse using the measurements in models is achieved *via* the theory of mechanical similarity.

The similarity principles for the study of partial and full-scale models are quite different.

In the full-scale model, the similarity is achieved *via* the equitable proportion between forces of gravity and elastic forces.

On the contrary, in the partial-scale models, the verification of similarity constants is achieved *via* the resemblance of eigenfrequencies between the real structure and the model.

Validation of Research Studies

The models used in research studies are usually full-scale models.

Depending to the factor under study, the model is manufactured to be sensitive to this factor. If for example we study bending vibrations of the structure, then we construct a model with weak moment of inertia about the horizontal axis, while if we study torsional vibrations, we construct a model with weak torsional constant and possibly the warping constant as well.

It should be noted that these models do not represent real scale bridges. In the tests conducted by Hirai [13] for instance, the models had the following characteristics:

$$\ell = 3{,}00\text{m}, \quad f/\ell = 1/10, \quad EI_y = 7.686*10^4\,\text{N}*\text{cm}^2, \quad 2\text{m} = 5{,}94\text{gr}/\text{cm}$$
$$C_M = 0{,}243\text{cm}^6 \quad , \quad \Theta_x/\ell = 0{,}00775\text{gr}\cdot\sec^2 \quad , \quad \omega_\varphi = 95{,}7\sec^{-1}$$

If however, the model corresponded to a bridge with opening *i.e.* 1.200m, then the scale should have been 1/n=1/400 and the real bridge would have an eigenfrequency $\omega_{bridge} = \dfrac{95{,}7}{\sqrt{400}} = 4{,}785\sec^{-1}$.

This eigenfrequency value is very high and does not correspond to a real bridge. In the same study, it was measured $\upsilon_{cr} = 11{,}90\text{m}/\sec$ for the model, while the critical velocity of the wind corresponding to the real bridge would be $\upsilon_{bridge} = 11{,}90\sqrt{400} = 238{,}00\text{m}/\sec$, *i.e.*, the 2/3 of the sound speed, which is higher even than the most extreme storm.

Modeling Theory – Similarity Principles

Partial-Scale Models

Regardless from any geometrical scale, the following nondimensional sizes for the real bridge and the model should be equal:

$$\frac{\omega b}{\upsilon_{cr}} \; , \; \frac{b^2}{m} \; , \; \frac{b}{r}$$

where: ω: is the eigenfrequency

υ_{cr}: is the critical velocity of the wind

b: is the bridge deck half width

m: is the mass of the bridge per unit length

$r = \sqrt{\dfrac{J}{F}}$: is the radius of gyration

If the geometrical scale is 1/n and the critical velocity for the bridge is n times bigger than the one of the model, then it will be:

$$\omega_{model} = \frac{n}{V}\omega_{bridge}$$
$$m_{model} = \frac{1}{n^2} m_{bridge}$$
$$c = \frac{1}{V^2} m_{bridge}\,\omega^2_{bend_{\,bridge}} \cdot \ell$$

where ℓ is the length of the partial scaled model and c is the constant of the springs (Fig. **10a**) required for the suspension of the deck.

Finally, we will also have:

$$e = \frac{1}{n} \cdot \frac{\omega_{tor_{bridge}}}{\omega_{bend_{bridge}}} \cdot r_{bridge}$$

where 2e is the distance between the springs, that will be placed at each end of the deck, $\omega_{tor_{bridge}}$ is the eigenfrequency of the bridge for lateral-torsional vibration, and $\omega_{bend_{bridge}}$ is the eigenfrequency for bending vibration.

Full-Scale Models

The similarity principles valid for construction of a full-scale model are the same with those of aeronautics.

The two systems, the model and the real structure, are the same from dynamical point of view if they have the same geometrical shape and their natural properties are such that, all acting forces and secondary actions are characterized by a single scale.

The following nondimensional sizes for the bridge and the model should be equal:

$$\frac{\sigma}{\rho}, \quad \frac{E}{\rho v^2}, \quad \delta_S, \quad \frac{g(2e)}{v^2}, \quad \frac{\upsilon(2e)}{v}$$

where:

σ: is the material density

E: is the modulus of elasticity

δ_S: is the elastic damping of the system

ρ: is the air density

v: is the viscoelastic coefficient

υ: is the wind velocity

g: is the gravitational constant

(2e): is the width of the bridge deck

If the geometrical scale is 1/n, then it will be:

$$\upsilon_{model} = \frac{1}{\sqrt{n}} \upsilon_{bridge}$$

and

$$\omega_{model} = \sqrt{n} \cdot \omega_{bridge}$$

where ω are the corresponding eigenfrequencies.

REFERENCES

[1] G. Airy "On the use of the suspension bridge with stiffened roadway for railway and other bridges of great span", 1986, Minutes of the proceedings, Institution of Civil Engineers, Vol. 26, 67.

[2] D. Dicker "Aerodynamic stability of H sections", 1966, Journal of the Engineering Mechanics Division, A.S.C.E., Vol. 92, No EM2, Proc. Paper 4764.

[3] F. Theodorsen "General theory of aerodynamic instability and the mechanism of flutter", 1934, National advisory committee for aeronautics. Technical Report 496, U.S. Government Printing Office, Washington P.C.

[4] Y.C. Fung "An Introduction to the theory of aeroelasticity", 1993, John Wiley and Sons Inc., New York.

[5] R.H. Scanlan, J.J. Tomko "Airfoil and bridge deck flutter derivatives", 1971, Journal of the Engineering Mechanics

Division, A.S.C.E., Vol. 97, No EM6, Proc. Paper 8609.

[6] R.H. Scanlan, K.S. Budlong "Flutter and aerodynamic response considerations for bluff objects in a smooth flow", 1972, Prceedings, International Union of Theoretical and Applied Mechanics, International Association for Hydraulic Research Symposium on Flow-Induced Structural Vibrations, Karlsruhe, Germany.

[7] R.H. Scallan, J.G. Beliveau, K.S. Budlong "Indicial aerodynamic function for bridge decks", 1974, Journal of Engineering Mechanics Division, A.S.C.E., EM4.

[8] R.H. Scanlan, N.P. Jones "Aeroelastic analysis of cable-stayed bridges", 1990, Journal of Structures Engineering, A.S.C.E. M6.

[9] A. Jain, N. Jones, R. Scanlan "Coupled aeroelastic and aerodynamic response analysis of long-span bridges", 1996, Journal of Wind Engineering and Industrial Aerodynamics, 60.

[10] E. Simin, R. Scanlan "Wind effects on structures", 1996, J.Wiley & Sons, New York.

[11] L. Singh "Experimental determination of aeroelastic and aerodynamic parameters of long-span bridges", 1997, PhD thesis, Johns Hopkins University, Baltimore.

[12] A. Jain "Multi-mode aeroelastic and aerodynamic analysis of long-span bridges", 1997, PhD thesis, John Hopkins University, Baltimore.

[13] A. Hirai "Aerodynamische Stabilität von Hängebrücken unter Wind Belastung", 1956, Baningenieur Heft 11.

CHAPTER 8

Dynamic Instability Problems

Abstract: This chapter deals with dynamic instability aspects of cable-stayed and suspension bridges. The technique for determining the instability regions of frequencies as well as the boundaries of critical frequencies based on the solution of Mathieu-Hill equations is presented in detail. In the cases examined, instability phenomena in pylons, bridge decks and cables are analytically presented.

GENERAL

Over the last years, a new field of the applied elasticity theory, called **theory of dynamic stability of elastic systems**, has been developed.

The main objectives studied in this field of elasticity are mostly related with vibration and stability problems of flexible systems. As it happens in a number of fields in mechanics, the theory of Dynamic Stability goes now through a period of particular growth.

On the other hand, bridges are structures that are mostly subjected to dynamic loads and it is evident that their stress and strain states as well as the problems rising from these states of instability have a dynamic character.

The main characteristic of the dynamic instability problems is that instability of a structure may occur even if the biggest stress values are smaller than the corresponding static ones, *i.e.* the buckling load for the same structure.

It has been proved that if some specific relations between the frequency θ of the applied loading and the natural frequency ω of the structure are valid, the structure becomes dynamically unstable. The relation expressing the frequencies for which dynamic instability occurs accompanied by rapid growth of big oscillation amplitudes is quite different than the one expressing the critical frequencies in classic resonance problems.

Many examples of structures or structural parts exist for which dynamic instability is possible to occur (see Fig. **1**).

As in the special case where a statically applied load may cause loss of stability of the structure, a dynamical load of same manner will cause loss of dynamic stability.

Such a dynamic load is characterized by the fact that it is usually treated as a parameter in the equation of dynamic equilibrium (or equation of motion). Such a loading is called parametric loading and this term is most suitable because it shows the relation to the phenomenon of parametric resonance (or parametric excitation).

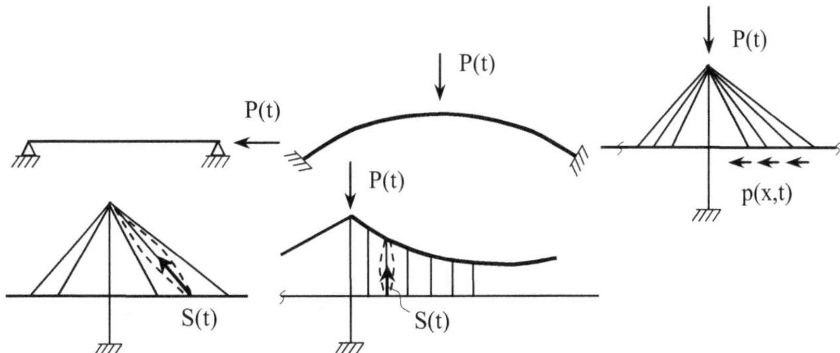

Figure 1: Structures where dynamic instability may occur

Introducing the significance of parametric loading, we may characterize the "theory of dynamic stability of elastic systems" as the study of oscillations caused by pulsating parametric loads.

George T. Michaltsos and Ioannis G. Raftoyiannis

Nevertheless, it would be more accurate to speak about loadings that are parametric for some deformation types instead of generally speaking about parametric loads. For example, an axially compressive force applied onto a straight bar is a parametric load regarding the transverse deformations of the bar and not the longitudinal ones.

Most of the problem cases of dynamic instability in bridges result to the so-called Mathieu - Hill equations, which will be examined directly in the following paragraph, or in systems of differential equations, examined later on.

DEFORMATIONS IN PLANE – MATHIEU-HILL EQUATIONS

For a better understanding of the problem let us consider the simple beam in Fig. **2**, which is subjected to the compressive load P(t), varying with respect to time as follows:

$$P(t) = P_\alpha + P_\beta \cdot \Phi(t) \tag{1}$$

This problem is one of the most usual and simple dynamic stability problems in bridge engineering, and was first analyzed in this manner by Beliaev [10].

The equation of dynamic equilibrium will be:

$$EI_y \frac{d^4 w}{dx^4} + P(t)\frac{d^2 w}{dx^2} + m\frac{d^2 w}{dt^2} = 0 \tag{2}$$

In order to solve eq (2), we will seek a solution in the form:

$$w(x,t) = W_n(x) \cdot f_n(t) \tag{3}$$

where $f_n(t)$ are time functions to be determined and $W_n(x) = \sin\frac{n\pi x}{\ell}$ (n=1,2,3,...) is a complete set of orthogonal functions that satisfy the boundary conditions of the simply supported beam in Fig. **2**, and as we observe these functions are the shape functions or otherwise the modal shapes of the freely vibrating beam in bending. Introducing eq (3) into eq (2), we obtain:

$$\left\{EI_y \frac{n^4\pi^4}{\ell^4} f_n(t) - \left[P_\alpha + P_\beta \Phi(t)\right] \cdot \frac{n^2\pi^2}{\ell^2} f_n(t) + m\frac{d^2 f_n}{dt^2}\right\} \sin\frac{n\pi x}{\ell} = 0 \tag{4}$$

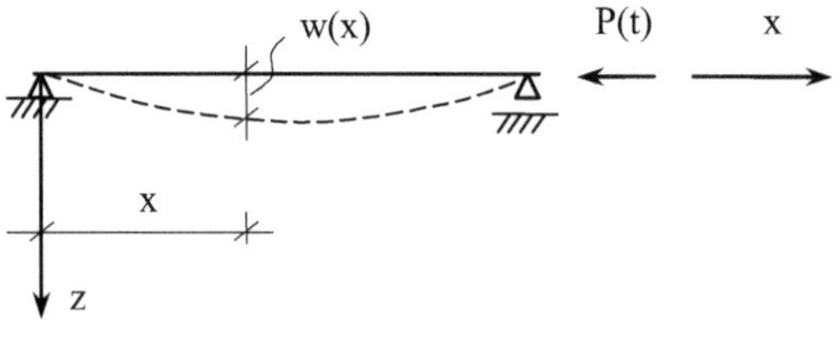

Figure 2: Simply supported beam under time varying axially compressive load

In order that the solution eq (3) satisfies eq (2), a necessary and sufficient condition is that the quantity inside the brackets of eq (4) must be zero for any time t. In other words:

$$\frac{d^2 f_n}{dt^2} + \frac{1}{m} \cdot \left[EI_y \frac{n^4\pi^4}{\ell^4} - \left[P_\alpha + P_\beta \Phi(t)\right]\frac{n^2\pi^2}{\ell^2}\right] f_n = 0 \quad \text{or} \quad \frac{d^2 f_n}{dt^2} + \frac{EI_y}{m} \cdot \frac{n^4\pi^4}{\ell^4} \cdot \left[1 - \frac{P_\alpha + P_\beta \Phi(t)}{EI_y \dfrac{n^2\pi^2}{\ell^2}}\right] f_n = 0$$

and since $\omega_n^2 = \frac{n^4\pi^4 EI_y}{m\ell^4}$ are the eigenfrequencies for the freely vibrating beam in Fig. **2**, we will have:

$$\left.\begin{aligned} &\frac{d^2 f_n}{dt^2} + \omega_n^2\left[1 - \frac{P_\alpha + P_\beta \Phi(t)}{P_{en}}\right] f_n = 0 \\ \text{where:}\qquad &P_{en} = \frac{n^2\pi^2 EI_y}{\ell^2} \end{aligned}\right\} \tag{5a,b}$$

P_{en} is the nth eigenvalue for buckling or the Euler buckling load, corresponding to the nth modal shape. The above relation eq (5a), can be also written as follows:

$$\left.\begin{aligned} &\frac{d^2 f_n}{dt^2} + \Omega_n^2\left(1 - 2\mu_n \Phi(t)\right) f_n = 0 \\ \text{where:}\qquad &\Omega_n = \omega_n\sqrt{1 - \frac{P_\alpha}{P_{en}}} \\ &\mu_n = \frac{P_\beta}{2(P_{en} - P_\alpha)} \end{aligned}\right\} \tag{6a,b,c}$$

Ω_n is the cyclic frequency of the freely vibrating beam shown in Fig. **2**, which is loaded axially by the force P_α, while μ_n is a non-dimensional quantity called **excitation parameter**. Since eq (6a) is valid for any value of n, we may disregard the index and hence, eq (6a) becomes:

$$\frac{d^2 f}{dt^2} + \Omega^2\left[1 - 2\mu\Phi(t)\right] f = 0 \tag{7}$$

Equation (7) is known as the Mathieu equation, while sometimes it is also met as Hill equation in the bibliography. Equations Mathieu-Hill are met in various problems of Physics and Engineering Science. Some problems of theoretical Physics result to such equations, and especially the problem of distribution of electromagnetic waves or the quantum theory of metals, the problem of movement of electrons to a crystal mesh and many others.

The Mathieu-Hill equation is also met in relative research on the theory of oscillations, the parametric excitation of electric oscillators, in problems of celestial Mechanics and especially in the theory of relative movement of Earth – Moon.

Rich bibliography exists related to these equations. An extensive presentation and investigation of various types of Mathieu – Hill equations, can be found in various textbooks, as for example in Mc-Lachlan [2].

One of the most interesting characteristics of these equations is that for some specific relations between their factors, they give unbounded solutions. Using as parameters the values of μ and Ω, one can derive the regions of dynamic instability for any natural problem.

The determination of the regions of dynamic instability is one of the most important problems in the theory of Dynamic Stability. It is worth, however, to study and define some interesting attributes of Mathieu-Hill equations.

The Basic Attributes of Mathieu – Hill Equations [3]

1st Attribute **(Floquet Solutions)**

Let us consider the general case of eq (7), where Φ(t) is a periodical function with period:

$$T = \frac{2\pi}{\theta} \tag{8}$$

We assume that function Φ(t) can be represented by a Fourier series in the form:

$$\Phi(t) = \sum_{n=1}^{\infty} \left(\gamma_n \cos n\theta t + \delta_n \sin n\theta t\right) \tag{9}$$

For a periodical function with period T, the following relation is valid:

$$\Phi(t+T) = \Phi(t)$$

Hence, if $f(t)$ is a solution of eq (7), then $f(t+T)$ will be also a solution of eq (7).

Let $f_1(t)$ and $f_2(t)$ be two linearly independent solutions of eq (7). Then, $\overset{*}{f}_1(t+T)$ and $\overset{*}{f}_2(t+T)$, will also be solutions which can be written as the summation of the initial solutions in the form:

$$\left.\begin{aligned} \overset{*}{f}_1(t+T) &= \alpha_{11} f_1(t) + \alpha_{12} f_2(t) \\ \overset{*}{f}_2(t+T) &= \alpha_{21} f_1(t) + \alpha_{22} f_2(t) \end{aligned}\right\} \tag{10}$$

where α_{ij} are constants of the above linear transformation. If $\alpha_{12} = \alpha_{21} = 0$, we conclude to a simpler linear transformation, also known as **"Floquet transformation"** or **"Floquet solutions"**. If we set: $\alpha_{11} = \rho_1$ and $\alpha_{22} = \rho_2$ we will obtain:

$$\left.\begin{aligned} \overset{*}{f}_1(t+T) &= \rho_1 f_1(t) \\ \overset{*}{f}_2(t+T) &= \rho_2 f_2(t) \end{aligned}\right\} \tag{11}$$

It is well known from the linear transformation theory that each transformation in the form of eq (10) takes the form of eq (11), if the so-called **"characteristic equation"** is valid:

$$\begin{vmatrix} (\alpha_{11} - \rho) & \alpha_{12} \\ \alpha_{21} & (\alpha_{22} - \rho) \end{vmatrix} = 0 \tag{12}$$

2nd Attribute

The characteristic equation (12) defines the character of the solution of a Mathieu-Hill equation. Let us examine now, how this characteristic equation is formulated.

We again consider $f_1(t)$ and $f_2(t)$, which are her two linearly independent solutions of eq (7), satisfying the initial conditions:

$$\left.\begin{aligned} f_1(0) &= 1 \quad , \quad f_1'(0) = 0 \\ f_2(0) &= 0 \quad , \quad f_2'(0) = 1 \end{aligned}\right\} \tag{13}$$

Differentiating eq (10) and setting t = 0, in both eq (10) as well as the resulting expressions, and employing eqs (13), we will get:

$$\left.\begin{aligned} \alpha_{11} &= f_1(T) \\ \alpha_{21} &= f_2(T) \\ \alpha_{12} &= f_1'(T) \\ \alpha_{22} &= f_2'(T) \end{aligned}\right\} \tag{14}$$

and hence the characteristic equation (12) will take the form:

$$\begin{vmatrix} f_1(T)-\rho & f_1'(T) \\ f_2(T) & f_2'(T)-\rho \end{vmatrix} = 0$$

or in a concise form:

$$\left.\begin{aligned} &\rho^2 - 2A\rho + B = 0 \\ \text{where:}\quad & A = \frac{f_1(T) + f_2'(T)}{2} \\ & B = f_1(T)\cdot f_2'(T) - f_2(T)\cdot f_1'(T) \end{aligned}\right\} \tag{15}$$

But since f_1 and f_2 satisfy eq (7), *i.e.*:

$$\begin{aligned} f_1'' + \Omega^2(1-2\mu\Phi)f_1 &= 0 \\ f_2'' + \Omega^2(1-2\mu\Phi)f_2 &= 0 \end{aligned}$$

we easily obtain that:

$$f_2 f_1'' - f_1 f_2'' = 0$$

or after integration:

$$f_2 f_1' - f_1 f_2' = \text{const.}$$

If we use the initial conditions eq (13), we will obtain:

$$B = f_1(T)f_2'(T) - f_2(T)f_1'(T) = 1$$

and hence, eq (15a) becomes:

$$\rho^2 - 2A\rho + 1 = 0 \tag{16}$$

where the roots ρ_1 and ρ_2 of the above are obviously related as follows:

$$\rho_1 \cdot \rho_2 = 1 \tag{17}$$

Thus, we observe that the roots of characteristic equation (and consequently its factors) do not depend from the choice of solutions $f_i(t)$.

3rd Attribute

In the 1st attribute it has been proven that between the special solutions of eq (7), two linearly independent solutions $f_\kappa(t)$ $(\kappa = 1,2)$ exist, that satisfy eq (11), *i.e.*:

$$\overset{*}{f}_\kappa(t+T) = \rho_\kappa f_\kappa(t)\ , \qquad (\kappa = 1,2)$$

These solutions can be written in the form:

$$f_\kappa(t) = X_\kappa(t)\cdot e^{\frac{t}{T}\cdot \ell n\rho_\kappa}\ , \quad (\kappa = 1,2) \tag{18}$$

where $X_\kappa(t)$ is a periodical function with period T. Provided however that the functions are periodical, the following relation will be also valid:

$$f_\kappa(t+T) = X_\kappa(t+T)e^{\frac{t+T}{T}\cdot \ell n\rho_\kappa} = X_\kappa(t)e^{\left(\frac{t}{T}+1\right)\cdot \ell n\rho_\kappa} = \rho_\kappa f_\kappa(t)$$

Since the solutions of eq (12) may also be complex, and based on the known attribute of complex functions: $\ell n\rho = \ell n|\rho| + i\cdot \arg\rho$, eq (18) can be also written in the form:

$$\left.\begin{aligned} &f_\kappa(t) = \varphi_\kappa(t)\cdot e^{\frac{t}{T}\cdot \ell n|\rho_\kappa|} \\ \text{where:}\quad &\varphi_\kappa(t) = X_\kappa(t)\cdot e^{\frac{i\cdot t}{T}\cdot \arg\rho_\kappa} \end{aligned}\right\} \qquad \textbf{(19a,b)}$$

If the root $|\rho_\kappa|$ is bigger than 1, then the solution of eq (19a) will be an unbounded exponential function. If $|\rho_\kappa|$ is less than 1, then the solution will decrease as t increases. If $|\rho_\kappa| = 1$, the solution will be a periodical (or periodical – like) function, *i.e.* bounded with respect to time.

Let: $|A| = \frac{1}{2}\cdot|f_1(T) + f_2'(T)| > 1$. Then, eq (16) will have real roots and the one of them, due to eq (17), will be bigger than 1. Hence, the general integral of eq (7) will be unbounded and increasing with respect to time:

$$f(t) = c_1X_1(t)\cdot e^{\frac{t}{T}\cdot \ell n\rho_1} + c_2X_2(t)\cdot e^{\frac{t}{T}\cdot \ell n\rho_2}$$

If instead: $|A| = \frac{1}{2}\cdot|f_1(T) + f_2'(T)| < 1$, then the characteristic of eq (16) will have complex conjugated roots which, since their product is equal to 1 their value will be also equal to 1.

The case of complex roots corresponds to the region of unbounded solutions. Curves separating the regions of bounded and unbounded solutions should obviously satisfy the condition: $|A| = 1$ or:

$$|f_1(T) + f_2'(T)| = 2 \qquad \textbf{(20)}$$

This equation can be used for the determination of the boundaries of dynamic instability regions.

Nevertheless, in order to employ the above equation, we should know the solutions f_1 and f_2 for at least the first period of oscillation. This method leads to cumbersome calculation difficulties. Only in certain special cases, it is possible to integrate an equation in the form of eq (7) in terms of simple functions.

Critical Eigenfrequencies Equations - Regions of Dynamic Instability

In this paragraph, we will present a method for the determination of the instability regions for the case of an arbitrary periodical function in equation (9).

In attribute 3, it has been proven, it has been proven (attribute 3) that the region of real characteristic roots coincides with the region of increasing unbounded solutions of the differential equation (7), while the region of complex characteristic roots of the factors ρ_i corresponds to bounded and (usually) periodical solutions. On the boundary lines that separate the regions of real and complex solutions, multiple roots exist. From the condition (17), we have concluded that such roots can be either $\rho_1 = \rho_2 = 1$ or $\rho_1 = \rho_2 = -1$.

In the first case, as concluded from eq (11), the solution of the differential equation will be periodical with period: $T = \frac{2\pi}{\theta}$, while in the second case with period: 2T.

Consequently, the regions of bounded solutions and the regions of unbounded solutions (regions of stability-instability, respectively) are separated by periodical solutions with periods T and 2T. More precisely, two solutions with identical periods define an instability region, while two solutions with different periods define a stability region.

The last attribute has been proven in a strictly mathematical way by Strutt [4]. If a region with real roots (instability region) lies between $\rho = 1$ and $\rho = -1$, then the root $\rho=0$ must be lying between those two roots, and since $\rho_1\rho_2=1$ the root $\rho = \infty$ must be lying between those two roots as well, due the continuous dependence of the real roots on the factors of the differential equation, which is impossible.

Hence, the roots $\rho=1$ and $\rho=-1$ enclose a region with complex roots, that corresponds to an instability region.

From the above investigation, we can conclude that the determination of the boundaries of dynamic instability regions is reduced in finding the conditions under which the given differential equation has periodical solutions with periods T or 2T. The fact that periodical solutions exist and that these can be analyzed into Fourier series is well known. This fact allows us to seek directly periodical solutions for eq (7) in the form of trigonometric series.

Equations for Boundary or Critical Eigenfrequencies

We shall apply this method in the case of a Mathieu differential equation:

$$f'' + \Omega^2\left(1 - 2\mu\cos\theta t\right) f = 0 \tag{21}$$

We seek a periodical solution with period 2T in the form:

$$f(t) = \sum_{\kappa=1,3,5,...}^{\infty}\left(\alpha_\kappa \cdot \sin\frac{\kappa\theta t}{2} + \beta_\kappa \cdot \cos\frac{\kappa\theta t}{2}\right) \tag{22a}$$

Introducing eq (22a) into eq (21) and setting equal the factors of same terms $\sin\frac{\kappa\theta t}{2}$ and $\cos\frac{\kappa\theta t}{2}$, we obtain the following linear homogenous system of equations with unknowns α_κ and β_κ:

$$\left.\begin{aligned}
&\left(1+\mu-\frac{\theta^2}{2\Omega^2}\right)\alpha_1 - \mu\alpha_3 = 0\\
&\left(1-\frac{\kappa^2\theta^2}{4\Omega^2}\right)\alpha_\kappa - \mu(\alpha_{\kappa-2}+\alpha_{\kappa+2}) = 0 \qquad (\kappa = 3,5,7,.....)\\
&\left(1-\mu-\frac{\theta^2}{2\Omega^2}\right)\beta_1 - \mu\beta_3 = 0\\
&\left(1-\frac{\kappa^2\theta^2}{4\Omega^2}\right)\beta_\kappa - \mu(\beta_{\kappa-2}+\beta_{\kappa+2}) = 0 \qquad (\kappa = 3,5,7,.....)
\end{aligned}\right\} \tag{23}$$

We observe that the first system contains only the factors α_κ, while the one second contains only the factors β_κ. These systems of equations are identical, except from the first equation in each system, where only the sign of μ differs. In order that a nontrivial solution exists, the determinant of the factors of the unknowns must be zero. By combining the two conditions from eq (23), we have:

$$\begin{vmatrix}
\left(1\pm\mu-\frac{\theta^2}{4\Omega^2}\right) & -\mu & 0 & \cdots\\
-\mu & \left(1-\frac{9\theta^2}{4\Omega^2}\right) & -\mu & \cdots\\
0 & -\mu & \left(1-\frac{25\theta^2}{4\Omega^2}\right) & \cdots\\
\cdots & \cdots & \cdots & \cdots
\end{vmatrix} = 0 \tag{24}$$

For the determination of periodical solutions with period T, we follow a similar method. Introducing into eq (21) a solution of the form:

$$f(t) = b_0 + \sum_{\kappa=2,4,6,...}^{\infty} \left(\alpha_\kappa \sin\frac{\kappa\theta t}{2} + \beta_\kappa \cos\frac{\kappa\theta t}{2} \right) \qquad \textbf{(22b)}$$

and setting again equal the factors of same terms $\sin\frac{\kappa\theta t}{2}$ and $\cos\frac{\kappa\theta t}{2}$, we obtain the following linear homogenous system of equations:

$$\left.\begin{aligned}
&\left(1-\frac{\theta^2}{\Omega^2}\right)\alpha_2 - \mu\alpha_4 = 0\\
&\left(1-\frac{\kappa^2\theta^2}{4\Omega^2}\right)\alpha_\kappa - \mu(\alpha_{\kappa-2}+\alpha_{\kappa+2}) = 0 \quad (\kappa = 2,4,6,.....)\\
&\beta_0 - \mu\beta_2 = 0\\
&\left(1-\frac{\theta^2}{\Omega^2}\right)\beta_2 - \mu(2\beta_0+\beta_4) = 0\\
&\left(1-\frac{\kappa^2\theta^2}{4\Omega^2}\right) - \mu(\beta_{\kappa-2}+\beta_{\kappa+2}) = 0 \qquad (\kappa = 4,6,.....)
\end{aligned}\right\} \qquad \textbf{(25)}$$

Setting again the determinant of the factors of the unknowns equal to zero for both the above systems, we obtain the following relations:

$$\begin{vmatrix}
\left(1-\frac{\theta^2}{\Omega^2}\right) & -\mu & 0 & \cdots\cdots \\
-\mu & \left(1-\frac{4\theta^2}{\Omega^2}\right) & -\mu & \cdots\cdots \\
0 & -\mu & \left(1-\frac{9\theta^2}{\Omega^2}\right) & \cdots\cdots \\
\cdots\cdots & \cdots\cdots & \cdots\cdots & \cdots\cdots
\end{vmatrix} = 0 \qquad \textbf{(26)}$$

$$\begin{vmatrix}
1 & -\mu & 0 & 0 & \cdots\cdots \\
-2\mu & \left(1-\frac{\theta^2}{\Omega^2}\right) & -\mu & 0 & \cdots\cdots \\
0 & -\mu & \left(1-\frac{4\theta^2}{\Omega^2}\right) & -\mu & \cdots\cdots \\
0 & 0 & -\mu & \left(1-\frac{9\theta^2}{\Omega^2}\right) & \cdots\cdots \\
\cdots\cdots & \cdots\cdots & \cdots\cdots & \cdots\cdots & \cdots\cdots
\end{vmatrix} = 0 \qquad \textbf{(27)}$$

The above equations (24), (26) and (27), that relate the frequencies of the applied load to the natural frequency of the beam and the magnitude of the applied force, are called **"equations of boundary or critical frequencies"**, where the boundary or critical frequencies are the frequencies θ of the external loads correspond to the boundaries of the instability regions.

Equation (24) gives the instability regions that are limited to periodical solutions with period 2T, while eqs (26) and (27) give the instability regions that are limited to periodical solutions with period T.

Convergence of the Determinant Terms

The determinants (24), (26) and (27) contain an infinite number of terms and consequently rises the question of their convergence [5]. It has been proven that these determinants fall into the known category of converging determinants and are called normal determinants.

A determinant:

$$\Delta = \begin{vmatrix} (1+d_{11}) & d_{12} & d_{13} & \cdots \\ d_{21} & (1+d_{22}) & d_{23} & \cdots \\ d_{31} & d_{32} & (1+d_{33}) & \cdots \\ \cdots & \cdots & \cdots & \cdots \end{vmatrix} \qquad \textbf{(28)}$$

is called normal when the double series $\sum_{i=1}^{\infty}\sum_{\kappa=1}^{\infty} d_{i\kappa}$ is fully converging. We shall examine for instance the determinant (24). Multiplying line κ (κ=1,2,3,...) by $\frac{-4\Omega^2}{(2\kappa-1)^2\theta^2}$ the expression takes the form of eq (28), where:

$$d_{\kappa\kappa} = \begin{cases} -\frac{4\Omega^2}{\theta^2}(1\pm\mu) & \text{for } \kappa = 1 \\ -\frac{4\Omega^2}{(2\kappa-1)^2\theta^2} & \text{for } \kappa \neq 1 \end{cases}$$

$$d_{i\kappa} = \begin{cases} -\frac{4\Omega^2}{(2\kappa-1)^2\theta^2} - \mu & \text{for } i = \kappa \pm 1 \\ 0 & \text{for } i \neq \kappa \pm 1 \end{cases}$$

It can be shown that the double series $\sum_{i=1}^{\infty}\sum_{\kappa=1}^{\infty} d_{i\kappa}$, is fully converging. Indeed, we obtain the inequality:

$$\sum_{i=1}^{\infty}\sum_{\kappa=1}^{\infty}\left|d_{i\kappa}\right| < \frac{4\Omega^2}{\theta^2}(1+2\mu)\sum_{\kappa=1}^{\infty}\frac{1}{(2\kappa-1)^2}$$

for which we know that the series of the right hand side is converging.

Dynamic Instability Regions

The determinants in eqs (24), (26) and (27) contain an infinite number of terms. Any of the above determinants (as well as their general forms that will be studied later on), can take the following form:

$$\begin{vmatrix} \alpha_1 & 1 & 0 & 0 & \cdots \\ 1 & \alpha_2 & 1 & 0 & \cdots \\ 0 & 1 & \alpha_3 & 1 & \cdots \\ \cdots & \cdots & \cdots & \cdots & \cdots \end{vmatrix} = 0$$

The first basic determinant (or first order determinant), is obviously the equation: $\alpha_1=0$.

The second order determinant is: $\begin{vmatrix} \alpha_1 & 1 \\ 1 & \alpha_2 \end{vmatrix} = 0$ or $\alpha_1\alpha_2 - 1 = 0$ or $\alpha_1 = \frac{1}{\alpha_2}$.

The third order determinant is: $\begin{vmatrix} \alpha_1 & 1 & 0 \\ 1 & \alpha_2 & 1 \\ 0 & 1 & \alpha_3 \end{vmatrix} = 0$ or $\alpha_1(\alpha_2\alpha_3 - 1) - \alpha_3 = 0$ or $\alpha_1 = \dfrac{1}{\alpha_2 - \dfrac{1}{\alpha_3}}$.

In general, we have:

$$\alpha_1 = \cfrac{1}{\alpha_2 - \cfrac{1}{\alpha_3 - \cfrac{1}{\alpha_4 - \ldots\ldots}}} \tag{29}$$

Considering for example the determinant in eq (24) for which:

$$\alpha_1 = -\frac{1}{\mu}\left(1 \pm \mu - \frac{\theta^2}{4\Omega^2}\right)$$

$$\alpha_\kappa = -\frac{1}{\mu}\left(1 - \frac{(2\kappa - 1)^2 \theta^2}{4\Omega^2}\right) \quad , \quad (\kappa \geq 2)$$

eq (29) gives: $\dfrac{\theta^2}{4\Omega^2} = 1 \pm \mu - \cfrac{\mu^2}{1 - \cfrac{9\theta^2}{4\Omega^2} - \cfrac{\mu^2}{1 - \cfrac{25\theta^2}{4\Omega^2} - \ldots\ldots}}$

We shall now estimate the error accumulating if we consider only parts of the determinant with infinite terms (*i.e.*, 1st order determinant, 2nd order determinant, 3rd order determinant, etc) in order to determine the corresponding regions of critical frequencies.

If we introduce the first (approximate) value that is computed from the basic determinant eq(24) of chapter 8 into the bottom diagonal term of the second order determinant and evaluate the resulting expression, we will obtain:

$$\theta = 2\Omega\sqrt{1 \pm \mu + \frac{\mu^2}{8 \pm 9\mu}}$$

The last term inside the root takes into account the correction due to the first order determinant. This correction is increasing as μ increases, it is though less than 1% [3] for values of $\mu \leq 0,5$ and it is always less than 3% for any value of $\mu \leq 1$.

We are henceforth in position to study the regions of dynamic instability that result from the determinants (24), (26) and (27) related to the Mathieu equation (21).

The main instability region is obtained if we set the top left term (basic determinant) in eq (24) equal to zero. Then, we will have: $1 \pm \mu - \dfrac{\theta^2}{4\Omega^2} = 0$ and consequently, the instability region boundaries will be given by:

$$\theta = 2\Omega\sqrt{1 \pm \mu} \tag{30}$$

In order to determine the boundaries of the second instability region, we consider equations (26) and (27), and the second order determinants give:

$$\begin{vmatrix} 1-\dfrac{\theta^2}{\Omega^2} & -\mu \\ -\mu & 1-\dfrac{4\theta^2}{\Omega^2} \end{vmatrix} = 0 \quad , \quad \begin{vmatrix} 1 & -\mu \\ -2\mu & 1-\dfrac{\theta^2}{\Omega^2} \end{vmatrix} = 0$$

These equations give the following expressions for the second instability region boundaries:

$$\left.\begin{aligned} \theta &= \Omega \cdot \sqrt{1+\frac{1}{3}\mu^2} \\ \theta &= \Omega \cdot \sqrt{1-2\mu^2} \end{aligned}\right\} \tag{31}$$

In order to determine the limits of the third instability region, we consider equation (24), and the second order determinant gives:

$$\begin{vmatrix} \left(1 \pm \mu - \dfrac{\theta^2}{4\Omega^2}\right) & -\mu \\ -\mu & \left(1-\dfrac{9\theta^2}{4\Omega^2}\right) \end{vmatrix} = 0$$

from which we obtain the following expressions for the third instability region boundaries:

$$\theta = 2\Omega \cdot \sqrt{\frac{(10 \pm 9\mu) - \sqrt{(10 \pm 9\mu)^2 + 36(\mu^2 \mp \mu - 1)}}{18}} \tag{32}$$

The first three instability regions on the plane $\left(\mu\ , \dfrac{\theta}{2\Omega}\right)$ are shown as shaded areas in Fig. **3**. We observe that the area of each dynamic instability region is decreasing as the order of the region increases, with the first (basic) instability region having the biggest area.

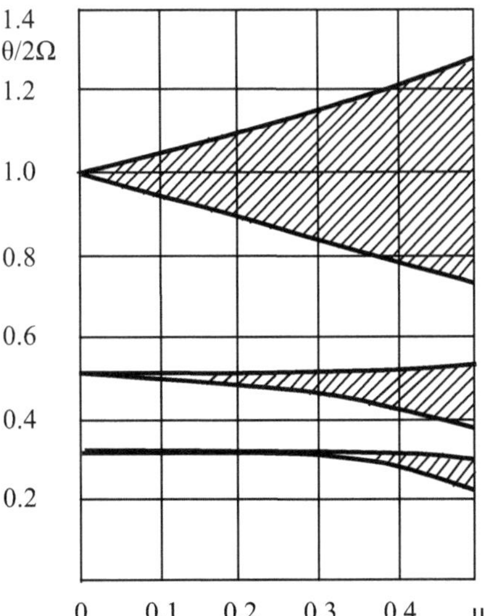

Figure 3: Dynamic instability regions

The Case of Damping

In the case of damping, equation (2) acquires an additional term, namely $c\dot{w}$ and hence, following the same procedure as in paragraph 2 of chapter 8, equation (7) becomes:

$$\left.\begin{aligned} &\frac{d^2 f}{dt^2}+\lambda\frac{df}{dt}+\Omega^2\left[1-2\mu\Phi(t)\right]f=0 \\ &\text{where: } \lambda=\frac{c}{m} \end{aligned}\right\} \tag{33}$$

Employing the transformation:

$$f(t)=e^{-\frac{\lambda \cdot t}{2}}\rho(t) \tag{34}$$

equation (33), after some manipulation, becomes:

$$\frac{d^2\rho}{dt^2}+\left[\Omega^2-\frac{\lambda^2}{4}-2\mu\Omega^2\Phi(t)\right]\rho=0$$

which can be finally written as:

$$\left.\begin{aligned} &\rho''+\bar{\Omega}^2[1-2\bar{\mu}\Phi(t)]\rho=0 \\ &\text{where: } \quad \bar{\Omega}^2=\Omega^2-\frac{\lambda^2}{4} \\ &\text{and: } \quad \bar{\mu}=\frac{4\Omega^2}{4\Omega^2-\lambda^2}\mu \end{aligned}\right\} \tag{35a,b,c}$$

We observe that equation (35a) can result from the initial eq (7) by replacing Ω by $\bar{\Omega}$ and μ by $\bar{\mu}$, as given in eqs (35b,c), respectively.

Then, the instability regions are also given by Fig. **3**, in which the values $\frac{\theta}{2\Omega}$ and μ are substituted by $\frac{\theta}{\sqrt{4\Omega^2-\lambda^2}}$ and $\left(\frac{4\Omega^2}{4\Omega^2-\lambda^2}\mu\right)$, respectively.

The Case of an Arbitrary Φ(t)

We shall now study the following equation:

$$f''+\Omega^2\left[1-2\mu\Phi(t)\right]f=0 \tag{36}$$

We again seek a periodical solution in the form of series:

$$f(t)=\sum_{1,3,5,\ldots}^{\infty}\left(\alpha_\kappa\sin\frac{\kappa\theta t}{2}+\beta_\kappa\cos\frac{\kappa\theta t}{2}\right) \tag{37}$$

From the general theory of orthocanonical functions, it is known that a function Φ(t) can be analyzed into series as follows:

$$\left.\begin{aligned}&\Phi(t)=\sum_{\kappa=1}^{\infty} h_\kappa\cdot\cos\kappa\theta t\\ \text{where:}\quad & h_\kappa=\frac{\int_0^T \Phi(t)\cdot\cos(\kappa\theta t)dt}{\int_0^T \cos^2(\kappa\theta t)dt}\\ & T=\frac{2\pi}{\theta}\end{aligned}\right\}\tag{38}$$

and hence, eq (36) becomes:

$$\left.f''+\Omega^2\left(1-2\sum_{\kappa=1}^{\infty}\mu_\kappa\cos\kappa\theta t\right)f=0\quad\text{where:}\quad \mu_\kappa=\mu h_\kappa\ ,\quad \mu=\frac{P_\beta}{2(P_e-P_\alpha)}\right\}\tag{39}$$

Introducing eq (37) into eq (39a) and following the above procedure, we obtain the following equation for the boundary frequencies:

$$\begin{vmatrix}\left(1\pm\mu_1-\dfrac{\theta^2}{4\Omega^2}\right) & -(\mu_1\pm\mu_2) & -(\mu_2\pm\mu_3)\ldots\ldots\ldots\ldots\\ -(\mu_1\pm\mu_2) & \left(1\pm\mu_3-\dfrac{9\theta^2}{4\Omega^2}\right) & -(\mu_1\pm\mu_4)\ldots\ldots\ldots\ldots\\ -(\mu_2\pm\mu_3) & -(\mu_1\pm\mu_4) & \left(1\pm\mu_5-\dfrac{25\theta^2}{4\Omega^2}\right)\ldots\ldots\\ \ldots\ldots\ldots\ldots & \ldots\ldots\ldots\ldots & \ldots\ldots\ldots\ldots\end{vmatrix}=0\tag{40}$$

The Existence of a Second Part

Many dynamic stability problems are described by the following general form:

$$\ddot f+\Omega^2\left[1+2\mu A(z)\Phi(t)\right]f=p_0+pB(z)\Phi(t)\tag{41}$$

where A(z) and B(z) are orthogonal functions or a combination of orthogonal functions in the range from 0 to b. Integrating eq (41) from 0 to b, we will obtain:

$$\ddot f+\Omega^2\left[1+2\mu\cdot\frac{\int_0^b A(z)dz}{b}\cdot\Phi(t)\right]f=p_0+p\cdot\frac{\int_0^b B(z)dz}{b}\cdot\Phi(t)$$

or in a more concise form:

$$\left.\begin{aligned}&\ddot f+\Omega^2\left[1+2\bar\mu\Phi(t)\right]f=p_0+\bar p\Phi(t)\\ \text{with:}\quad &\bar\mu=\mu\cdot\frac{\int_0^b A(z)dz}{b}\\ &\bar p=p\cdot\frac{\int_0^b B(z)dz}{b}\end{aligned}\right\}\tag{42}$$

which is the same form as eq (36), but it also contains a second member.

Introducing again the solution eq (37) into eq(41) and analyzing $\Phi(t)$ into series according to eq(38), we conclude to the system of eq(40), which however has a second member as well. Consequently, the stability of the system, which is governed by eq (42), will also depend on the existence or non-existence of a solution of the above system, that is from matrices (40), the vanishing of which leads to the sought instability regions. We note finally, that equation (42), can also be studied *via* numerical integration, using a suitable model with a small initial value $\frac{df}{dt}$, so that the process starts.

Plane Deformations – The Most General Case

Let us consider the beam shown in Fig. **4**, with the following assumptions:

1. The beam has an initial constructional curvature $w_0(x)$ prior to loading imposition.
2. The beam in Fig. **4** is loaded by p, q_1, P with common time function, which is generally related to the dynamic deformation w of the beam, while q_2 can be an external loading with a different time function. Thus, the following relations are valid:

$$\left.\begin{aligned}&\overline{P}(t)=P_0+P\Phi(t)\\&\overline{p}(x,t)=p_0(x)+p(x)\Phi(t)\\&\overline{q}_1(x,t)=q_{01}(x)+q_1(x)\Phi(t)\\&\overline{q}_2(x,t)=q_{02}(x)+q_2(x)F(t)\end{aligned}\right\}\tag{43}$$

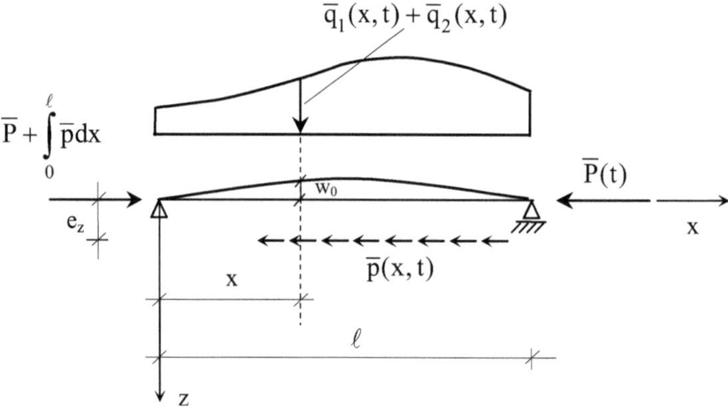

Figure 4: Beam under time varying axial and transverse loads

3. The load $\overline{p}(x,t)$ is eccentrically applied at distance e_z from axis x.

Under the above assumptions, the general equation of motion will be [6]:

$$\left.\begin{aligned}&(EI_y w'')''+(\overline{P}_x w')'+(\overline{P}_x w_0')'+(\overline{p}e_z)'+m\ddot{w}=\overline{q}_1+\overline{q}_2\\&\text{where: }\ \overline{P}_x=\overline{P}+\int_x^\ell \overline{p}(\lambda,t)d\lambda\end{aligned}\right\}\tag{44}$$

For a beam with constant cross-section, eq (44a) can be written as:

$$EI_y w''''+\overline{P}_x'(w+w_0)'+\overline{P}_x(w+w_0)''+e_z\overline{p}'+m\ddot{w}=\overline{q}_1+\overline{q}_2\tag{45}$$

Applying the differentiation general formula by Leibnitz, eq (44b) gives:

$$\overline{P}'_x = \overline{P}' + \frac{d}{dx}\int_x^{\ell} \overline{p}(\lambda,t)\, d\lambda = \frac{d}{dx}\int_x^{\ell} p(\lambda)\Phi(t)d\lambda =$$

$$= \Phi(t)\left\{ p(\ell)\frac{d\ell}{dx} - p(x)\frac{dx}{dx} + \int_x^{\ell}\frac{\partial p(\lambda)}{\partial x}d\lambda \right\} = -p(x)\Phi(t) = -\overline{p}(x,t)$$

and hence, eq (45) takes the form:

$$EI_y w'''' - \overline{p}(x,t)(w+w_0)' + \left(\overline{P} + \int_0^{\ell}\overline{p}(\lambda,t)d\lambda \right)\cdot (w+w_0)'' + e_z\overline{p}'(x,t) + m\ddot{w} = \overline{q}_1 + \overline{q}_2 \qquad \textbf{(46)}$$

The last equation, due to eq (43), becomes:

$$\left.\begin{aligned} & EI_y w'''' - \left[p_0(x) + p(x)\Phi(t)\right]\cdot (w+w_0)' + \\ & \quad + \left\{ P_0 + p_0(x) + \left(P + \int_x^{\ell} p(\lambda)d\lambda \right)\Phi(t) \right\}\cdot (w+w_0)'' + \\ & \quad + e_z\left[p_0'(x) + p'(x)\Phi(t)\right] - \left[q_{01}(x) + q_1(x)\Phi(t)\right] + m\ddot{w} = q_{02}(x) + q_2(x)F(t) \end{aligned}\right\} \qquad \textbf{(47)}$$

We seek a solution in the form:

$$w(x,t) = X_n(x) f_n(t)$$

where $X_n(x)$ is the shape function of the beam in Fig. **4** for bending.

Introducing the above into eq (47) and taking into account that X_n satisfies the equation of motion for free vibration, *i.e.*: $EI_y X'''' - m\omega_n^2 X_n = 0$ we finally get:

$$\left.\begin{aligned} & m\omega_n^2 X_n f_n - \left[p_0(x) + p(x)\Phi(x)\right]\left(X_n' f_n + w_0'\right) + \\ & \quad + \left\{ P_0 + p_0(x) + \left(P + \int_x^{\ell} p(\lambda)d\lambda \right)\Phi(t) \right\}\cdot \left(X_n'' f_n + w_0''\right) + \\ & \quad + e_z\left[p_0'(x) + p'(x)\Phi(t)\right] - \left[q_{01}(x) + q_1(x)\Phi(t)\right] + mX_n\ddot{f} = q_{02}(x) + q_2(x)F(t) \end{aligned}\right\} \qquad \textbf{(48)}$$

Multiplying both parts of the equation by X_n and integrating from 0 to ℓ, we obtain:

$$\left.\begin{aligned} & \ddot{f}_n + \left\{ \omega_n^2 + A + B\Phi(t) \right\} f_n = \Gamma + \Delta\Phi(t) + EF(t) \\ & \text{where:} \quad A = \frac{\int_0^{\ell}\left[P_0 + p(x)\right] X_n'' X_n dx - \int_0^{\ell} p_0(x) X_n' X_n dx}{m\int_0^{\ell} X_n^2 dx} \\ & \qquad\qquad B = \frac{\int_0^{\ell}\left[P + \int_x^{\ell} p(\lambda)d\lambda \right] X_n'' X_n dx - \int_0^{\ell} p(x) X_n' X_n dx}{m\int_0^{\ell} X_n^2 dx} \end{aligned}\right\} \qquad \textbf{(49)}$$

$$\left.\begin{aligned}
&\Gamma = \frac{\int_0^\ell p_0(x)w_0'X_n dx - \int_0^\ell [P_0 + p_0(x)]w_0''X_n dx - e_z\int_0^\ell p_0'(\lambda)X_n dx + \int_0^\ell q_{01}(x)X_n dx + \int_0^\ell q_{02}(x)X_n dx}{m\int_0^\ell X_n^2 dx}\\
&\Delta = \frac{\int_0^\ell p_0(x)w_0'X_n dx - \int_0^\ell\left[P + \int_x^\ell p(\lambda)d\lambda\right]w_0''X_n dx - e_z\int_0^\ell p'(x)X_n dx + \int_0^\ell q_1(x)X_n dx}{m\int_0^\ell X_n^2 dx}\\
&E = \frac{\int_0^\ell q_2(x)X_n dx}{m\int_0^\ell X_n^2 dx}
\end{aligned}\right\} \quad \textbf{(49)}$$

Omitting the indexes n, we will have:

$$\left.\begin{aligned}
&\ddot{f} + \Omega_\kappa^2\left[1 + 2\mu\Phi(t)\right]f = \Gamma + \Delta\Phi(t) + EF(t)\\
&\text{where:} \quad \Omega_\kappa^2 = \omega_\kappa^2 + A \ , \quad \mu = \frac{B}{2\Omega_\kappa^2}
\end{aligned}\right\} \quad \textbf{(50)}$$

The above equation has the form of eq (41).

DEFORMATION IN SPACE – SYSTEMS OF DIFFERENTIAL EQUATIONS

Many stability problems, where the structure is deformed in space, may lead to systems of differential equations, which can be written in matrix form as follows:

$$\Omega\frac{d^2 f}{dt^2} + \left[E - \alpha A - \beta\Phi(t)B\right]f = 0 \quad \textbf{(51)}$$

where Ω, A, B are known matrices with constant terms, E is the unit matrix, α and β are known constant factors, f is the unknown time function, and Φ(t) is a periodical function of time with period T, for which it is obviously valid that:

$$\Phi(t + T) = \Phi(t) \quad \textbf{(52)}$$

and can be expanded into a Fourier series.

The ith equation of the above system of differential equations, can be written in the following form:

$$\left.\begin{aligned}
&\frac{d^2 f_i}{dt^2} + \sum_{\kappa=1}^{n}\Phi_{i\kappa}(t)f_\kappa = 0 \ , \quad (i = 1,2,....,n)\\
&\text{where:} \quad (\Phi_{i\kappa}) = \Omega^{-1}\left[E - \alpha A - \beta\Phi(t)B\right]
\end{aligned}\right\} \quad \textbf{(53a,b)}$$

Introducing the following variables:

$$\left.\begin{aligned}
&x_j = f_j \quad (j = 1,2,....n)\\
&x_j = \frac{df_{j-n}}{dt} \quad (j = n+1, n+2,....2n)
\end{aligned}\right\} \quad \textbf{(54a,b)}$$

the above system of differential equations (53) together with eqs (54b), form a new system with 2n first order differential equations, *i.e.*:

$$\left.\begin{aligned} \frac{dx_i}{dt} - x_{n+i} = 0 \quad (i = 1,2,...,n) \\ \frac{dx_i}{dt} + \sum_{\kappa=1}^{\infty} \Phi_{i\kappa}(t)x_\kappa = 0 \quad (i = n+1, n+2,....,2n) \end{aligned}\right\} \qquad \textbf{(55a,b)}$$

or in matrix form:

$$\frac{d\overline{x}}{dt} + \Phi(t)\overline{x} = 0 \qquad \textbf{(56)}$$

where $\overline{x}$ is a vector with terms x_i (i=1,2,…,2n) and $\Phi(t)$ is a matrix with dimensions 2n by 2n constructed as shown in eq (55b).

We assume now that we know $2n = m$ independent to each other solutions, which are called **"fundamental solutions"**:

$$x_{1\kappa}(t), \quad x_{2\kappa}(t),....., x_{m\kappa}(t) \qquad k = 1,2,3,..., \ m = 2n)$$

Thus, we have the matrix:

$$X(t) = \begin{pmatrix} x_{11}(t) & x_{12}(t) & .... & x_{1m}(t) \\ x_{21}(t) & x_{22}(t) & .... & x_{2m}(t) \\ .... & .... & .... & .... \\ x_{m1}(t) & x_{m2}(t) & .... & x_{mm}(t) \end{pmatrix}$$

where the first index is related to the number of function and the second index to the number of solution. Moreover, it is obvious that X(t) satisfies the differential equation:

$$\frac{d^2X}{dt^2} + X\Phi(t) = 0 \qquad \textbf{(57)}$$

which is obviously satisfied also by X(t+T).

Following a similar procedure as in previous paragraph, we may consider the following linear transformation:

$$X(t+T) = RX(t) \qquad \textbf{(58)}$$

where R is a matrix with terms $r_{i\kappa}$ that can be determined as follows.

We have assumed that the fundamental system $x_{i\kappa}$ satisfies the initial conditions. Then, eq (58) for t=0 gives: $X(T) = R$ or:

$$r_{i\kappa} = x_{i\kappa}(T) \qquad \textbf{(59)}$$

Then, the characteristic equation corresponding to eq (12) can be formulated as follows:

$$\left|R - \rho E\right| = 0 \qquad \textbf{(60)}$$

Equation (60) has m solutions corresponding to m linearly independent solutions of the differential equations system (56).

Let ρ_1, ρ_2, ..., ρ_m be the roots of the characteristic equation of (60), where we have assumed that multiple roots do not exist. Then, *via* a suitable selection of the initial functions system, the matrix R can be reduced in a diagonal form as follows:

$$R = \begin{pmatrix} \rho_1 & 0 & ..... & 0 \\ 0 & \rho_2 & ..... & 0 \\ .... & .... & ..... & ..... \\ 0 & 0 & ..... & \rho_m \end{pmatrix}$$

In other words, a fundamental system of solutions exists, for which the linear transformation is valid:

$$x_{i\kappa}(t+T) = \rho_\kappa x_{i\kappa}(t) \quad , \quad (i, \kappa = 1,2,.....,m)$$

or in vectorial form:

$$\overline{x}_{i\kappa}(t+T) = \rho_\kappa \overline{x}_\kappa(t) \quad , \quad (\kappa = 1,2,....,m) \tag{61}$$

Here, $\overline{x}_\kappa\left(x_{i\kappa}, x_{2\kappa}, ..., x_{m\kappa}\right)$ is the vector for the κth solution, *i.e.* it is the κth column of the matrix X(t).

The solutions of the system can take the form of eq (19a):

$$\overline{x}_\kappa(t) = e^{\frac{t}{T}\ell n\rho_\kappa} y_\kappa(t) \tag{62}$$

where $y_\kappa(t)$ is a periodical vector with period T. We can also write:

$$\left.\begin{aligned} &\overline{x}_\kappa(t) = e^{\frac{t}{T}\cdot\ell n|\rho_\kappa|} \varphi_\kappa(t) \\ \text{with:}\quad &\varphi_\kappa(t) = y_\kappa(t)\cdot\exp\left(\frac{i\cdot t}{T}\cdot\arg\rho_\kappa\right) \end{aligned}\right\} \tag{63a,b}$$

where $\varphi_\kappa(t)$ also a periodical vector.

We call the exponential terms:

$$\lambda_\kappa = \frac{1}{T}\ell n\rho_\kappa \tag{64}$$

as characteristic exponents of the non trivial solutions (62).

From the form of the above solutions (62), it follows that if all the characteristic exponents λ_κ have a negative real part, then the general solution (56) is decreasing with time. That is, the initial equilibrium state (initial movement) is stable. If however, among the characteristic exponents even one with positive real part exists, then the system will have increasing unbounded solutions, *i.e.* the initial equilibrium state will be unstable.

Given that $\ell n\rho = \ell n|\rho| + i\cdot\alpha rg\rho$, the following are valid:

a) If all the roots of the characteristic equation have absolute values smaller than 1, then the system is dynamically stable. b) If between the characteristic roots it exists even one with absolute value bigger than 1, then the system is dynamically unstable.

In characteristic roots with absolute value equal to 1, it obviously corresponds a clearly imaginary exponent. In this case, it is possible that either stability or instability exists.

Equation for the Determination of Characteristic Exponents [7]

The vector φ(t) of solution (63a) contains periodic terms with period T. This solution is continuous and hence, the terms in φ(t) can be expanded in a Fourier series. Consequently, we can determine the characteristic exponents λ.

Let us consider for instance the system of second order differential equations:

$$\Omega\frac{d^2 f}{dt^2}+\left[E-\alpha A-\beta B\cos\theta t\right]f=0 \tag{65}$$

We seek a solution of eq (65) in the form:

$$f(t)=e^{\lambda t}\left[\frac{1}{2}b_0+\sum_{\kappa=1}^{\infty}\left(\alpha_\kappa\sin\kappa\theta\,t+b_\kappa\cos\kappa\theta\,t\right)\right] \tag{66}$$

where α_κ and b_κ are vectors independent of time. The expansion (66) obviously refers to all terms of $f(t)$ in the form:

$$f_i(t)=e^{\lambda t}\left[\frac{1}{2}b_{i0}+\sum_{\kappa=1}^{\infty}\left(\alpha_{i\kappa}\sin\kappa\theta\,t+b_{i\kappa}\cos\kappa\theta\,t\right)\right]$$

where $\alpha_{i\kappa}$ and $b_{i\kappa}$ are time dependent coefficients.

Introducing eq (66) into eq (65) and setting equal the coefficients of the terms $e^{\lambda t}\sin\kappa\theta\,t$ and $e^{\lambda t}\cos\kappa\theta\,t$, respectively, we obtain the following algebraic homogenous system:

$$\begin{aligned}
&\lambda^2\Omega b_0+(E-\alpha A)b_0+\beta B b_1=0\\
&(\lambda^2-\kappa^2\theta^2)\,\Omega\alpha_\kappa+2\lambda\kappa\theta\Omega b_\kappa+(E-\alpha A)\alpha_\kappa-\frac{1}{2}\beta B(\alpha_{\kappa-1}+\alpha_{\kappa+1})=0\\
&(\lambda^2-\kappa^2\theta^2)\,\Omega b_\kappa+2\lambda\kappa\theta\Omega\alpha_\kappa+(E-\alpha A)b_\kappa-\frac{1}{2}\beta B(b_{\kappa-1}+b_{\kappa+1})=0\\
&(\kappa=1,2,3,.....,\quad \alpha_0=0)
\end{aligned}$$

In order for this system to have nontrivial solutions, the determinant of the coefficients of the unknowns must be zero, that is:

$$\begin{vmatrix}
\left((\lambda^2-\theta^2)\Omega+E-\alpha A\right) & -\frac{1}{2}\beta B & 2\lambda\theta\Omega\\
-\beta B & (\lambda^2\Omega+E-\alpha A) & 0\\
-2\lambda\theta\Omega & 0 & \left((\lambda^2-\theta^2)\Omega+E-\alpha A\right)
\end{vmatrix}=0 \tag{67}$$

Each term in the above expression represents n^2 subterms.

However, a direct application of eq (67) is cumbersome.

Determination of the Equations of Critical Frequencies

The characteristic equation resulting from eq (60) has the form:

$$\begin{aligned} &\rho^m + \alpha_1\rho^{m-1} + \alpha_2\rho^{m-2} + \ldots + \alpha_{m-2}\rho^2 + \alpha_{m-1}\rho + \alpha_m = 0 \\ \text{where:}\quad &\alpha_\kappa = \alpha_{m-\kappa} \end{aligned} \tag{68}$$

This is a reverse equation or in other words, if a root ρ exists, also the root $\frac{1}{\rho}$ of eq (68) will exist.

This can be proved by the well-known Liapunov theorem, according to which: "If the differential system is regular, that is to say if $\frac{dp_\kappa}{dt} = -\frac{\partial H}{\partial q_\kappa}$, $\frac{dq_\kappa}{dt} = \frac{\partial H}{\partial p_\kappa}$ or if it can be transformed in a regular form *via* a linear transformation with constant or periodical factors, then the characteristic equation for this system is a reverse equation ". (Here, by H(p,q,t) we denote a Hamiltonian function.)

We consider a pair of particular solutions that correspond to a pair of inverse characteristic roots:

$$\left.\begin{aligned} f_\kappa(t) &= y_\kappa(t)\cdot\exp\left(\frac{t}{T}\ell n\rho_\kappa\right) \\ f_{n+\kappa}(t) &= y_{n+\kappa}(t)\cdot\exp\left(-\frac{t}{T}\ell n\rho_\kappa\right) \end{aligned}\right\} \tag{69}$$

We also consider that ρ_κ is a real number different than 1. Then, one of the particular solutions will increase with time without limit (unbounded). Consequently, the region of real roots ρ will be the region of unbounded solutions that increase with time (instability region).

By altering the factors of the system, one can determine the condition for which the characteristic number will remain $\rho_\kappa = 1$ or $\rho_\kappa = -1$.

In this case, the solution will be periodical with period T or 2T. Altering further the factors of the system, the considered pair of characteristic roots will become complex and conjugated:

$$\begin{aligned} \rho_\kappa &= \alpha + ib \\ \rho_{n+\kappa} &= \alpha - ib \end{aligned}$$

that since the relation $\rho_\kappa \cdot \rho_{n+\kappa} = 1$ is valid, they will have an absolute value equal to unit. The region of complex roots is the region of bounded solutions (stability region).

Consequently, at the boundaries of the instability regions, the system of differential equations has periodical solutions with period T or 2T.

More precisely, two solutions with the same period enclose an instability region, while two solutions with different periods enclose a stability region.

Based on the above, the determination of the instability regions is reduced to the determination of the conditions, for which the system of differential equations (51) has periodical solutions with period T or 2T.

Let us consider the following system of differential equations:

$$\Omega\frac{d^2 f}{dt^2} + \left[E - \alpha A - \beta B\cos\theta t\right] f = 0 \tag{70}$$

We seek the solution of eq (70) in the form of series:

$$f(t)=\sum_{1,3,5,...}\left(\alpha_\kappa \sin\frac{\kappa\theta\, t}{2}+b_\kappa \cos\frac{\kappa\theta\, t}{2}\right) \quad \textbf{(71)}$$

where α_κ and b_κ are vectors with respect to time t. These series are converging, since the periodical solutions of the differential system (70) satisfy the Dirichlet convergence condition in all cases.

Introducing eq (71) into eq (70) and setting equal the coefficients of the terms $\sin\frac{\kappa\theta t}{2}$ and $\cos\frac{\kappa\theta t}{2}$, respectively, we obtain the following system in matrix form:

$$\left(E-\alpha A+\frac{1}{2}\beta B-\frac{1}{4}\theta^2\Omega\right)\alpha_1-\frac{1}{2}\beta B\alpha_3=0 \qquad \left(E-\alpha A-\frac{1}{2}\beta B-\frac{1}{4}\theta^2\Omega\right)b_1-\frac{1}{2}\beta Bb_3=0$$

$$\left(E-\alpha A-\frac{1}{4}\kappa\theta^2\Omega\right)\alpha_\kappa-\frac{1}{2}\beta B\left(\alpha_{\kappa-2}+\alpha_{\kappa+2}\right)=0 \qquad \left(E-\alpha A-\frac{1}{4}\kappa\theta^2\Omega\right)b_\kappa-\frac{1}{2}\beta B\left(b_{\kappa-2}+b_{\kappa+2}\right)=0$$

$(\kappa=3,5,.......)$

Consequently, the condition for the existence of solutions with period $\frac{4\pi}{\theta}$ will be:

$$\begin{vmatrix} \left(E-\alpha A\pm\frac{1}{2}\beta B-\frac{1}{4}\theta^2\Omega\right) & -\frac{1}{2}\beta B & 0 & \cdots \\ -\frac{1}{2}\beta B & \left(E-\alpha A-\frac{9}{4}\theta^2\Omega\right) & -\frac{1}{2}\beta B & \cdots \\ 0 & -\frac{1}{2}\beta B & \left(E-\alpha A-\frac{25}{4}\theta^2\Omega\right) & \cdots \\ \cdots & \cdots & \cdots & \cdots \end{vmatrix}=0 \quad \textbf{(72)}$$

Seeking now a solution in the form:

$$f(t)=\frac{1}{2}b_0+\sum_{2,4,6}\left(\alpha_\kappa \sin\frac{\kappa\theta t}{2}+b_\kappa \cos\frac{\kappa\theta t}{2}\right) \quad \textbf{(73)}$$

and following the same process, we obtain the following conditions for solutions with period $\frac{2\pi}{\theta}$:

$$\left.\begin{aligned} &\begin{vmatrix} (E-\alpha A-\theta^2\Omega) & -\frac{1}{2}\beta B & 0 & \cdots \\ -\frac{1}{2}\beta B & (E-\alpha A-4\theta^2\Omega) & -\frac{1}{2}\beta B & \cdots \\ 0 & -\frac{1}{2}\beta B & (E-\alpha A-16\theta^2\Omega) & \cdots \\ \cdots & \cdots & \cdots & \cdots \end{vmatrix}=0 \\ &\begin{vmatrix} (E-\alpha A) & -\beta B & 0 & 0 & \cdots \\ -\frac{1}{2}\beta B & (E-\alpha A-\theta^2\Omega) & -\frac{1}{2}\beta B & 0 & \cdots \\ 0 & -\frac{1}{2}\beta B & (E-\alpha A-4\theta^2\Omega) & -\frac{1}{2}\beta B & \cdots \\ 0 & 0 & -\frac{1}{2}\beta B & (E-\alpha A-16\theta^2\Omega) & \cdots \\ \cdots & \cdots & \cdots & \cdots & \cdots \end{vmatrix}=0 \end{aligned}\right\} \quad \textbf{(74a,b)}$$

It can be proven that the above determinants are regular, smooth and converging functions. The above conditions (72) and (74) are fully proportional to the corresponding conditions determined for the case of the equivalent Mathieu-Hill equation. From the above, by replacing the matrices by their eigenvalues, that is by setting:

$\Omega \rightarrow \frac{1}{\omega^2}$, $A \rightarrow \frac{1}{\alpha_\kappa}$, $B \rightarrow \frac{1}{\beta_\kappa}$, $E = 1$, we obtain the conditions for the Mathieu-Hill equation. This proportionality is not limited to their form only.

Let us consider now some extreme cases:

1. Let us assume that α=0 and $\beta \rightarrow 0$. Then, eq (72) takes the form:

$$\begin{vmatrix} E-\frac{1}{4}\theta^2\Omega & 0 & 0 & \ldots \\ 0 & E-\frac{9}{4}\theta^2\Omega & 0 & \ldots \\ 0 & 0 & E-\frac{25}{4}\theta^2\Omega & \ldots \\ \ldots & \ldots & \ldots & \ldots \end{vmatrix} = 0$$

and similarly eq (74) becomes:

$$\begin{vmatrix} E-\theta^2\Omega & 0 & 0 & \ldots \\ 0 & E-4\theta^2\Omega & 0 & \ldots \\ 0 & 0 & E-16\theta^2\Omega & \ldots \\ \ldots & \ldots & \ldots & \ldots \end{vmatrix} = 0$$

That is the equations that determine the instability regions are coincident by pairs, *i.e.* the instability regions are degenerated to curves. These regions are determined from:

$$\left| E - \frac{\kappa^2\theta^2}{4}\Omega \right| = 0 \quad (\kappa = 1,2,3,\ldots) \tag{75}$$

and since $\frac{1}{\omega_\kappa^2}$ are the characteristic values of matrix Ω, we will have:

$$\overline{\theta} = \frac{2\omega}{\kappa} \qquad (\kappa = 1,2,3,\ldots)$$

2. If $\alpha \neq 0$, then we obtain by a similar method:

$$\left| E - \alpha A - \frac{1}{4}\kappa^2\theta^2\Omega \right| = 0 \quad (\kappa = 0,1,2,3,\ldots) \tag{76}$$

The equations that determine the eigenfrequencies for free vibration of the system, are:

$$\left| E - \alpha A - \overline{\omega}^2\Omega \right| = 0 \tag{77}$$

From the above, we can easily determine:

$$\bar{\theta} = \frac{2\bar{\omega}}{\kappa} \quad (\kappa = 1,2,3,...) \tag{78}$$

Using the condition (76) with κ = 0, we can easily determine as a special case, the equation that gives the boundaries for the static stability region:

$$|E - \alpha A| = 0$$

From the above, it is obvious that the spectrum for resonance frequencies of the system (70) is fully proportional to the spectrum of the Mathieu-Hill equation. In problems where coupled equations can be reduced to simple independent equations, these frequencies are determined by the formula:

$$\bar{\omega}_\kappa = \omega_\kappa \cdot \sqrt{1 - \frac{\alpha}{\alpha_\kappa}} \;, \quad (\kappa = 1,2,3,...)$$

or in the most general case, in order to determine the natural frequencies, eq (77) should be solved.

Next, for an accurate calculation of the instability regions, one must solve equations (72) and (74). Consequently, based on the above analysis, we determine the main instability regions corresponding to κ =1 in eq (78).

An approximate expression for the boundaries of the main instability regions can be found by setting the first term in matrix (72) equal to zero:

$$\left| E - \alpha A \pm \frac{1}{2}\beta B - \frac{1}{4}\theta^2 \Omega \right| = 0 \tag{79}$$

This approximation results from the assumption that the periodical solutions at the boundaries of the instability regions are:

$$f(t) = \alpha \sin\frac{\theta t}{2} + b\cos\frac{\theta t}{2}$$

If we consider now the equations from which the natural frequencies are determined for a system with load parameters α, β/2, and α, -β/2, we will have:

$$\left| E - \alpha A - \frac{1}{2}\beta B - \bar{\omega}^2_{(\alpha + \frac{1}{2}\beta)} \cdot \Omega \right| = 0$$

$$\left| E - \alpha A + \frac{1}{2}\beta B - \bar{\omega}^2_{(\alpha - \frac{1}{2}\beta)} \cdot \Omega \right| = 0$$

or comparing them with eq (79) we obtain:

$$\bar{\theta} = 2\bar{\omega}_{(\alpha \pm \frac{1}{2}\beta)}$$

Hence, in the first approach, the frequencies that correspond to the boundaries of the main instability regions, are double the natural frequency of system, when loaded by a constant load with parameters α, β/2 and α, -β/2 respectively.

The Usual Case

One of the most usual cases of these equations that result from problems of dynamic instability of bridges is the following:

$$\left.\Omega\frac{d^2 f}{dt^2}+\left[E-(\alpha+\beta\cos\theta t)A\right]f=0 \quad \text{where:} \quad \Omega=\begin{pmatrix}\frac{1}{\omega_1^2} & 0\\ 0 & \frac{1}{\omega_2^2}\end{pmatrix}, \quad A=\begin{pmatrix}0 & \alpha_{12}\\ \alpha_{21} & 0\end{pmatrix}\right\} \quad \textbf{(80a,b)}$$

For the above case, we will derive the equations giving the instability regions. The equation for the boundaries of static stability is:

$$|E-\alpha A|=0$$

which, due to eq (80b), has the following solution:

$$\bar{\alpha}=\pm\frac{1}{\sqrt{\alpha_{12}\alpha_{21}}} \quad \textbf{(81)}$$

As a first approximation, the boundaries of the main instability regions can be found from eq (79): $\left|E-\left(\alpha\pm\frac{1}{2}\beta\right)A-\frac{1}{4}\theta^2\Omega\right|=0$, which due to eq (80b) is written as:

$$\begin{vmatrix}1-\frac{\theta^2}{4\omega_1^2} & -\left(\alpha\pm\frac{1}{2}\beta\right)\alpha_{12}\\ -\left(\alpha\pm\frac{1}{2}\beta\right)\alpha_{21} & 1-\frac{\theta^2}{4\omega_2^2}\end{vmatrix}=0$$

From the above equation, and using also eq (81), we obtain:

$$\left.\begin{aligned}&\left(1-\frac{\theta^2}{4\omega_1^2}\right)\cdot\left(1-\frac{\theta^2}{4\omega_2^2}\right)-K=0\\ \text{where:}\quad &K=\frac{\left(\alpha\pm\frac{1}{2}\beta\right)^2}{\bar{\alpha}^2}\end{aligned}\right\} \quad \textbf{(82a,b)}$$

Solution of eq (82a) results the expressions for the boundaries of the main two instability regions:

$$\left.\begin{aligned}&\bar{\theta}=\frac{2\omega_1}{\sqrt{2\gamma}}\cdot\sqrt{1+\gamma-\sqrt{(1-\gamma)^2+4\gamma K}}\\ &\bar{\theta}=\frac{2\omega_2}{\sqrt{2}}\cdot\sqrt{1+\gamma+\sqrt{(1-\gamma)^2+4\gamma K}}\\ \text{where:}\quad &\gamma=\frac{\omega_1^2}{\omega_2^2}\end{aligned}\right\} \quad \textbf{(83a,b,c)}$$

and since in the usual cases met in practice it is $\gamma << 1$, we will have:

$$\left.\begin{aligned}\overline{\theta} &= 2\omega_1\sqrt{1-\frac{1}{1-\gamma}K} \\ \overline{\theta} &= 2\omega_2\sqrt{1+\frac{\gamma}{1-\gamma}K}\end{aligned}\right\} \quad \textbf{(84a,b)}$$

If we set:

$$\frac{\alpha}{\overline{\alpha}} = \mu \quad , \quad \frac{\beta}{2\overline{\alpha}} = \nu \quad \textbf{(85)}$$

equations (84) can be written as follows:

$$\left.\begin{aligned}\overline{\theta} &= 2\omega_1\sqrt{1-\frac{1}{1-\gamma}(\mu\pm\nu)^2} \\ \overline{\theta} &= 2\omega_2\sqrt{1+\frac{\gamma}{1-\gamma}(\mu\pm\nu)^2}\end{aligned}\right\} \quad \textbf{(86)}$$

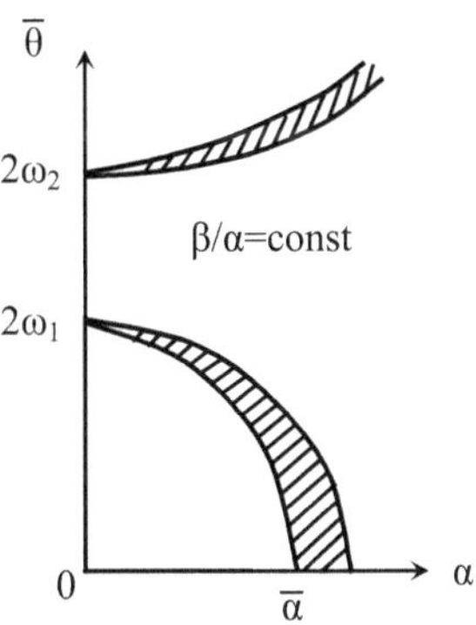

Figure 5: The main two instability regions

For a constant ratio $\frac{\beta}{\alpha}$, we obtain the diagram in Fig. **5** that gives the main two instability regions. We observe that the first region is found neighboring to frequency $2\omega_1$ while the second one is neighboring to $2\omega_2$.

For the second instability region, that corresponds to value κ=2 in eq (78), we consider the equations:

$$\left.\begin{aligned}&\left|E-\alpha A-\theta^2\Omega\right| = 0 \\ &\begin{vmatrix}(E-\alpha A) & -\beta A \\ -\frac{1}{2}\beta A & (E-\alpha A-\theta^2\Omega)\end{vmatrix} = 0\end{aligned}\right\} \quad \textbf{(87a,b)}$$

that correspond to the following approximate expression:

$$f(t) = \frac{1}{2}b_0 + \alpha_2 \sin\theta t + b_2 \cos\theta t$$

Expanding the first of eq (87), we obtain:

$$\begin{vmatrix} 1-\dfrac{\theta^2}{\omega_1^2} & -\alpha\alpha_{12} \\ -\alpha\alpha_{21} & 1-\dfrac{\theta^2}{\omega_2^2} \end{vmatrix} = 0 \quad \text{or} \quad \left(1-\frac{\theta^2}{\omega_1^2}\right)\cdot\left(1-\frac{\theta^2}{\omega_2^2}\right)-\mu^2=0$$

which for $\gamma << 1$ gives:

$$\left.\begin{array}{ll} \overline{\theta}=\omega_1\sqrt{1-\dfrac{1}{1-\gamma}\mu^2} & \overline{\theta}=\omega_2\sqrt{1+\dfrac{\gamma}{1-\gamma}\mu^2} \end{array}\right\} \quad \textbf{(88a,b)}$$

Equations (88) provide one boundary curve for each of the two instability regions. The second boundary curve of each region is obtained from eqs (87a,b), which can be written as:

$$\begin{vmatrix} 1 & -\alpha\alpha_{12} & 0 & -\beta\alpha_{12} \\ -\alpha\alpha_{21} & 1 & -\beta\alpha_{12} & 0 \\ 0 & -\dfrac{1}{2}\beta\alpha_{12} & 1-\dfrac{\theta^2}{\omega_1^2} & -\alpha\alpha_{12} \\ -\dfrac{1}{2}\beta\alpha_{21} & 0 & -\alpha\alpha_{21} & 1-\dfrac{\theta^2}{\omega_2^2} \end{vmatrix}=0$$

which for $\gamma <<1$, gives:

$$\left.\begin{array}{ll} \overline{\theta}=\omega_1\sqrt{1-\dfrac{\mu^2-2(1-\gamma)v^2-(\mu^2-2v^2)^2}{(1-\gamma)(1-\mu^2)-2v^2}} & \overline{\theta}=\omega_2\sqrt{1+\dfrac{\gamma[\mu^2-(\mu^2-2v^2)^2]-2(1-\gamma)v^2}{(1-\gamma)(1-\mu^2)-2\gamma v^2}} \end{array}\right\} \quad \textbf{(89a,b)}$$

For the third instability region, that corresponds to value $\kappa = 3$ in eq (78), we consider the equation:

$$\begin{vmatrix} \left(E-\left(\alpha\pm\dfrac{1}{2}\beta\right)A-\dfrac{1}{4}\theta^2\Omega\right) & -\dfrac{1}{2}\beta A \\ -\dfrac{1}{2}\beta A & \left(E-\alpha A-\dfrac{9}{4}\theta^2\Omega\right) \end{vmatrix}=0$$

The above equation can be written in expanded form as follows:

$$\begin{vmatrix} 1-\dfrac{\theta^2}{4\omega_1^2} & -\left(\alpha\pm\dfrac{1}{2}\beta\right)\alpha_{12} & 0 & -\dfrac{1}{2}\beta\alpha_{12} \\ -\left(\alpha\pm\dfrac{1}{2}\beta\right)\alpha_{21} & 1-\dfrac{\theta^2}{4\omega_2^2} & -\dfrac{1}{2}\beta\alpha_{21} & 0 \\ 0 & -\dfrac{1}{2}\beta\alpha_{12} & 1-\dfrac{9\theta^2}{4\omega_1^2} & -\alpha\alpha_{12} \\ -\dfrac{1}{2}\beta\alpha_{21} & 0 & -\alpha\alpha_{21} & 1-\dfrac{9\theta^2}{4\omega_2^2} \end{vmatrix}=0 \quad \textbf{(90)}$$

This expanded form of the above determinant is a reverse equation of degree higher than two. We can, however, simplify this problem according to the following justification by Bolotin [3]. The instability region that we seek is found near the values $\theta = \frac{2}{3}\omega_1$. We set this value of θ to all the terms of the determinant except the one in position 33. Then, eq (90) becomes:

$$\begin{vmatrix} \frac{8}{9} & -\left(\alpha \pm \frac{1}{2}\beta\right)\alpha_{12} & 0 & -\frac{1}{2}\beta\alpha_{12} \\ -\left(\alpha \pm \frac{1}{2}\beta\right)\alpha_{21} & 1-\frac{\gamma}{9} & -\frac{1}{2}\beta\alpha_{21} & 0 \\ 0 & -\frac{1}{2}\beta\alpha_{12} & 1-\frac{9\theta^2}{4\omega_1^2} & -\alpha\alpha_{12} \\ -\frac{1}{2}\beta\alpha_{21} & 0 & -\alpha\alpha_{21} & 1-\gamma \end{vmatrix} = 0$$

The solution of the above equation gives the expression for the critical frequencies:

$$\bar{\theta} = \frac{2}{3}\omega_1\sqrt{1 - \frac{\frac{8}{9}[\mu^2 + v^2(1-\gamma)] - [\mu(\mu \pm v) + v^2]^2}{(1-\gamma)[\frac{8}{9} - (\mu \pm v)^2] + v^2}} \qquad \textbf{(91a)}$$

In a similar way, we can find the corresponding expression for the region $\theta = \frac{2}{3}\omega_2$:

$$\bar{\theta} = \frac{2}{3}\omega_2\sqrt{1 + \frac{\frac{8}{9}[\gamma\mu^2 + v^2(1-\gamma) - \gamma[\mu(\mu \pm v) + v^2]^2}{(1-\gamma)[\frac{8}{9} - (\mu \pm v)^2] + \gamma v^2}} \qquad \textbf{(91b)}$$

Finally, the frequencies of a vibrating system can be determined by the relations:

$$\begin{aligned} \bar{\omega}_1 &= \omega_1\sqrt{1 - \frac{\alpha^2}{(1-\gamma)\bar{\alpha}^2}} \\ \bar{\omega}_2 &= \omega_2\sqrt{1 + \frac{\gamma\alpha^2}{(1-\gamma)\bar{\alpha}^2}} \end{aligned} \qquad \textbf{(92a,b)}$$

DYNAMIC INSTABILITY PROBLEMS IN CABLE - STAYED BRIDGES

From the analysis in paragraph of page 156, it is clear that the loads q(x) and p(x) due to the cables (Fig. **6**) can be given from the following relations:

$$p(x) = F_1(x)w(x) - F_2(x)\int_{\alpha_1}^{\alpha_2} F_3(x)w(x)dx$$

$$q(x) = q(x)\tan\varphi = q(x)F_4(x)$$

where $F_1(x)$, $F_2(x)$, $F_3(x)$, $F_4(x)$ are known functions of x, while $F_4(x)$ - specifically in the parallel wiring - is a constant number. Consequently, the above relations can be written as follows:

$$\left.\begin{aligned} p(x) &= F_1(x)w(x) - F_2(x)\int_{\alpha_1}^{\alpha_2} F_3(x)w(x)dx \\ q(x) &= \overline{F}_1(x)w(x) - \overline{F}_2(x)\int_{\alpha_1}^{\alpha_2} F_3(x)w(x)dx \end{aligned}\right\} \qquad \textbf{(93a,b)}$$

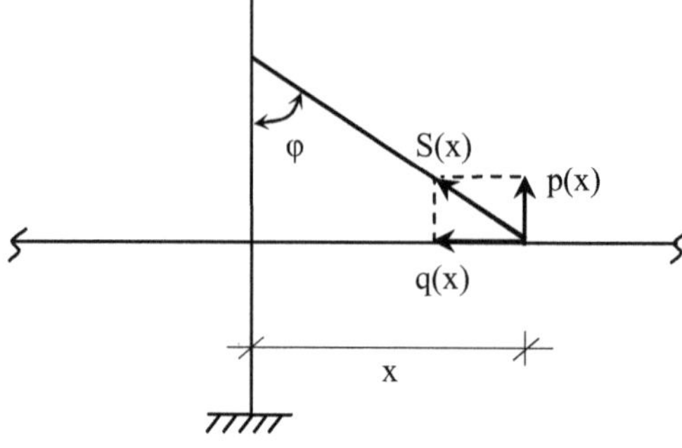

Figure 6: Loads imposed to the bridge due to the cable system

Dynamic Instability of the Pylons

The Fan System [8]

In this case (see also Fig. 7) we will have:

$$P(t) = \int_{\alpha_1}^{\alpha_2} p(x)dx$$

$$S(t) = \int_{\alpha_1}^{\alpha_2} p(x)\tan\varphi dx = \frac{1}{h-h_0}\cdot\int_{\alpha_1}^{\alpha_2} xp(x)dx$$

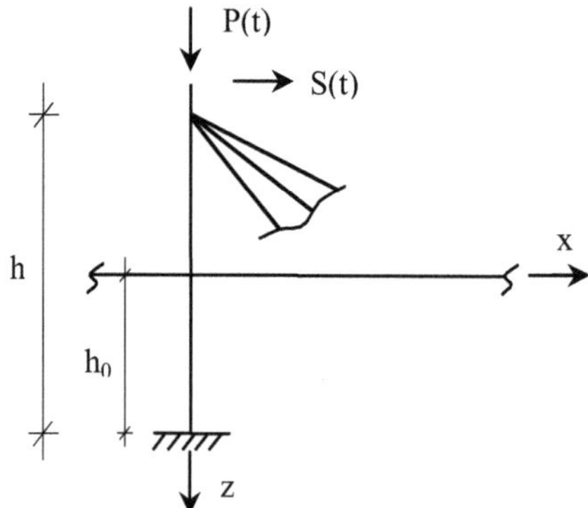

Figure 7: Fan arrangement of cables

a. On a bridge at rest, without external loading, we will have w=0 and $p(x)=p_0(x)$ being the known prestressing forces of cables, each one depending on the initial constructional curvature w_0 (the deformations w, are measured from the curve w_0). Thus, the force $P_0 = \int_{\alpha_1}^{\alpha_2} p_0(x)dx$ will be constant and also independent from the vibrations of the bridge.

b. For a vibrating bridge, the following equation is valid:

$$p(x,t) = p_0(x) + p_d(x,t)$$

where $p_d(x,t)$ depends on the dynamic deformations $w_d(x,t)=w(x,t)$ of the bridge deck.

Due to the dynamic character of the loading, the deformations of the deck can be written in the form:

$$w(x,t) = \sum_n X_n(x)\Phi_n(t) \quad \textbf{(94)}$$

where $X_n(x)$ are the known shape functions of the bridge deck and $\Phi_n(t)$ are known time functions. Hence, we can write:

$$\left.\begin{aligned} & P(t) = P_0 + \sum_n A_n\Phi_n(t) \\ & S(t) = S_0 + \sum_n B_n\Phi_n(t) \\ \text{where:}\quad & P_0 = \int_{\alpha_1}^{\alpha_2} p_0(x)dx \quad , \quad S_0 = \frac{1}{h-h_0}\int_{\alpha_1}^{\alpha_2} xp_0(x)dx \\ & A_n = \int_{\alpha_1}^{\alpha_2}\left[F_1(x)X_n(x) - F_2(x)\int_{\alpha_1}^{\alpha_2} F_3(x)X_n(x)dx\right]dx \\ & B_n = \frac{1}{h-h_0}\int_{\alpha_1}^{\alpha_2} x\left[F_1(x)X_n(x) - F_2(x)\int_{\alpha_1}^{\alpha_2} F_3(x)X_n(x)dx\right]dx \end{aligned}\right\} \quad \textbf{(95a,b,c,d,e,f)}$$

The equation of motion of the vibrating pylon is:

$$\frac{d^2}{dz^2}\left[E_pI_p(z)u''(z,t)\right] + P(t)u''(z,t) + m_p(x)\ddot{u}(z,t) = S(t)\delta(z-h) \quad \textbf{(96)}$$

where E_p, $I_p(z)$, $m_p(z)$ are the elasticity modulus, the moment of inertia and the mass per unit length of the pylon, respectively.

In order to solve eq (96), we seek a solution in the form:

$$u(z,t) = Z_\kappa(z)f_\kappa(t) \quad \textbf{(97)}$$

where $Z_\kappa(z)$ are the shape functions of a cantilever beam (that constitute a complete orthocanonical system).

Introducing eq (97) into eq (96) and taking into account that Z_κ fulfills the equation for free vibration:

$$\frac{d^2}{dz^2}\left[E_pI_p(z)Z''_\kappa(z)\right] - m_p\omega_\kappa^2 Z_\kappa = 0$$

we finally obtain:

$$m_p\omega_\kappa^2 Z_\kappa f_\kappa + PZ''_\kappa f_\kappa + m_p Z_\kappa \ddot{f}_\kappa = S\delta(x-h) \quad \textbf{(98)}$$

Multiplying the above equation by Z_κ, integrating the outcome from 0 to h and employing the boundary conditions, eq (98) becomes:

$$\ddot{f}_\kappa + \left[\omega_\kappa^2 + \frac{\int_0^h Z_\kappa'^2 dz}{\int_0^h m_p Z_\kappa^2 dz} \cdot P_0 + \frac{\int_0^h Z_\kappa'^2 dz}{\int_0^h m_p Z_\kappa^2 dz} \cdot \sum_n A_n \Phi_n \right] \cdot f_\kappa = \frac{S_0 Z_\kappa(h)}{\int_0^h m_p Z_\kappa^2 dz} + \frac{Z_\kappa(h)}{\int_0^h m_p Z_\kappa^2 dz} \cdot \sum_n B_n \Phi_n \qquad \textbf{(99)}$$

Considering that the vibration is influenced mainly by the first eigenmode and the first time function and omitting the indicator κ, since eq (99) is valid for any κ, we obtain:

$$\left.\begin{array}{l} \ddot{f} + \Omega^2 \left[1 + 2\mu\Phi_1(t)\right] f(t) = p_0 + p\Phi_1(t) \\ \text{where :} \quad \Omega^2 = \omega_\kappa^2 + \dfrac{\int_0^h Z_\kappa'^2 dz}{\int_0^h m_p Z_\kappa^2 dz} \cdot P_0 \\ \mu = \dfrac{\int_0^h Z_\kappa'^2 dz}{2\Omega^2 \int_0^h m_p Z_\kappa^2 dz} \cdot A_1(x) \\ p_0 = \dfrac{S_0 Z_\kappa(h)}{\int_0^h m_p Z_\kappa^2 dz} \\ p = \dfrac{Z_\kappa(h)}{\int_0^h m_p Z_\kappa^2 dz} B_1(x) \end{array}\right\} \qquad \textbf{(100a,b,c,d,e)}$$

The above equation (100a), falls into the general case of paragraph in page 228 and can be solved as presented above. Depending however on the values of p_0, p and $\Phi_1(t)$, it can be also reduced to simpler forms.

The Harp System

In this case, the pylon is subjected to the loads shown in Fig. **8** and are given by eq (93a,b). Thus, if p_0 and q_0 are the loads acting on the deck of the bridge at rest, then the loads for the vibrating bridge will be:

$$p(x,t) = p_0(x) + p_d(x,t)$$
$$q(x,t) = q_0(x) + q_d(x,t)$$

Due to the dynamic character of the loading, the deformations of the deck can be expressed in the form of equations (94) and, consequently, we can write:

$$\left.\begin{aligned} p(x,t) &= p_0(x) + \sum_n A_n(x)\Phi_n(t) \\ q(x,t) &= q_0(x) + \sum_n B_n(x)\Phi_n(t) \\ \text{where:}\quad A_n(x) &= F_1(x)X_n(x) - F_2(x)\int_{\alpha_1}^{\alpha_2} F_3(x)X_n(x)dx \\ B_n(x) &= \overline{F}_1(x)X_n(x) - \overline{F}_2(x)\int_{\alpha_1}^{\alpha_2} F_3(x)X_n(x)dx \end{aligned}\right\} \qquad \textbf{(101a,b,c,d)}$$

and p_0, q_0 are the components of the cable prestressing forces.

From the geometry in Fig. **16** of chapter 6, we can easily determine:

$$\left.\begin{aligned} \overline{p}(z,t) &= \frac{\delta_r}{\gamma}p_0(z-h_0) + \frac{\delta_r}{\gamma}p_d\left[(z-h_0),t\right] = \overline{p}_0(z) + \sum_n \overline{A}_n(z)\Phi_n(t) \\ \overline{q}(z,t) &= \frac{\delta_r}{\gamma}q_0(z-h_0) + \frac{\delta_r}{\gamma}q_d\left[(z-h_0),t\right] = \overline{q}_0(z) + \sum_n \overline{B}_n(z)\Phi_n(t) \\ \text{where:}\quad \overline{p}_0(z) &= \frac{\delta r}{\gamma}\cdot p_0(z-h_0) \\ \overline{q}_0(z) &= \frac{\delta r}{\gamma}\cdot q_0(z-h_0) \\ \overline{A}_n(z) &= \frac{\delta r}{\gamma}\cdot A_n(z-h_0) \\ \overline{B}_n(z) &= \frac{\delta r}{\gamma}\cdot B_n(z-h_0) \end{aligned}\right\} \qquad \textbf{(102a,b,c,d,e,f)}$$

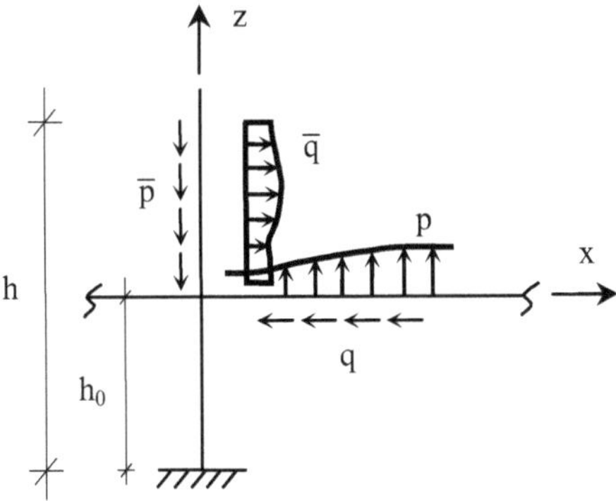

Figure 8: Loads imposed to the bridge due to the harp cable system

The pylons equation of motion is (see also page 230):

$$\begin{aligned} &\frac{d^2}{dz^2}\left[E_pI_pu''(z,t)\right] - \left[\overline{p}_0 + \sum_n \overline{A}_n\Phi_n\right]u' + \\ &+\left[\overline{p}_0 + \int_z^h \sum_n \overline{A}_n(\lambda)d\lambda\cdot\Phi_n\right]u'' - \left[\overline{q}_0 + \sum_n \overline{B}_n\Phi_n\right] + m\ddot{u} = 0 \end{aligned} \qquad \textbf{(103)}$$

We seek a solution in the form of eq (97) and hence, eq (103) becomes:

$$\frac{d^2}{dz^2}\left[E_pI_pZ''_\kappa f_\kappa\right]-\left[\overline{p}_0+\sum_n\overline{A}_n\Phi_n\right]\cdot Z'_\kappa f_\kappa+$$

$$+\left[\overline{p}_0+\int_z^h\sum_n\overline{A}_n(\lambda)d\lambda\cdot\Phi_n\right]\cdot Z''_\kappa f_\kappa-\left[\overline{q}_0+\sum_n\overline{B}_n\Phi_n\right]+mZ_\kappa\ddot{f}_\kappa=0$$

Multiplying by Z_κ, integrating the outcome from 0 to h and taking into account that Z_κ fulfils the equation for free vibration:

$$\frac{d^2}{dz^2}\left[E_pI_p(z)Z''_\kappa\right]-m_p(z)\omega_\kappa^2Z_\kappa=0$$

we finally obtain:

$$\begin{aligned}&m_p\omega_\kappa^2Z_\kappa f_\kappa-\left[\overline{p}_0+\sum_n\overline{A}_n\Phi_n\right]Z'_\kappa f_\kappa+\\&+\left[\overline{p}_0+\int_z^h\sum_n\overline{A}_n(\lambda)d\lambda\cdot\Phi_n\right]\cdot Z''_\kappa f_\kappa-\left[\overline{q}_0+\sum_n\overline{B}_n\Phi_n\right]+m_pZ_\kappa\ddot{f}_\kappa=0\end{aligned}\tag{104}$$

Multiplying by Z_κ, integrating the outcome from 0 to h and employing the boundary conditions, eq (104) becomes:

$$\omega_\kappa^2f_\kappa-\frac{\int_0^h\left[\overline{p}_0+\sum_n\overline{A}_n\Phi_n\right]Z'_\kappa Z_\kappa dz}{\int_0^h m_pZ_\kappa^2dz}f_\kappa+\frac{\int_0^h\left[\overline{p}_0+\int_z^h\sum_n\overline{A}_n(\lambda)d\lambda\cdot\Phi_n\right]Z''_\kappa Z_\kappa dz}{\int_0^h m_pZ_\kappa^2dz}f_\kappa-\frac{\int_0^h\left[\overline{q}_0+\sum_n\overline{B}_n\Phi_n\right]Z_\kappa dz}{\int_0^h m_pZ_\kappa^2dz}+\ddot{f}_\kappa=0\tag{105}$$

Considering also that the vibration is influenced mainly by the first eigenmode and the first time function and omitting the indicator κ, since eq (105) valid for any κ, we obtain:

$$\left.\begin{aligned}&\ddot{f}+\Omega^2\left[1+2\mu\Phi_1(t)\right]f=s_0+s\Phi_1(t)\\&\text{where:}\quad\Omega^2=\omega_\kappa^2+\frac{\int_0^h\overline{p}_0Z''Z_\kappa dz-\int_0^h\overline{p}_0Z'_\kappa Z_\kappa dz}{\int_0^h m_pZ_\kappa^2dz}\\&\mu=\frac{\int_0^h\left[\int_z^h\overline{A}_1(\lambda)d\lambda\right]Z''_\kappa Z_\kappa dz-\int_0^h\overline{A}_1Z'_\kappa Z_\kappa dz}{\int_0^h m_pZ_\kappa^2dz}\\&s_0=\frac{\int_0^h\overline{q}_0Z_\kappa dz}{\int_0^h m_pZ_\kappa^2dz}\quad,\quad s=\frac{\int_0^h\overline{B}_1Z_\kappa dz}{\int_0^h m_pZ_\kappa^2dz}\end{aligned}\right\}\tag{106a,b,c,d,e}$$

The last one has the form of eq (53) and can be solved in a similar manner.

Let us consider for example a three-span cable-stayed bridge with span lengths $\ell_1=\ell_3=80\text{m}$, and $\ell_2=200\text{m}$, and mass per unit length m_b=600kg/m. The first three eigenfrequencies are determined using the formulas in Chapter 6 and are: $\omega_{b1} = 2.91$, $\omega_{b2} = 7.22$, and $\omega_{b3} = 11.19 \sec^{-1}$. The cross-sectional area of the cables A(x) varies along axis x according to the law proposed by Bruno-Golotti [11]. The two pylons have a height H=90 m with h_2=20 m each. Moreover, by considering p_o=p=0 (which corresponds to vertical loads only) we find: S_o=1,080,000 dN and A_1=950,000 dN.

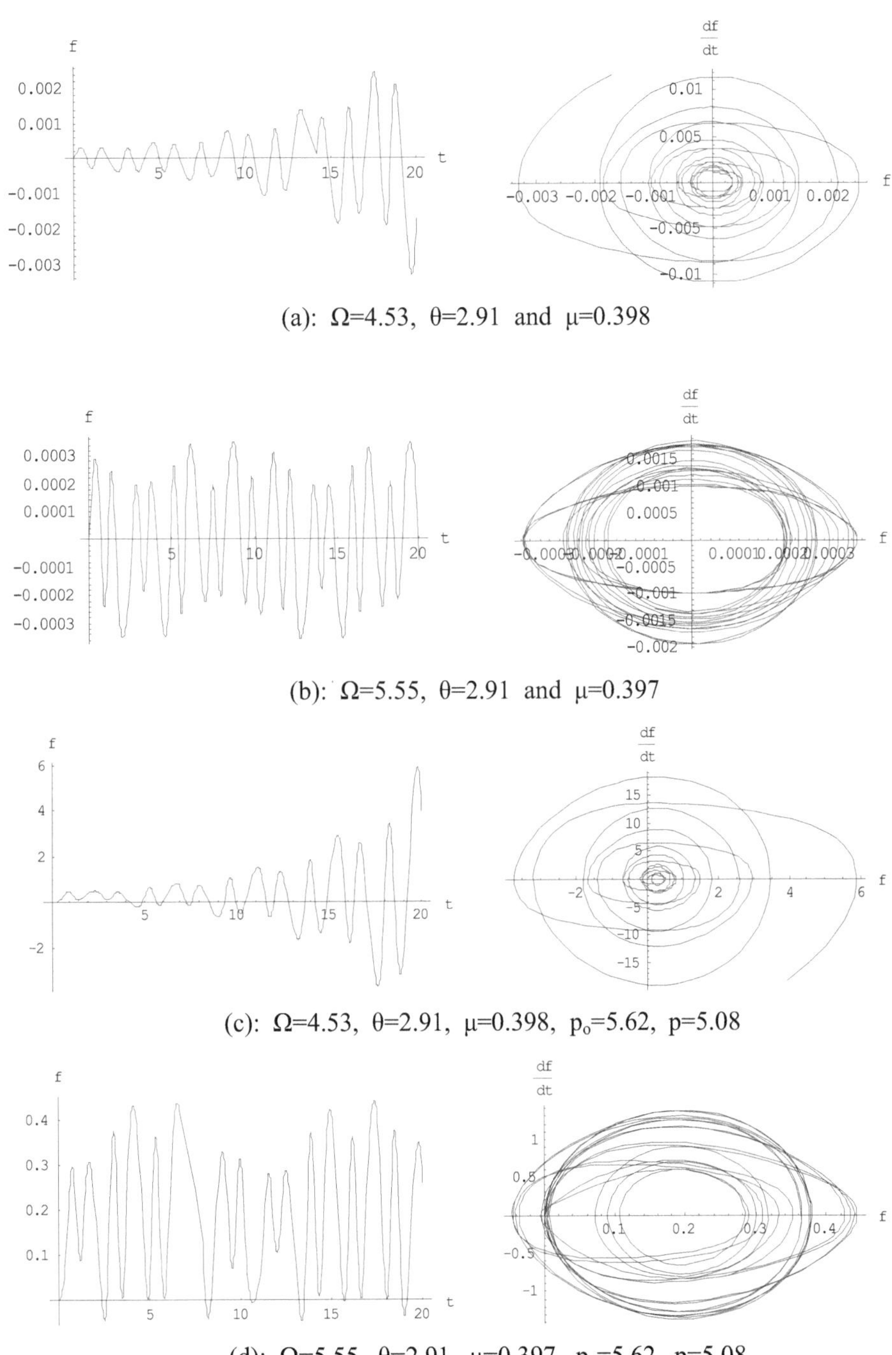

(a): Ω=4.53, θ=2.91 and μ=0.398

(b): Ω=5.55, θ=2.91 and μ=0.397

(c): Ω=4.53, θ=2.91, μ=0.398, p_o=5.62, p=5.08

(d): Ω=5.55, θ=2.91, μ=0.397, p_o=5.62, p=5.08

Figure 9: Time functions and phase-plane portraits for a pylon with various properties

As an example of bad design, we assume that each pylon has the following characteristic properties: m_p=3000kg/m and I_p=1.50m^4 from which we get: $\omega_{p1} = 1.406$, $\omega_{p2} = 8.800$, $\omega_{p3} = 24.671 \sec^{-1}$. From the shape functions of the pylon (modal shapes of a cantilever) we obtain: $\int_0^H Z_1'^2 dz = 0.05161$. Using the above, we get Ω_1=4.53, μ_1=0.398, and also $\frac{\theta_1}{\Omega_1} = 0.642$. Note that point $(\theta_1/\Omega_1, \mu_1)$ lies within the unstable region of figure **(3)**. Via numerical integration of eq (7) we obtain the diagrams in Fig. **9a** showing that the pylon finally collapses.

An improved design of the pylon with: m_p=2000kg/m, and I_p=1.50m^4 (corresponding to a better distribution of mass) will give: ω_1=1.772, Ω_1=5.55, μ_1=0.397 and $\frac{\theta_1}{\Omega_1} = 0.52$. In this case, point ($\theta_1/\Omega_1$, μ_1) lies within the stable region of figure **(3)**. The resulting vibration is bounded as shown in Fig. **9b** . Next, we consider $p_o \neq 0$ and $p \neq 0$ (which corresponds to the existence of transverse loads as well). Due to these loads we will have: P_o=1,350,000 dN and hence B_1=1,220,000 dN and $Z_1(H)$=1.999, $\int_0^H Z_1^2 dz = 160.02$. Then for the pylon corresponding to the bad design we will compute p_o=5.62 and p=5.08. In this case, we obtain the time function and phase-plane diagrams shown in Fig. **9c**, which show that the pylon is unstable.

For the improved design case pylon we have: p_o=8.43 and p=7.62, and from the corresponding diagrams in Fig. **9d**, we conclude that the pylon is stable.

Instability of a Vibrating Wire

Parametric instability of a vibrating wire may occur when some of the parameters that influence its natural oscillation (in this case the axial stresses) are altered with respect to time due to external causes, in which even the change of its characteristics is included, e.g. frost or ice cover.

Let us consider the cable shown in Fig. **10**, which is an elastic and weightless wire with length ℓ, cross-sectional area A, modulus of elasticity E and mass per unit length m.

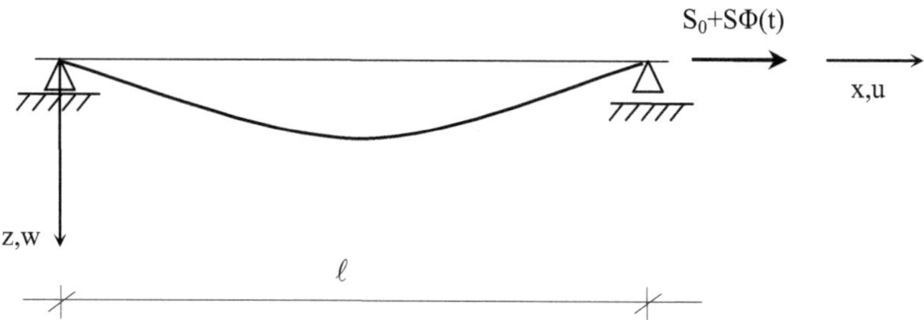

Figure 10: Cable subjected to axial tensile force

The cable constitutes a part of a cable-stayed bridge and is loaded in tension with the force $S_0 + S \cdot \Phi(t)$, where S_0 is a constant force due to self-weight and other permanent loads, and $S \cdot \Phi(t)$ is a force due to the vibration of the deck. We also assume that $S_0 > S \cdot \Phi(t)$, so that the cable is always under tension.

In order to determine the axial deformation of the cable due to its vibration, we consider the infinitesimal element dx shown in Fig. **11**. If at time instant t the length AB=dx goes to position A′B′=ds, we will have:

$$ds = \sqrt{(dx+du)^2 + dw^2} = dx\sqrt{\left(1+\frac{du}{dx}\right)^2 + \left(\frac{dw}{dx}\right)^2} = dx\sqrt{1+2\frac{du}{dx}+\left(\frac{du}{dx}\right)^2+\left(\frac{dw}{dx}\right)^2}$$

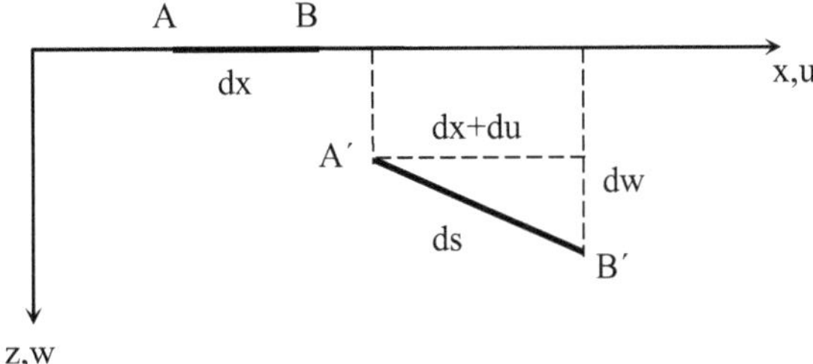

Figure 11: Initial and deformed state of a cable element

Since the deformation u is very small and also: $\left(\frac{du}{dx}\right)^2 << 1$, we will have:

$$ds = dx\sqrt{1 + 2\frac{du}{dx} + \left(\frac{dw}{dx}\right)^2} \cong dx\left(1 + \frac{du}{dx} + \frac{1}{2}\left(\frac{dw}{dx}\right)^2\right)$$

Hence, since it is $ds = (1+\varepsilon)dx$ we finally have:

$$\varepsilon = \frac{du}{dx} + \frac{1}{2}\left(\frac{dw}{dx}\right)^2 \quad \textbf{(107)}$$

We compute next the energy of the cable element dx. We have:

a) Kinetic energy: $K = \frac{1}{2}\int_0^\ell m(\dot{u}^2 + \dot{w}^2)dx = \frac{m}{2}\int_0^\ell (\dot{u}^2 + \dot{w}^2)dx$

b) Strain energy: $V = \frac{EA}{2}\left(u' + \frac{1}{2}w'^2\right)^2 dx$

c) Load potential: $\Omega = (S_0 + S\cdot\Phi)u(\ell)$, but since $u(\ell) = \int_0^\ell (ds - dx) = \int_0^\ell \left(u' + \frac{1}{2}w'^2\right)dx$ it follows that:

$$\Omega = [S_0 + S\cdot\Phi(t)]\int_0^\ell \left(u' + \frac{1}{2}w'^2\right)dx$$

Employing the boundary conditions $u(0) = w(0) = w(\ell) = 0$, and applying the Hamilton principle:

$$\int_{t_1}^{t_2} \delta(K - V - \Omega)dt = 0 \quad \textbf{(108)}$$

we will have:

1) $$\int_{t_1}^{t_2} \delta K dt = \int_{t_1}^{t_2}\int_0^\ell \frac{m}{2}(2\dot{u}\delta\dot{u} + 2\dot{w}\delta\dot{w})dxdt = m\int_0^\ell (\dot{u}\delta u + \dot{w}\delta w)_{t_1}^{t_2} dx - m\int_{t_1}^{t_2}\int_0^\ell (\ddot{u}\delta u + 2\ddot{w}\delta w)dxdt$$

and omitting the axial mass inertia forces (as being very small), we get:

$$\int_{t_1}^{t_2}\delta K dt = m\int_0^{\ell}(\dot{u}\delta u + 2\dot{w}\delta w)_{t_1}^{t_2} dx = m\int_{t_1}^{t_2}\int_0^{\ell}\ddot{w}\delta w dx dt$$

2) $$\int_{t_1}^{t_2}\delta V dt = EA\int_{t_1}^{t_2}\int_0^{\ell}(u'\delta u' + \frac{1}{2}w'^2\delta u' + u'w'\delta w' + \frac{1}{2}w'^3\delta w')dxdt$$

Neglecting the term w'^3 as being of higher order, we finally get:

$$\int_{t_1}^{t_2}\delta V dt = EA\int_{t_1}^{t_2}\left\{(u'\delta u\Big|_0^{\ell} - \int_0^{\ell}u''\delta u dx + \frac{1}{2}w'^2\delta u\Big|_0^{\ell} - \int_0^{\ell}w'w''\delta u dx + u'w'\delta w\Big|_0^{\ell} - \int_0^{\ell}(u'w'' + u''w')\delta w dx\right\}dt$$

3) $$\int_{t_1}^{t_3}\delta\Omega dt = (S_0 + S\Phi)\cdot\int_{t_1}^{t_2}\int_0^{\ell}(\delta u' + w'\delta w')dxdt = (S_0 + S\Phi)\cdot\int_{t_1}^{t_2}(w'\delta w\Big|_0^{\ell} - \int_0^{\ell}w''\delta w dx)dt$$

Finally, eq (108) provides the following governing equations of the problem:

$$EA(u'' + w'w'') = 0$$

$$-m\ddot{w} + (u'w'' + u'w')EA + EA\frac{3}{2}w'^2w'' + (S_0 + S\cdot\Phi)w'' = 0$$

The above equations can be also written as follows:

$$\left.\begin{aligned} EA\frac{\partial}{\partial x}\left(u' + \frac{1}{2}w'^2\right) = 0 \\ -m\ddot{w} + (S_0 + S\Phi)w'' + EA\frac{\partial}{\partial x}\left[w'\left(u' + \frac{1}{2}w'^2\right)\right] = 0 \end{aligned}\right\} \quad \textbf{(109a,b)}$$

Moreover, from eq (108) we also obtain the following time condition:

$$EA\left[u(\ell,t) + \frac{1}{2}w'^2(\ell,t)\right] = S_0 + S\Phi(t) \quad \textbf{(110)}$$

Since the derivative with respect to x from eq (109a) is zero, then the term $u' + \frac{1}{2}w'^2$ is an exclusive function with respect to time t, which due to eq (110) will be equal to $S_0 + S\Phi(t)$. Hence, eq (109b) can be written as:

$$-m\ddot{w} + 2[S_0 + S\Phi(t)]w'' = 0 \quad \textbf{(111)}$$

Next, we will determine the shape function for the freely vibrating cable under the tensile force S_0. Given that the bending stiffness of the cable is almost zero - the bending stiffness of a cable is actually very small, see also a relative study in Ref [10] - the governing equation for free vibration will be:

$$S_0 w'' - m\ddot{w} = 0 \quad \textbf{(112)}$$

Seeking a solution in the form of separate variables $w = W(x)\cdot T(t)$ and following the well-known procedure, the above equation will result the following two uncoupled equations:

$$\left.\begin{aligned} S_0 W'' + \omega^2 m W = 0 \\ \ddot{T} + \omega^2 T = 0 \end{aligned}\right\} \qquad \textbf{(113)}$$

The solution to eq (113a) is:

$$W = c_1 \sin\lambda x + c_2 \cos\lambda x$$
$$\text{where: } \lambda^2 = \frac{m\omega^2}{S_0}$$

Employing the boundary conditions $w(0) = w(\ell) = 0$, we obtain the shape functions and corresponding eigenfrequencies for the vibrating cable:

$$\left.\begin{aligned} & W_n(x) = \sin\frac{n\pi x}{\ell} \;,\quad (n = 1,2,...) \\ & \omega_n^2 = \frac{n^2\pi^2 S_0}{m\ell^2} \end{aligned}\right\} \qquad \textbf{(114a,b)}$$

For eq (111), we seek a solution in the form: $w(x,t) = W_\kappa f_\kappa$ which introduced into eq (111) gives: $mW_\kappa \ddot{f}_\kappa - 2[S_0 + S\Phi(t)]W_\kappa'' f_\kappa = 0$ or due to eq (113a), it is:

$$\ddot{f}_\kappa + \frac{2\omega_\kappa^2}{S_0}[S_0 + S\Phi(t)]f_\kappa = 0$$

or, introducing ω_κ from eq (114b), we get:

$$\ddot{f}_\kappa + \frac{2n^2\pi^2}{m\ell^2}\cdot[S_0 + S\Phi(t)]f_\kappa = 0$$

that can be also written as:

$$\left.\begin{aligned} & \ddot{f}_\kappa + \Omega_\kappa^2\,[1 - 2\mu\Phi(t)]f_\kappa = 0 \\ & \text{where: } \Omega_\kappa = \frac{2n^2\pi^2 S_0}{m\ell^2} \;,\quad \mu = -\frac{S}{2\Omega_\kappa} \end{aligned}\right\} \qquad \textbf{(115)}$$

This is the classical form of eq (7).

DYNAMIC INSTABILITY PROBLEMS IN SUSPENSION BRIDGES

General

After the collapse of the Tacoma bridge in 1940, numerous studies have been conducted and published.

Shortly after the collapse of the Tacoma bridge, Th. Karman has published an article [9], where an attempt to explain the collapse of the deck due to static stability loss is made, caused by action of the aerodynamic moments.

In almost all studies that followed, the problem of aerodynamic stability of the deck has been studied based on the principles of the flutter theory. This theory (see also Chapter 7) is used to obtain approximate solutions but in the same manner as in similar analyses of airplanes wings. Thus, in the majority of studies, the deck of bridge is considered as a thin plate with infinite length. The actual fact that the characteristics of a deck are essentially different than those of a plate and thus, they should be determined experimentally, is completely ignored.

Experimental studies brought a new light in the problem of aerodynamic stability of decks. Experiments in aerodynamic tunnels showed that the excitation of the structure in oscillation has a small resemblance to the phenomenon of flutter and in particular to fluttering that takes place in airplanes wings.

On the contrary, serious clues exist that these phenomena are more similar to the parametric resonance problems. It is consequently obvious, that for the excitation in parametric resonance, the presence of external periodical excitations is essential.

Since at the dynamic stability loss of a suspension bridge large deformations are developed, we are compelled to take into account their influence on the change of intensive sizes.

Let us consider the simply supported beam shown in Fig. **12a**, where at position x the bending moment $M_y(x)$ is developed and if we assume a deformed state as shown in Fig. **12b,c**, then the following intensity sizes are developed at x:

$$\left.\begin{aligned} \bar{M}_x &= M_y \sin(\arctan \upsilon') \cong M_y \upsilon' \\ \bar{M}_y &= M_y \cos\theta \cong M_y \\ \bar{M}_z &= M_y \sin\theta \cong M_y \theta \end{aligned}\right\} \quad \textbf{(116)}$$

We also assume, without loss of generality and in order to simplify the derived expressions, that the cross-section is doubly-symmetric, and hence $y_M = z_M = 0$. Finally, we assume that moment M_y is caused by the load:

$$p_z = p_0 + p_t \cos\varphi t \quad \textbf{(117)}$$

and hence:

$$M_y = \frac{x(\ell - x)}{2} \cdot (p_0 + p_t \cos\varphi t) \quad \textbf{(118)}$$

Then, using the notation in chapter 6 we will have:

$$\begin{aligned} \Sigma P_z &= g + p_z + 2(H_g + H_p)\cdot(z + w)'' - m\ddot{w} = \\ &= g + p_z + 2H_g z'' + 2H_p z'' + 2H_g\left(1 + \frac{p_z}{g}\right)w'' - m\ddot{w} = p_z + 2H_p z'' + 2H_g\left(1 + \frac{p_z}{g}\right)w'' - m\ddot{w} \end{aligned}$$

where H_g is the tensile force of the cable. Thus, eq(168) of chapter 6 can be written as:

$$\left.\begin{aligned} & EI_z w'''' - 2H_g\left(1 + \frac{p_z}{g}\right)w'' + m\ddot{w} = 2H_p z'' + p_0 + p_t \cos\varphi t \\ & EI_z \upsilon'''' + M_y \theta'' + m\ddot{\upsilon} = 0 \\ & EC_S \theta'''' - GJ_d \theta'' + M_y \upsilon'' + \Theta_x \ddot{\theta} = ez''\cdot(H_{p_1} - H_{p_2}) + e^2\theta''(H_1 + H_2) \end{aligned}\right\} \quad \textbf{(119a,b,c)}$$

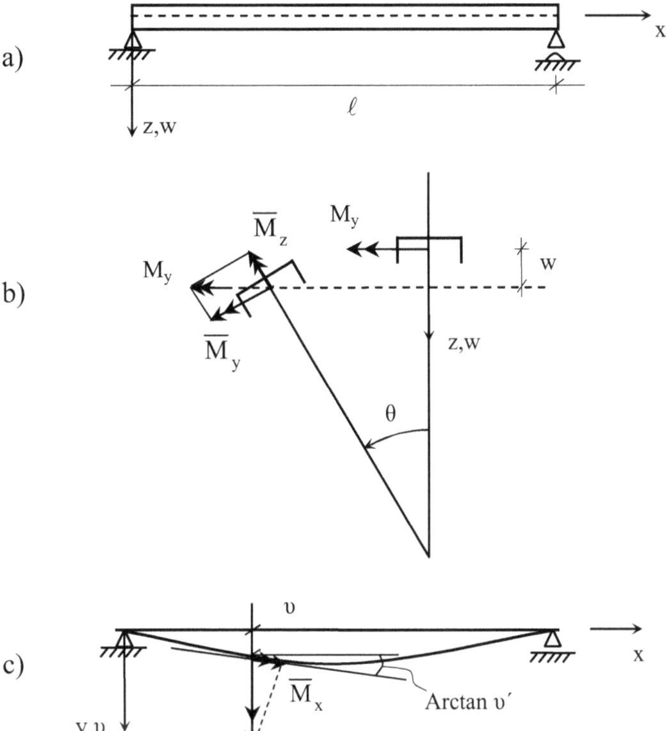

Figure 12: Deformed state of a bridge

Dynamic Instability in Bending

In this case, it will be $\upsilon = \theta = 0$ and hence, eq (119a) as in chapter 6 will take the form:

$$EI_y w'''' - 2H_g\left(1+\frac{p_0}{g}+\frac{p_t}{g}\cos\varphi t\right)w'' + m\ddot{w} = -\frac{2z''^2}{\dfrac{L_c}{E_c F_c}}\cdot\int_0^{\ell} w dx + p_0 + p_t\cos\varphi t$$

We seek a solution in the form: $w(x,t) = W_\kappa f_\kappa(t)$, where $W_\kappa(x)$ is the shape function given by eq(176) of chapter 6 and satisfies eq(174a) of the same chapter for free vibration. Thus, we easily conclude to the following relation:

$$mW\ddot{f} + 2\omega_\kappa^2 W f - 2H_g\frac{p_0}{g}W'' f - 2H_g\frac{p_t}{g}W''\cos\varphi t\cdot f = p_0 + p_t\cos\varphi t$$

where ω_κ are the eigenfrequencies of the bridge for bending vibrations from eq(178) of chapter 6. Then, multiplying by W and integrating from 0 to ℓ we will get:

$$\ddot{f} + \left(\frac{2\omega_\kappa^2}{m} - 2H_g\frac{p_0}{gm}\cdot\frac{\int_0^\ell W'^2 dx}{\int_0^\ell W^2 dx} - 2H_g\cdot\frac{p_t}{gm}\cdot\frac{\int_0^\ell W'^2 dx}{\int_0^\ell W^2 dx}\cos\varphi t\right)\cdot f = \frac{p_0}{m\int_0^\ell W^2 dx} + \frac{p_t}{m\int_0^\ell W^2 dx}\cos\varphi t$$

or finally:

$$\left.\begin{aligned}
&\ddot{f}+\Omega_{\kappa}^{2}\left(1-2\mu\cos\varphi t\right)\cdot f=A+B\cos\varphi t\\
&\text{where:}\qquad \Omega_{\kappa}^{2}=\frac{2\omega_{\kappa}^{2}}{m}-2H_{g}\frac{p_{0}}{g\cdot m}\cdot\frac{\int_{0}^{\ell}W'^{2}dx}{\int_{0}^{\ell}W^{2}dx}\\
&\mu=\frac{H_{g}}{\Omega_{\kappa}^{2}}\frac{p_{t}}{mg}\cdot\frac{\int_{0}^{\ell}W'^{2}dx}{\int_{0}^{\ell}W^{2}dx}\\
&A=\frac{p_{0}}{m\int_{0}^{\ell}W^{2}dx}\quad,\quad B=\frac{p_{0}}{m\int_{0}^{\ell}W^{2}dx}
\end{aligned}\right\}\qquad(120)$$

which has the form of eq (41).

Lateral-Torsional Dynamic Instability

Considering the case of pure lateral-torsional dynamic instability and due to eq(180) of chapter 6 we obtain the following relations:

$$EI_{z}\upsilon''''+M_{y}\theta''+m\ddot{\upsilon}=0$$

$$EC_{S}\theta''''-(GJ_{d}+e^{2}H_{g})\theta''+M_{y}\upsilon''+\Theta_{x}\ddot{\theta}=B\int_{0}^{\ell}\theta dx$$

or due to eq (118):

$$\left.\begin{aligned}
&EI_{z}\upsilon''''+\frac{x(\ell-x)}{2}\cdot(p_{0}+p_{t}\cos\varphi t)\theta''+m\ddot{\upsilon}=0\\
&EC_{S}\theta''''-(GJ_{d}+e^{2}H_{g})\theta''+\frac{x(\ell-x)}{2}\cdot(p_{0}+p_{t}\cos\varphi t)\upsilon''+\Theta_{x}\ddot{\theta}=B\int_{0}^{\ell}\theta dx
\end{aligned}\right\}\qquad(121)$$

We seek solution in the form:

$$\left.\begin{aligned}
\upsilon(x,t)&=V(x)f_{\upsilon}(t)\\
\theta(x,t)&=\Phi(x)f_{\theta}(t)
\end{aligned}\right\}\qquad(122)$$

where V and Φ are the κ shape functions, given by eq(188b) and eq(191a) of chapter 6, respectively, and satisfy also eq(182a,b) of the same chapter for free vibration. Thus, we easily conclude to the relations:

$$mV\cdot\ddot{f}_{\upsilon}+\frac{x(\ell-x)}{2}\cdot(p_{0}+p_{t}\cdot\cos\varphi t)\Phi''f_{\theta}+m\omega_{y}^{2}Vf_{\upsilon}=0$$

$$\Theta_{x}\Phi\cdot\ddot{f}_{\theta}+\frac{x(\ell-x)}{2}\cdot(p_{0}+p_{t}\cdot\cos\varphi t)V''f_{\upsilon}+\Theta_{x}\omega_{\theta}^{2}\Phi f_{\theta}=0$$

Multiplying the first by V and integrating from 0 to ℓ and then, multiplying the second by Φ and integrating from 0 to ℓ, we obtain:

$$\ddot{f}_\upsilon + \omega_y^2 f_\upsilon + \frac{\int_0^\ell \frac{x(\ell - x)}{2}\Phi''V dx}{m\int_0^\ell V^2 dx}\cdot(p_0 + p_t\cos\varphi t) f_\theta = 0$$

or

$$\ddot{f}_\theta + \omega_\theta^2 f_\theta + \frac{\int_0^\ell \frac{x(\ell - x)}{2}V''\Phi dx}{\Theta_x\int_0^\ell \Phi^2 dx}\cdot(p_0 + p_t\cos\varphi t) f_\upsilon = 0$$

$$\left.\begin{aligned}
&\frac{1}{\omega_y^2}\ddot{f}_\upsilon + f_\upsilon - \alpha_1(p_0 + p_t\cos\varphi t) f_\theta = 0\\
&\frac{1}{\omega_\theta^2}\ddot{f}_\theta + f_\theta - \alpha_2(p_0 + p_t\cos\varphi t) f_\upsilon = 0\\
&\text{where:}\quad \alpha_1 = -\frac{\int_0^\ell \frac{x(\ell - x)}{2}\Phi''V dx}{m\omega_y^2\int_0^\ell V^2 dx}\\
&\alpha_2 = -\frac{\int_0^\ell \frac{x(\ell - x)}{2}V''\Phi dx}{\Theta_x\omega_\theta^2\int_0^\ell \Phi^2 dx}
\end{aligned}\right\}\qquad \textbf{(123a,b,c,d)}$$

The above equations (123a,b) in matrix form are:

$$\left.\begin{aligned}
&\Omega\ddot{F} + [E - p_0 A - p_f A\cdot\cos\varphi t)\cdot F = 0\\
&\text{where:}\quad F = \begin{bmatrix} f_\upsilon \\ f_\theta \end{bmatrix}\\
&\Omega = \begin{pmatrix} \frac{1}{\omega_y^2} & 0 \\ 0 & \frac{1}{\omega_\theta^2} \end{pmatrix}\\
&A = \begin{bmatrix} 0 & \alpha_1 \\ \alpha_2 & 0 \end{bmatrix}
\end{aligned}\right\}\qquad \textbf{(124a,b,c,d)}$$

The above equation has the same form as eq (51) and can be solved in a similar manner.

REFERENCES

[1] N.M. Beliaev *"Stability of prismatic rods subjected to variable longitudinal forces"*, 1924, Collection of papers: Engineering Construction and Structural Mechanics (Inzheneruye sooruzhoniia I stroitel' naia mekhanika), Put Leningrand.

[2] N.W. McLachlan *"Theory and applications of Mathieu functions"*, 1964, Dover Publications Inc. New York.

[3] V.V. Bolotin *"The dynamic stability of elastic systems"*, 1964, Holden – Day Inc. London.

[4] M.J.O. Strutt *"Lamesche, Matheusche und verwandte Functionen in Physic and Technik"*, 1932, Springer Verlag, Berlin.

[5] E.T. Whittaker, G.N. Watson *"A course of modern analysis"*, 1952, University Press, Cambridge.

[6] G.T. Michaltsos *"Thin-walled Metal Structures – Calculation Procedures"*, 2008, Symeon Publ., Athens (in Greek).
[7] V.V. Bolotin *"Dynamic stability of plane bending forms"*, 1953, Iuzh. St. 14.
[8] G.T. Michaltsos, J.X. Raftoyiannis *"Stability of a cable - stayed bridge's pylon under time-depended loading"*, 7th Nat. Congress on Mechanics, June 2004, Chania, Greece.
[9] Th. Karman *"Aerodynamic stability of suspension bridges"*, 1940, Engineering News, Record 125.
[10] C. Kollbrunner, N. Hajdin, B. Stipanić *"Contribution to the Analysis of cable-stayed Bridges"*, 1980, Verlag Schulthess A.G., Zürich.
[11] D. Bruno, V. Golotti *"Vibration analysis of cable-stayed bridges"*, 1994, Journal Structure Ing. International, 1.

CHAPTER 9

Damping Systems

Abstract: In this chapter, the most characteristic cases of damping systems in bridges are presented. Damping systems are categorized into external damping systems (such as special bearings) and internal damping systems (such as masses and dampers in the bridge) and their modeling aspects are presented in detail.

INTRODUCTION [1]

The growth of materials technology, production and construction methods, but also the exceptional improvement of numerical methods and computers, resulted to lighter, very flexible and architecturally bolder constructions. Simultaneously, however, the requirements for safety, comfort and functionalism were increased. Thus, the efforts of designers for control of structural vibrations caused mainly by dangerous or not dynamic loads were also intensified. These efforts have led to the growth of the so-called structural control systems. The control of structural vibrations can be achieved via a suitable combination of mass and stiffness that influences the basic dynamic characteristics of the structure. Recently, however, it became popular to achieve the control of structural vibrations via the improvement of their ability to dissipate the dynamic energy.

For convenience, the damping factor is expressed in a non-dimensional form, as the percentage of the dissipated energy over the total energy, also known as **damping ratio**. Note that for steel structures, the desirable value for the damping ratio is near 1%, while for concrete structures near 2%. Practically, we can notice that a damping ratio 1% corresponds to a reduction of the initial amplitude of oscillation to half after roughly 11 cycles.

The improvement of the structural ability to dissipate energy is achieved via the above systems, which aim to absorb the main percentage of the energy introduced into the structure by dynamic loads.

Finally, it should be also noted that while the bearing elements of a structure suffer the main damages due to dynamic loads, in a structure with damping systems the damage is allocated to these systems.

In general, the repair of damages in a damping system has much less cost than the structure itself and can be achieved more easily. This happens because the damping systems are not designed to undertake gravity loads and hence, they can be removed and repaired without risk and without interrupting the service of the structure.

Damping systems can be placed on the structure either externally or internally.

In the first case (external damping type), systems of seismic insulation placed in the foundation is the most usual case, while in second case (internal non - insulation type) the energy absorption systems (energy dissipation devices) are the most representative ones.

Finally, damping systems are mainly distinguished according to their dynamic characteristics and functionality as follows:

- systems causing change of stiffness in the structure, in order to shift the fundamental period of the structure away from the dominating periods of the dynamic loads.
- systems aiming to the increase of the damping ratio of the structure, which results to displacements and accelerations control.
- finally, systems aiming to the re-distribution of the structural mass and as a consequence, to the change of the fundamental period of the structure, as in the first category.

DAMPING SYSTEMS CATEGORIES

It is established to classify the damping control systems into the following three categories:

George T. Michaltsos and Ioannis G. Raftoyiannis

- Passive control devices. These systems function autonomously without external energy offer. They are very simple and have relatively small cost.
- Active or semi - active control devices. These systems require external energy offer in order to operate. They constitute the so-called "smart" systems and their operation is based on electronic means of particularly advanced technology.
- Hybrid control devices. These systems constitute a combination of the above two categories systems, since the external energy offer is not essential for their operation but strengthens their efficiency.

The Passive Control Systems

As already reported, passive control systems do not require external energy for their operation. They constitute the so-called "inactive" systems. These systems mainly fall into the following three sub-categories:

Systems of Seismic Insulation [2]

Seismic insulation in bridges is achieved with the use of dampers placed between the deck and the piers or abutments.

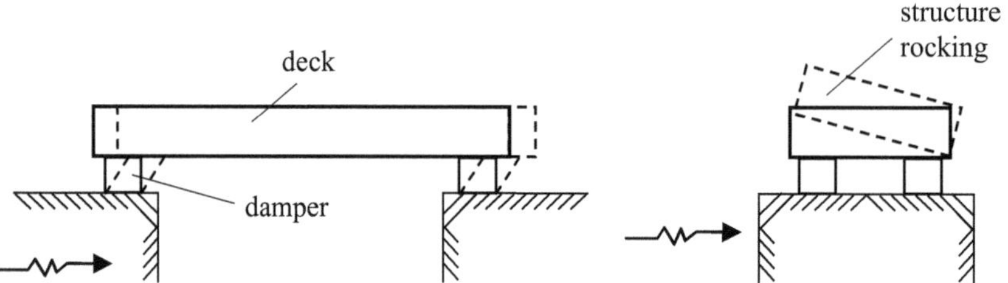

Figure 1: Structural system with seismic insulation

Dampers receive the main part of dynamic displacements due to their small horizontal stiffness, while the structure itself experiences very small relative deformations and hence, small strains. On the contrary, the vertical stiffness of the dampers is very high, thus contributing to the improvement of the structural behavior against rocking.

The most characteristic types of dampers are the following:

1. **Elastomeric dampers** (Fig. **2**). Into this category fall the common bearings made from natural or synthetic flexible materials with low damping (ELB – Elastometallic or Elastomeric Bearings).

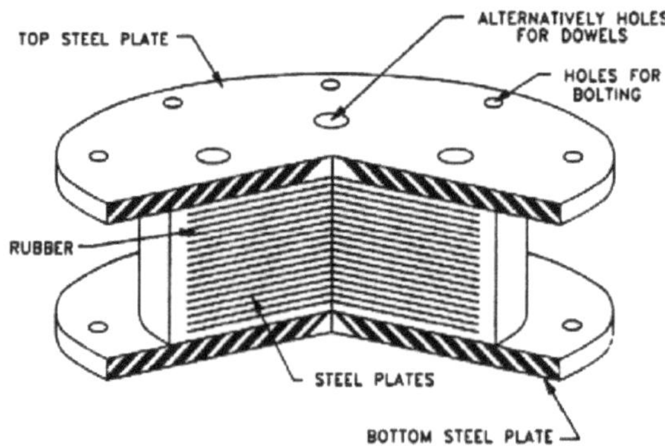

Figure 2: Elastomeric bearing

2. **Slipping dampers** (Fig. **3**). These dampers are based on slipping. Characteristic examples constitute the friction systems based on the pendulum principle (FPS – Friction Pendulum Systems). It is noted that these systems have a negligible influence on the increase of damping in the superstructure and hence, they are considered to be not very efficient.

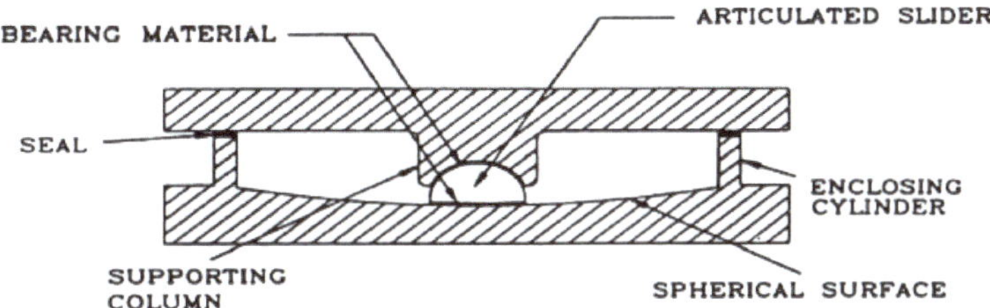

Figure 3: Slipping bearing

3. **Highly efficient dampers** (Fig. **4**). In this category of dampers fall the elastomeric bearings with core of lead for the increase of the damping ratio (LRB – Lead Rubber Bearings). These systems combine the reduction of stiffness with the increase of damping in the superstructure in the order of 10% to 20%.

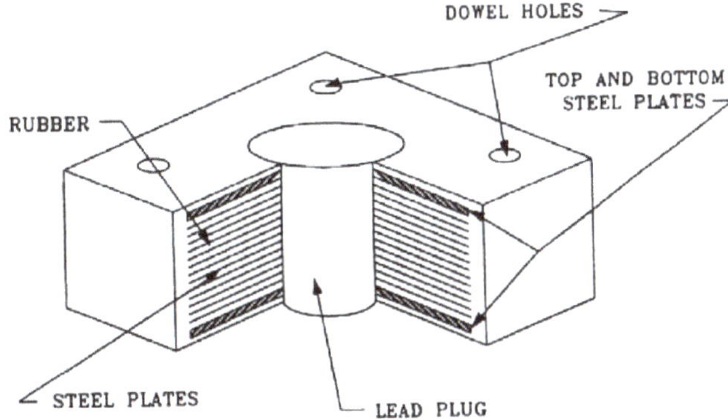

Figure 4: Elastomeric bearing

Energy Absorption Devices [3]

These devices are used mostly in stiffness systems of tall buildings and less in bridges (mostly in truss bridges) and even less in the foundation of bridges.

They are relatively small devices that are placed in the structure aiming at the increase of dissipated energy and consequently, the reduction of stresses and strains in the structure. The most characteristic types are the following:

1. **Metal dampers** (Fig. **5a**). Metal Dampers dissipate energy via the viscous behavior their metal parts when they are deformed in the plastic region (Fig. **5b**). It should be noted that these dampers have the capability to increase also the stiffness of the structure and thus, they are often called A.D.A.S. – Added Damping and Stiffness.

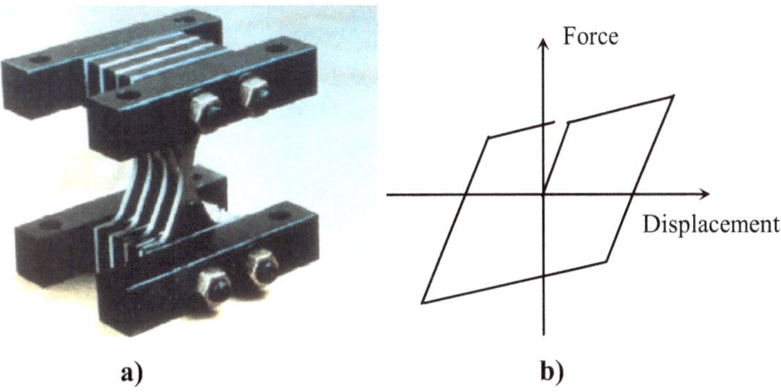

Figure 5: Metal dampers with elastoplastic behavior

2. **Friction dampers** (Fig. **6a**). This type of dampers dissipates energy via the developing friction. They have the capability of increasing the ability of the structure to dissipate energy in the range of 10% to 20% without particular change of the fundamental period. These dampers correspond to an almost orthogonal hysteretic loop (Fig. **6b**). The most usual type has also longitudinal holes, where bolts are placed (S.B.S. – Slotted Bolted Connection).

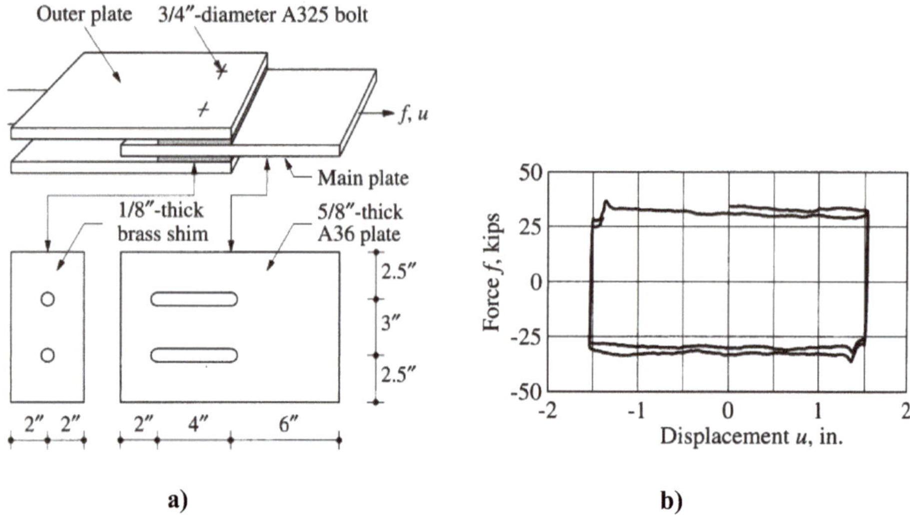

Figure 6: Friction dampers and corresponding hysteretic loop

3. **Viscous and viscoelastic dampers** (Figs. **7** & **8**). This type of dampers is based on the ability of various polymers to dissipate energy via shearing deformation. They have the capability of increasing the ability of the structure to dissipate energy in the range of 10% to 20%. Polymers can be in fluid (Fig. **7a**) or solid state (Fig. **8a**).

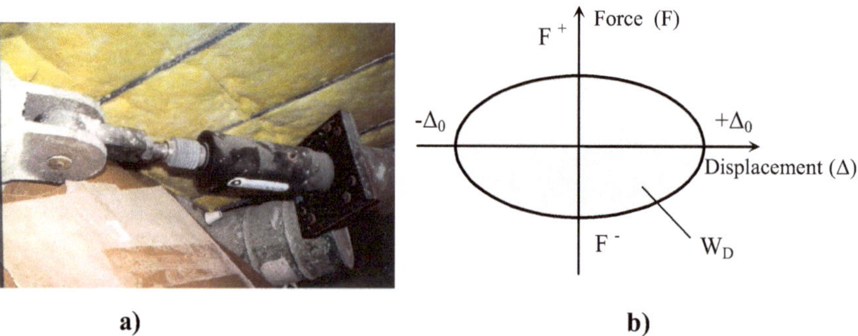

Figure 7: Viscous dampers in fluid state

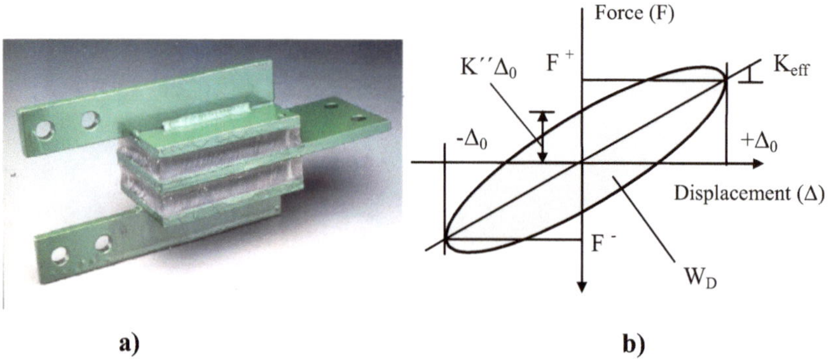

Figure 8: Viscous dampers in solid state

Mass Dampers [4]

Mass dampers are appliances with particularly big mass, which are connected elastically to the structure via springs and/or dampers occasionally. Their operation is based on developing particularly high inertia forces that are opposed to the inertia forces of the structure developed by its oscillation. In order that mass dampers are effective, their eigenfrequency should approach one of the structure eigenfrequencies. For this reason, mass dampers are often called also tuned mass dampers (TMD). TMD have been proved particularly effective in flexible structures and constitute, in various forms, the most usually applied type.

In the following Fig. **10**, one can see various types of mass dampers. From these types, a, b, c and d types are based on the pendulum principle. This can be simple (rare case) or it can be connected to the structure via a damper or a damper and a spring. This system, in the form of fret (for height savings), is commonly used in tall buildings, while in bridges the most common type is the pendulum with damping and springs.

One of the most usually preferred systems is that of a mass with rotational springs (e), but mainly the freely slipping mass with spring (f) or with spring and dampers (g).

Characteristic examples of using such a damping system are the two pedestrian bridges in Bellagio - Las Vegas.

The two pedestrian bridges have openings 46m and 38m. Architectural requirements but also local restrictions, have led to the design of structures with relatively small height 2m. The eigenfrequencies of the bridges were inside the interval of frequencies that may result from the movement of crowd of pedestrians on the bridges. For the energy absorption that results in such cases from pedestrians, a system of 6 TMDs (Fig. **11**) with weight 1.330kgr each have been placed in each bridge. After their placement, the dampers were tuned on site to the eigenfrequencies of the bridges. According to the measurements and checks performed on the bridges, it was realized that their damping ratio became ten times higher. More specifically, the damping ratio was increased from 0,6% to 8%. In order that the dampers are not in public view, they were placed in the interior of the bridge elements. The final cost of dampers installation and maintenance was smaller than the cost required to strengthen the bridges structure.

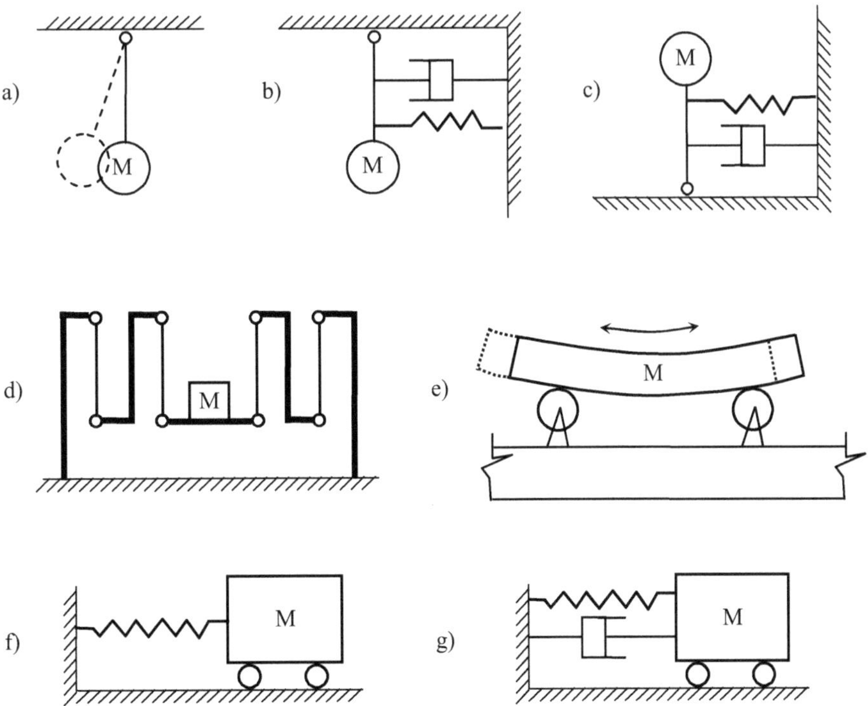

Figure 10: Tuned mass damping systems

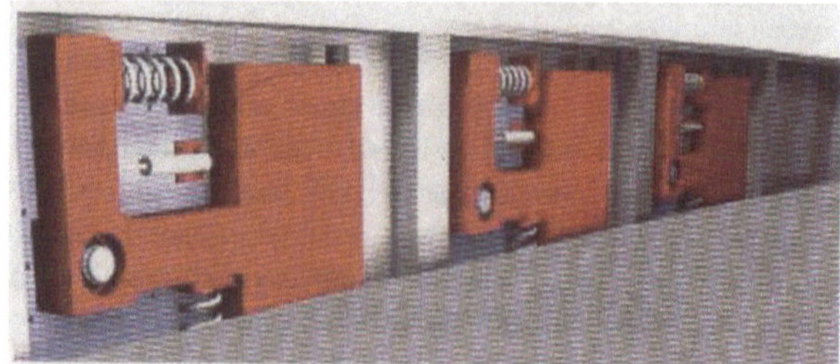

Figure 11: Tuned mass dampers (TMD) placed in a bridge (detail photo)

Active and Semi-Active Control Systems [5]

Active Control Systems

The various existing active control systems have the following basic differences with regard to passive control systems:

1. Contrary to passive systems, that are inactive appliances, active systems are active appliances that require particularly big quantities of energy, since they also apply big forces to the structure.
2. Part of a typical active control system constitute the sensors which are placed in selected positions on the structure in order to measure in real time sizes of deformations or forces that are developing into the structure.
3. Information collected by the sensors is processed also in real time by a controller, that is practically a computer that constitutes the "brain" of the system. The controller, based on the incoming information, determines with precision the forces that should be applied via special mechanisms (actuators) to the structure in order to minimize its response.

Active control systems are usually functioning automatically, that is with feedback information. The operation principle of such a system is presented in the diagram of Fig. **12**.

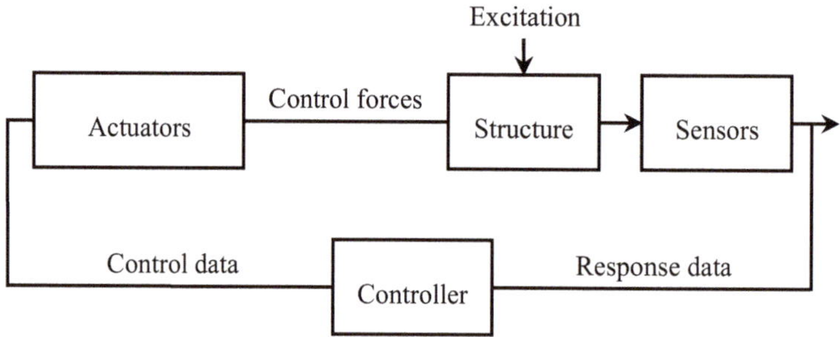

Figure 12: Active control system operation diagram

Semi-Active Control Systems

The semi-active systems have the same operation principle with the active systems, but require exceptionally less energy for their operation. Consequently, they are not in position to apply forces capable to maintain the structure motionless, however they are capable to stop its movement. In other words, semi-active systems can dissipate energy, but are unable to import energy to the manufacture. These systems are simpler, robust, henceforth reliable and more economical with respect to active systems, while their effectiveness is less than that of active systems.

The most characteristic types of this category are dampers with fluids (semi - active fluid dampers). Their operation is based on the transformation of one freely flowing viscous fluid to semi-solid state with controlled escape force and vice-versa. This process is takes place in minimal time (in the order of millisecond), when fluids are exposed to magnetic field, as in the case of magneto-rheological fluids (MR fluids) or to electric field, as in the case of electro-rheological fluids (ER fluids). In Fig. **13**, one can see a 12 tons magneto-rheological fluid damper.

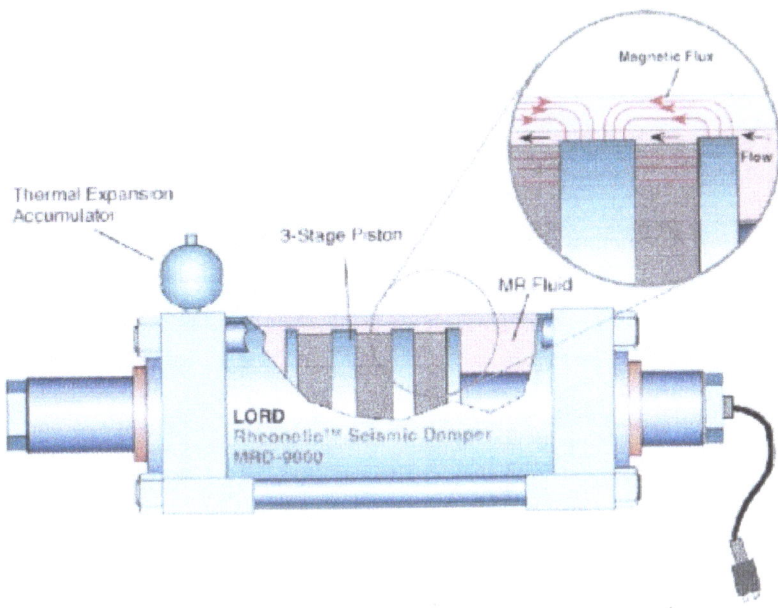

Figure 13: Magneto-reological (MR) fluid damper

Advantages of Active and Semi-Active Dampers [6]

It is generally useful to draw the advantages of active and semi-active control systems.

1. They present increased effectiveness to the control of structural response. The degree of their effectiveness depends almost exclusively on the number of control appliances that has been placed into the structure.
2. They can be applied in order to confront many dangers. The systems of active control are effective against almost all dynamic loadings, as for example seismic or wind loads.
3. They offer big flexibility and possibilities for satisfaction of the design parameters. For example, one can give priority to the functionality and comfort for small wind loads or to the safety against powerful seismic excitations.
4. A major disadvantage of these systems is their dependence on external energy sources that unfortunately are not always offered, particularly during powerful excitations, where these systems are supposed to function. This disadvantage is waived in the case of semi-active systems, which require of course less energy usually supplied by batteries.

Hybrid Control Systems

Hybrid systems constitute combinations of active and semi-active control systems. The main objective of hybrid systems is, obviously, to combine and exploit the advantages offered separately by each system.

In this way, the main part of the control forces can be applied by the passive part of the system, while the active part proceeds to the essential small corrections, aiming at the most optimal response of the structure.

It is obvious that there is still one more advantage that allocates the hybrid systems and this is that in case of loss of their catering in the required energy, they maintain a big part of their effectiveness, in as far as acting as pure passive control systems.

MODELING – BEHAVIOR OF DAMPING SYSTEMS

General – Deformation of Bridges

Both passive and semi-active or hybrid systems function in precisely the same way at the time interval when the intervention of a control system is not required to modify their characteristics or otherwise in as far as their characteristics remain unchanged.

Hence, their modeling and behavior is common. Consequently, we will examine the following systems:

a) Seismic insulation or energy absorption systems

b) Mass damping systems

For better understanding and further analysis, it is advisable at this point to study the deformations at an arbitrary point of a bridge, where we intend to place a damping system and in general the dynamic response of the bridge under dynamic, seismic or wind loading.

a) In Fig. **14a**, one can see the deformed state of a bridge under the action of horizontal dynamic loads. We observe that point Γ will move to position Γ′′, with the following displacement components:

1. The horizontal displacement ($\upsilon_{st}+\upsilon$), where υ_{st} is the dynamic displacement of point Γ to Γ′ resulting from the movement of the bridge as a rigid body, due to ground movement and υ is the elastic deformation component along y-axis of the total elastic deformation of point Γ′ of the bridge. In the case of wind loading, it is of course υ_{st}=0. The case where supports A and B suffer rocking, that is already presented in the end of Chapter 5, may also exist.
2. The vertical displacement component w along z-axis of the total elastic deformation of point Γ′ of the bridge, which in these cases is small.
3. The rotation θ of the cross-section at point Γ′.

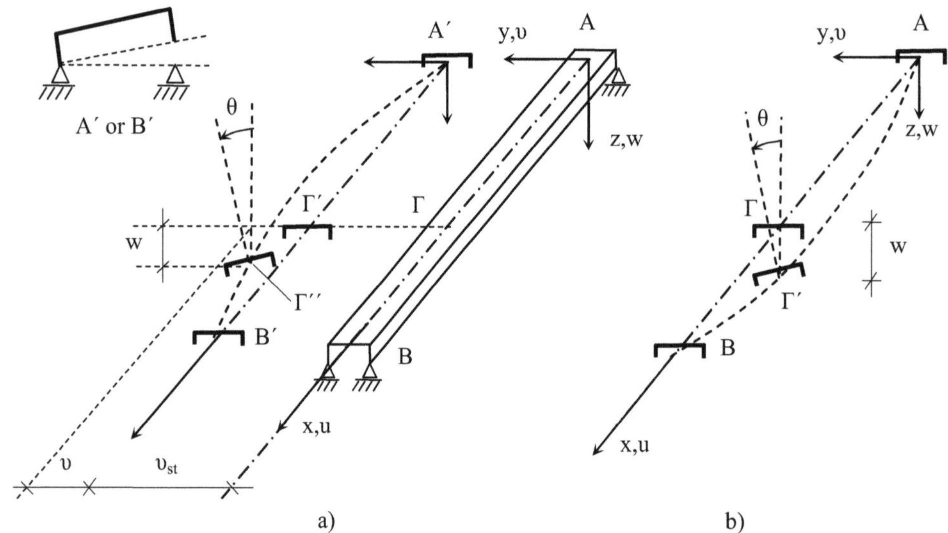

Figure 14: Deformation of a bridge under dynamic loads

b) In Fig. **14b**, one can see the deformations of the same bridge due to dynamic loads acting vertically, that appear when the bridge is found in the neighborhood of an earthquake epicenter with small or big fault depth, and which is associated with particularly intense vertical seismic component, or in the case of crowd resonance in pedestrian bridges. Then, the corresponding w_{st} can be taken as zero.

One DOF Systems without Friction

In this paragraph, we will examine the free oscillation of the basic types of one degree-of-freedom (1-DOF) systems usually employed for modeling the basic damping systems.

The derivation of the corresponding differential equations of motion requires not only good knowledge of mechanics, but also good judgment for the system modeling and right attribute of its characteristics in our effort to conclude to a system, the differential equations of which is possible to handle mathematically.

The Gravity Pendulum

The gravity or mathematical pendulum is schematically shown in Fig. **15**. The total mass of the system is placed at a distance ℓ from the suspension point 0 via a weightless bar. The pendulum movement is assumed to remain in plane, and consequently, it is fully described by the angle of deviation θ from the axis 0S of static equilibrium.

If γ is the acceleration of mass M, the moments of the above forces with respect to point 0 must be in equilibrium, i.e.:

$$M\gamma\ell = -Mg\ell\sin\theta$$

and since $\gamma = \ell\ddot{\theta}$, we will have $M\ell^2\ddot{\theta} = -Mg\sin\theta$ or:

$$J\ddot{\theta} + Mg\ell\sin\theta = 0 \qquad \textbf{(1)}$$

where $J = M\cdot\ell^2$ is the mass inertia of the system.

Expanding the expression for sinθ into Taylor series, we have: $\sin\theta = \theta - \frac{\theta^3}{3!} + \frac{\theta^5}{5!} - .....$

Maintaining only the first term of the series, the error for $\theta=5^0$ is 0,12% for $\theta=10^0$ is 0,51% for $\theta=15^0$ is 1,14% and for $\theta=20^0$ is 2,03%.

Consequently, we observe that for small angles of deviation ($\theta < 15^0$) the resulting error allows us to set : $\sin\theta \cong \theta$.

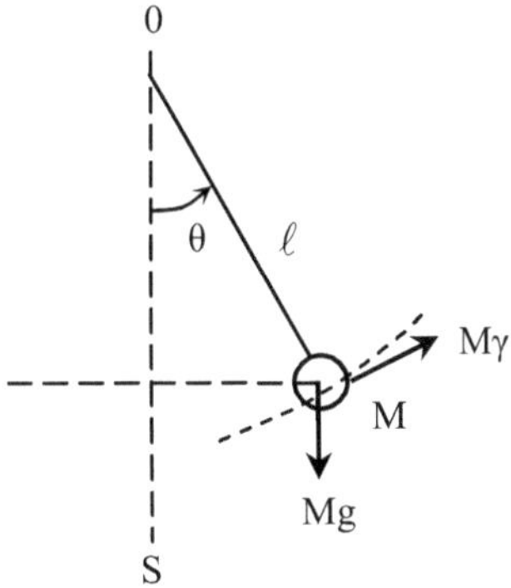

Figure 15: Gravity pendulum

The above eq (1) can then be written in the following form:

$$\ddot{\theta} + \frac{g}{\ell}\cdot\theta = 0 \qquad (2)$$

which constitutes the differential equation for free oscillations of the mathematical pendulum.

The Circular Disc Pendulum

If the mathematical pendulum is replaced by a natural pendulum, its mass will not be concentrated in one point. In Fig. **16**, one can see a circular disk with radius R and total mass M, suspended at point 0 and oscillating freely on its plane.

The initial equation of motion eq (1) is also valid, but J has a different value than the one in the previous paragraph. The rotary inertia of a circular disk with respect to its center K is: $J_K = \dfrac{MR^2}{2}$ and hence, we will have for pole 0 according to Steiner:

$$J_0 = J_K + M\ell^2 = M\left(\frac{R^2}{2} + \ell^2\right)$$

Thus, eq (1) can be written as:

$$\left.\begin{aligned} &\ddot{\theta} + \frac{g}{\lambda}\cdot\theta = 0 \\ &\text{where: } \lambda = \frac{\rho_0^2 + \ell^2}{\ell}\ ,\quad \rho_0 = \frac{R}{\sqrt{2}} \end{aligned}\right\} \tag{3}$$

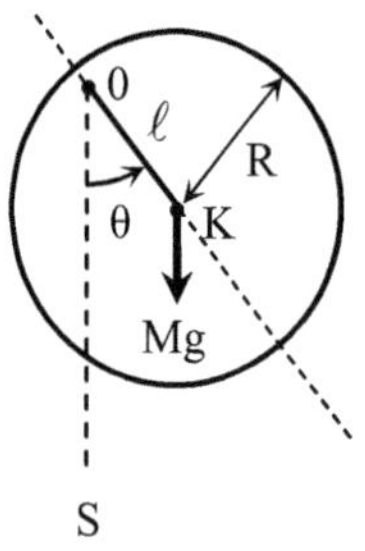

Figure 16: Circular disc pendulum

where ρ_0 is obviously the gyration radius of the disc, while the length λ is the **"length of the equivalent mathematical pendulum"** and corresponds to the length at which the total mass m of the pendulum can be considered being concentrated.

The Spring Pendulum

Let us consider the inclined plane shown in Fig. **17**, with angle $\hat{\alpha}$ with respect to the vertical axis S.

On this plane, a mass M that is suspended from point 0 through a weightless spring with constant k is sliding without friction. If the spring in the unstressed state has an initial length ℓ_0, then after releasing the mass M the spring will deform by $\delta = \dfrac{Mg}{k}\cdot\cos\alpha$ and will be resting at the position β-β, which is the static equilibrium position $(\ell_0 + \delta)$.

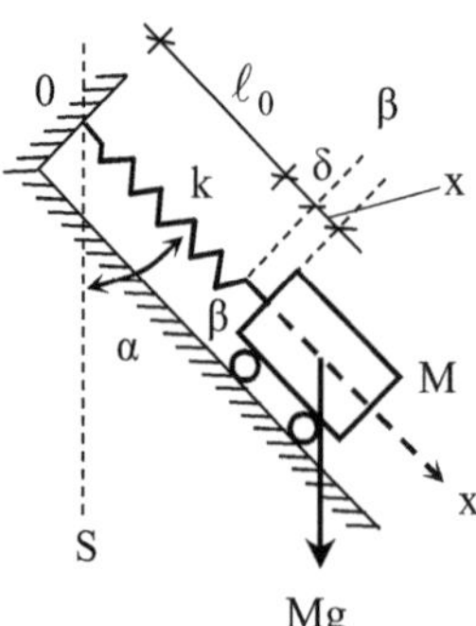

Figure 17: Spring pendulum

Consequently, the position of static equilibrium is altered with respect to angle α. If the mass M is displaced dynamically by x, it will oscillate due to the spring and inertia forces energy, and the equation of motion will be:

$$\ddot{x} + \frac{k}{M} x = 0 \tag{4}$$

that is independent from the angle of inclination plane.

Nevertheless, the biggest force of the spring will develop at the time when the oscillation amplitude becomes maximum and depends on angle α as follows:

$$k(\delta + x_{max}) = k\left(\frac{Mg}{k} \cdot \cos\alpha + x_{max}\right) \tag{5}$$

For $\alpha = 90^0$ it will be: $k \cdot (\delta + x_{max}) = k \cdot x_{max}$ and we will have the case of an oscillating mass on a horizontal plane as shown in Fig. **18**.

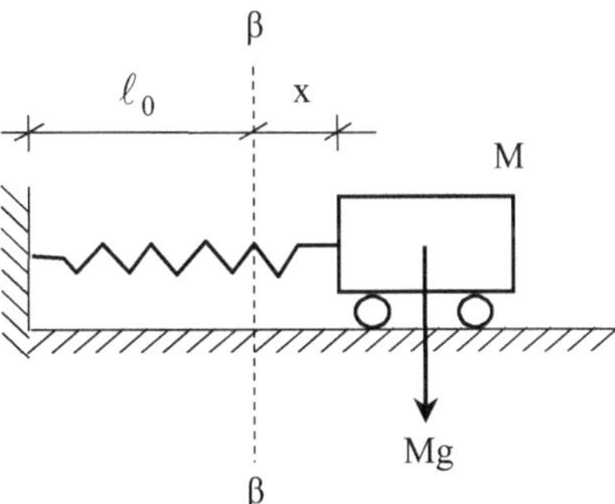

Figure 18: Spring oscillating mass

The Complex or Restrained Pendulum

The complex or restrained pendulum system is shown in Figs **(19a)** and **(b)** and is a combination of the mathematical pendulum and the spring pendulum, i.e. the mathematical pendulum restrained by a tangential spring. We assume that the static equilibrium position is on axis 0S that is determined by the angle α and also the direction on which the spring acts remains always tangential to the orbit of mass M (Fig. **19a**).

Then, the static deformation of the spring will be:

$$\Delta = \frac{Mg}{k} \sin\alpha \tag{6}$$

The deviation angle θ, used to measure the deformed configuration with respect to the axis of static equilibrium 0S, describes the motion of system. Hence, the restitution moment, due to the gravity forces will be:

$$Mg\ell \cdot [\sin(\alpha + \theta) - \sin\alpha] \cong Mg\ell\theta\cos\alpha$$

and the restitution moment due to the spring will be $k\ell^2\theta$ (for small oscillations about 0S). Thus, the differential equation of motion will take the form:

$$\ddot{\theta} + \left(\frac{g}{2} \cdot \cos\alpha + \frac{k}{M}\right) \cdot \theta = 0 \tag{7}$$

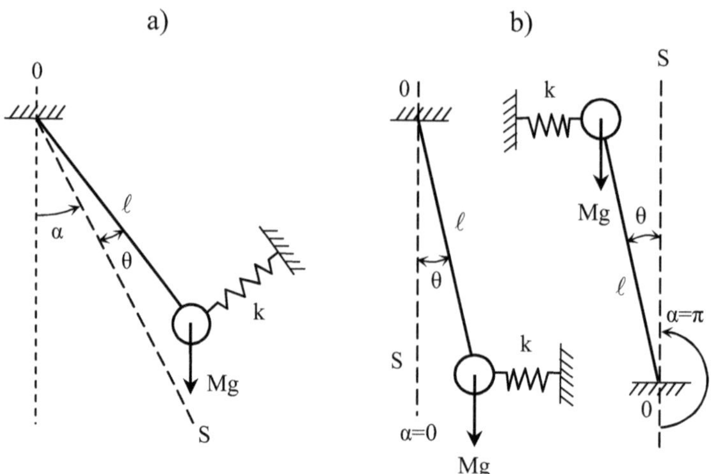

Figure 19: Complex pendulum

Easily, we can write the following equation of motion for the reverse pendulum with $\alpha = \pi$:

$$\ddot{\theta}+\left(-\frac{g}{2}+\frac{k}{M}\right)\theta=0 \tag{8}$$

Finally, if we assume that bar $\overline{0A}$ in Fig. **20** is rigid and weightless as the spring is acted by ℓ_0 from 0, the equation of motion will be the one of the following three:

$$\left.\begin{aligned}
&\ddot{\theta}+\left(\frac{g}{\ell}\cos\alpha+\frac{k\ell_0^2}{M\ell^2}\right)\theta=0 \quad \text{for } \alpha\neq 0\\
&\ddot{\theta}+\left(\frac{g}{\ell}+\frac{k\ell_0^2}{M\ell^2}\right)\theta=0 \quad \text{for } \alpha=0\\
&\ddot{\theta}+\left(-\frac{g}{\ell}+\frac{k\ell_0^2}{M\ell^2}\right)\theta=0 \quad \text{for } \alpha=\pi
\end{aligned}\right\} \tag{9a,b,c}$$

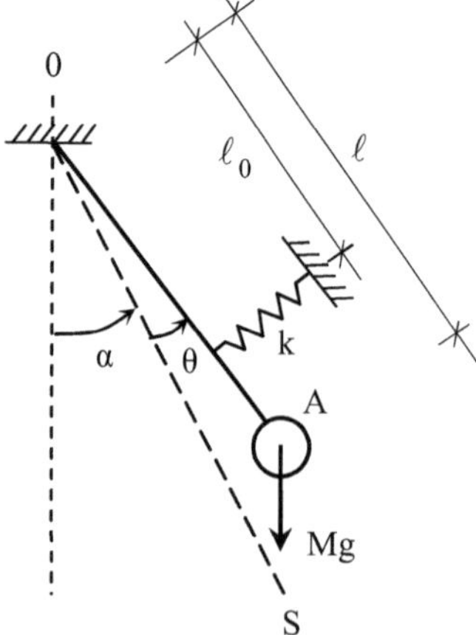

Figure 20: Complex pendulum

Rolling Pendulum

One part of a circular ring (Fig. **21**) is sliding on points Γ_1 and Γ_2 without friction. Following the motion of point K, we notice that it can be defined by angle θ. Then the restitution moment due to gravity forces will be $MgR\sin\theta$ and the general eq (1) can be written as:

$$J\ddot{\theta} + MgR\sin\theta = 0 \tag{10}$$

If ρ_0 is the gyration radius of AB with respect to the center of gravity, we will have: $J_0 = M\rho_0^2$ and hence, it will be: $J = J_0 + MR^2$.

Thus, eq (10) becomes:

$$M(\rho_0^2 + R^2)\ddot{\theta} + MgR\sin\theta = 0$$

or since $\sin\theta \cong \theta$:

$$\left.\begin{aligned} &\ddot{\theta} + \frac{g}{\lambda}\theta = 0 \\ \text{where: } &\lambda = \frac{R^2 + \rho_0^2}{R} \end{aligned}\right\} \tag{11a,b}$$

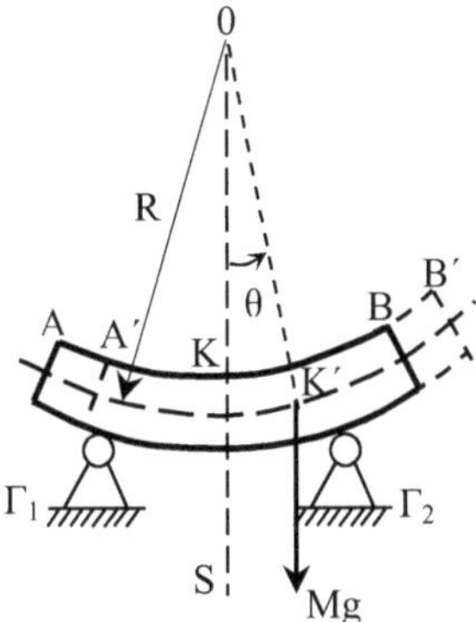

Figure 21: Rolling pendulum

The Behavior of a Damper on a Bridge

In this chapter, we are studying exclusively the behavior of a damper on a vibrating bridge and we determine the governing equation that describes its motion. We conclude that these equations have a common form, so that a single solution can satisfy them all.

Systems of Seismic Insulation or Energy Absorption

As described in the previous paragraphs, these systems are placed in the foundation of bridges. All these systems can be modeled in the same way. In the seismic insulation systems, the spring k is usually of significant importance, while in energy absorption systems is the damping coefficient c. That is the first ones are acting mostly as springs, while the second ones are acting mostly as dampers.

In Fig. **22a**, one can see the displacement of point A of the bridge in Fig. **14**, due to the ground motion υ_{st} and the displacement υ_0 due to bearing deformation. However, the deviation forces (i.e. the inertia forces) result from the deformation of the whole bridge. Consequently, the inertia forces will be:

$$\int_0^\ell M(\ddot{\upsilon}_A + \ddot{\upsilon})dx = \int_0^\ell M(\ddot{\upsilon}_{st} + \ddot{\upsilon}_0 + \ddot{\upsilon})dx$$

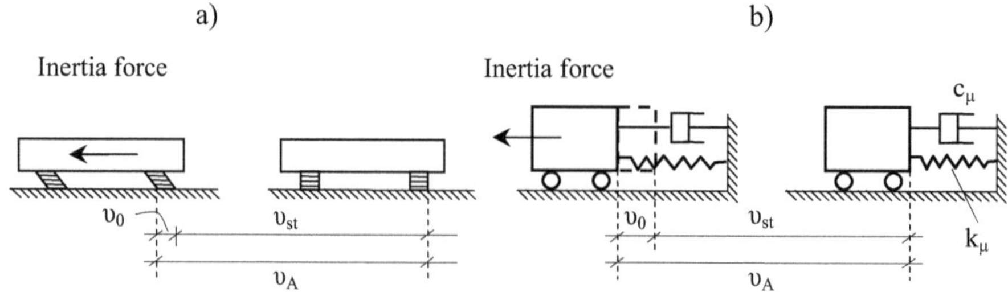

Figure 22: Modeling of a bridge on elastic bearings

These forces must be in equilibrium with the restitution forces:

$$k_\mu \cdot \upsilon_0 + c_\mu \cdot \dot{\upsilon}_0$$

and hence, the following equation is valid:

$$\ddot{\upsilon}_0 + \frac{c_\mu}{M\ell}\dot{\upsilon}_0 + \frac{k_\mu}{M\ell}\upsilon_0 = -\ddot{\upsilon}_{st}(t) - \frac{1}{\ell}\int_0^\ell \ddot{\upsilon}(x,t)\cdot dx \quad \textbf{(12)}$$

The above equation of motion is solved as usually and gives:

$$\left.\begin{aligned} &\upsilon_0(t) = -\frac{1}{\bar{\omega}_\mu}\cdot\int_0^t [\ddot{\upsilon}_{st}(\tau) + \frac{1}{\ell}\int_0^\ell \ddot{\upsilon}(x,\tau)dx]\cdot e^{-\beta\cdot(t-\tau)}\sin\bar{\omega}_\mu(t-\tau)\cdot d\tau \\ &\text{where:}\quad \frac{c_\mu}{M\ell} = 2\beta\ ,\ \ \omega_\mu^2 = \frac{k_\mu}{M\ell}\ ,\ \ \bar{\omega}_\mu = \sqrt{\omega_\mu^2 - \beta^2} \end{aligned}\right\} \quad \textbf{(13a,b)}$$

The above expression connects the elastic-dynamic deformations of the bridge with the dynamic deformations of bearings. In the above equations, c_μ and k_μ correspond to the damping and stiffness of all bearings.

Mass Dampers

a. The Simple Pendulum

Let us consider point Γ on the bridge shown in Fig. **14a**, on which a simple pendulum has been placed to serve as a damper.

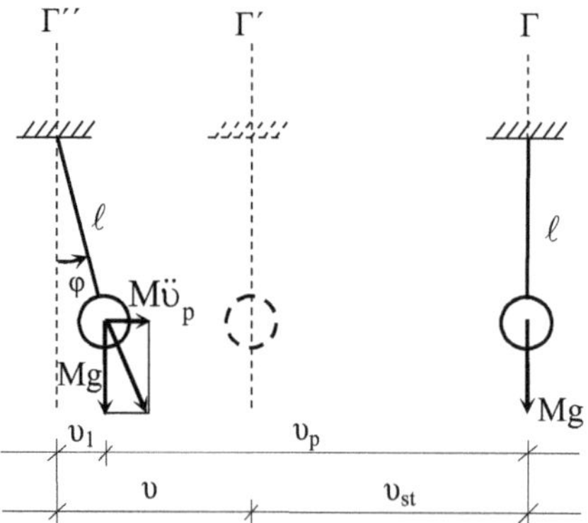

Figure 23: Simple mass damper

Assuming small deflections, the mass acceleration Mγ is almost horizontal and hence, the resulting inertia force can be considered horizontal as well. Then, the total inertia force will be $M\ddot{\upsilon}_p$, while the restitution force is Mg. From equilibrium of the above forces, it results that:

$$\tan\varphi = \frac{M\ddot{\upsilon}_p}{Mg} = \frac{\ddot{\upsilon}_p}{g}$$

and thus:

$$\upsilon_p = \upsilon_{st} + \upsilon - \upsilon_1 = \upsilon + \upsilon_{st} - \ell \sin\varphi = \upsilon + \upsilon_{st} - \frac{\ell}{g}\ddot{\upsilon}_p$$

since φ has been assumed very small and hence, $\varphi \cong \sin\varphi \cong \tan\varphi$.

Thus, the above equation of motion is written as follows:

$$\left.\begin{aligned} &\ddot{\upsilon}_p + \omega_p^2 \upsilon_p = \omega_p^2(\upsilon_{st} + \upsilon) \\ &\text{where:} \quad \omega_p^2 = \frac{g}{\ell} \end{aligned}\right\} \qquad \textbf{(14a,b)}$$

and finally:

$$\upsilon_p(x,t) = \omega_p^2 \int_0^t [\upsilon_{st}(\tau) + \upsilon(x,\tau)] \sin\omega_p(t-\tau)\cdot d\tau \qquad \textbf{(15)}$$

b. The Restrained Pendulum

For the case of the restrained pendulum in Fig. **24**, we assume that mass M is suspended from point Γ through a weightless rigid bar. Then, following the same methodology, the equation of motion will be:

$$\ddot{\upsilon}_p + \frac{c_p}{M}\dot{\upsilon}_p + \left(\frac{g}{\ell} + \frac{k_p \ell_0^2}{M\ell^2}\right)\upsilon_p = \frac{g}{\ell}(\upsilon_{st} + \upsilon) \qquad \textbf{(16)}$$

The above equation has the following solution:

$$\left.\begin{aligned} &\upsilon_p(x,t) = \frac{g}{\ell\bar{\omega}_p} \int_0^t [\upsilon_{st}(\tau) + \upsilon(x,\tau)] e^{-\beta(t-\tau)} \sin\bar{\omega}_p(t-\tau)\cdot d\tau \\ &\text{where:} \quad \frac{c_p}{2M} = \beta \ , \quad \frac{g}{\ell} + \frac{k_p \ell_0^2}{M\ell^2} = \omega_p^2 \ , \quad \bar{\omega}_p = \sqrt{\omega_p^2 - \beta^2} \end{aligned}\right\} \qquad \textbf{(17a,b)}$$

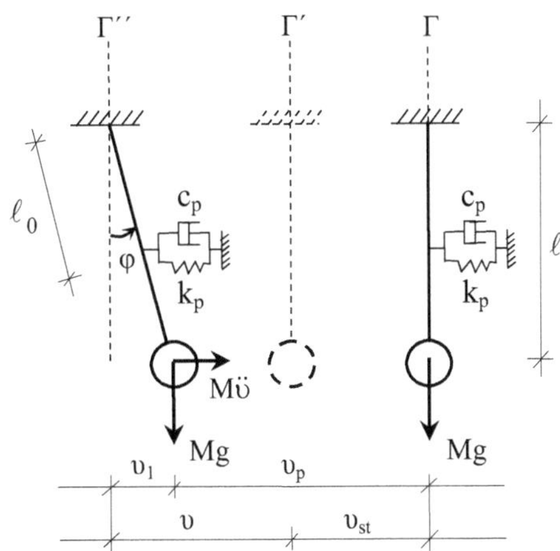

Figure 24: Restrained mass damper

For the reverse pendulum case the above equations are also valid but with:

$$\omega_p^2 = -\frac{g}{\ell} + \frac{k_p \ell_0^2}{M\ell^2} \tag{18}$$

Particular attention should be paid to the choice of spring constant k_p, so that ω_p^2 is positive in eq (18), otherwise the system will be unstable.

c. The Oscillating Mass

For the case of an oscillating mass damper attached at point Γ of the vibrating bridge shown in Fig. **14**, we will have (see Fig. **25**):

$$\upsilon_p = \upsilon_{st} + \upsilon - \upsilon_l$$

and since the deviation forces are equal to the restitution forces, it should be: $M\ddot{\upsilon}_p = \upsilon_l k_p + \dot{\upsilon}_l c_p$, which according to the above will give:

$$\ddot{\upsilon}_p + \frac{c_p}{M}\dot{\upsilon}_p + \frac{k_p}{M}\upsilon_p = \frac{k_p}{M}(\upsilon_{st} + \upsilon) + \frac{c_p}{M}(\dot{\upsilon}_{st} + \dot{\upsilon}) \tag{19}$$

Equation (19) has the solution:

$$\left.\begin{array}{l} \upsilon_p(x,t) = \dfrac{1}{\bar{\omega}_p}\displaystyle\int_0^t [\omega_p^2(\upsilon_{st}(\tau) + \upsilon(x,\tau)) + 2\beta(\dot{\upsilon}_{st}(\tau) + \dot{\upsilon}(x,\tau))]\, e^{-\beta(t-\tau)} \sin\bar{\omega}_p(t-\tau)d\tau \\ \text{where: } \dfrac{c_p}{M} = 2\beta \ , \quad \dfrac{k_p}{M} = \omega_p^2 \ , \quad \bar{\omega}_p = \sqrt{\omega_p^2 - \beta^2} \end{array}\right\} \tag{20a,b}$$

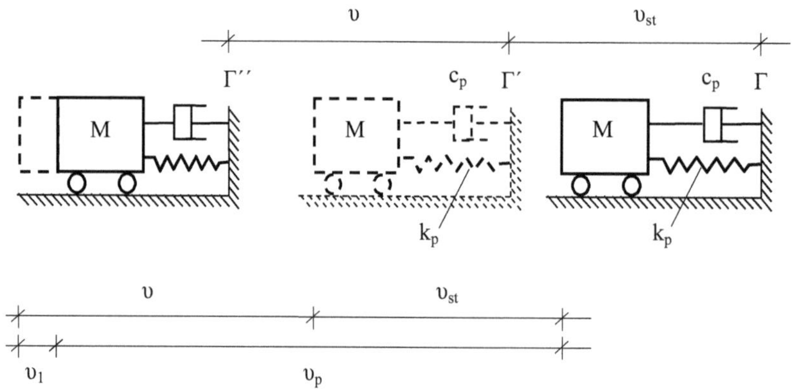

Figure 25: Oscillating mass damper

d. The Rolling Mass Pendulum

For the case of a rolling mass type damper attached to point Γ of the oscillating bridge in Fig. **14**, we will have (see also Fig. **26**): $\upsilon_p = \upsilon_{st} + \upsilon - \upsilon_l$. Since the deviation forces are equal to the restitution forces, we will have:

$$M\frac{\rho_0^2 + R^2}{R}\ddot{\upsilon}_p + Mg\upsilon_p = c_p\dot{\upsilon}_l + k_p\upsilon_l$$

which finally gives:

$$\ddot{\upsilon}_p + \frac{c_p}{M\lambda}\dot{\upsilon}_p + \left(\frac{g}{\lambda} + \frac{k_p}{M\lambda}\right)\upsilon_p = \frac{c_p}{M\lambda}(\dot{\upsilon}_{st} + \dot{\upsilon}) + \frac{k_p}{M\lambda}(\upsilon_{st} + \upsilon) \tag{21}$$

where : $\lambda = \dfrac{\rho_0^2 + R^2}{R}$ and ρ_0 is the gyration radius of the circular ring sector

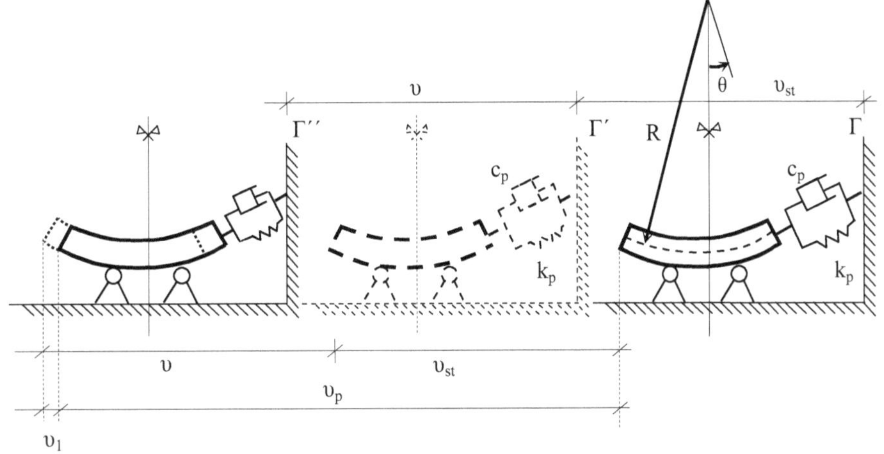

Figure 26: Rolling mass damper

Equation (21) has the following solution:

$$\upsilon_p(x,t) = \frac{1}{\bar{\omega}_p}\int_0^t \left\{ 2\beta\,[\dot{\upsilon}_{st}(\tau) + \dot{\upsilon}(x,\tau)] + \omega_p^2[\upsilon_{st}(\tau) + \upsilon(x,\tau)] \right\} e^{-\beta(t-\tau)} \sin\bar{\omega}_p(t-\tau)d\tau \tag{22}$$

where: $2\beta = \dfrac{c_p}{M\lambda}$, $\omega_p^2 = \dfrac{g}{\lambda} + \dfrac{k_p}{M\lambda}$, $\bar{\omega}_p = \sqrt{\omega_p^2 - \beta^2}$

The Bridge - Dampers System

Design Principles

In this paragraph, we shall study the application of damping systems on a bridge with respect to the phenomenon and the deformation that we are dealing with.

It is obvious that we can apply more the one dampers at the same cross-section of the bridge and/or in various positions along its length.

We should, however, take into account in the design of a bridge some basic principles regarding the use of dampers:

1. In general, the effect of dampers is efficient when they are "in resonance" with some of the bridge eigenfrequencies. With the term "efficient" it is meant the control (reduction of size) of the motion amplitudes but also the intensity sizes.
2. Dampers are functioning more efficiently in the region of frequencies of the external loading near the frequency of the dampers that is called "Operating Frequency Range".
3. Increase of the damping ratio of dampers causes as a rule a reduction of the motion amplitudes and intensity sizes and most important, an increase of operating frequencies range (region of influence

frequencies of dampers). Still, an important result is the elimination of peaks in sizes of the bridge response resulting from resonance to the new eigenfrequencies of the bridge-dampers system.

4. The increase of mass M of the damper leads to increased effectiveness, but also to a wider operating frequencies range. Note that due to the significant change of the eigenfrequencies, it is possible to have an increase in the response sizes, but for frequencies outside the region of the dampers output.

5. In order that the dampers are efficient, they should be placed at positions with big modal amplitudes that correspond to eigenfrequencies of the bridge, to which the dampers are coordinated to.

6. For the control of more than one loading and provided that we use simple passive systems, the use of particular dampers for each loading is recommended.

Loading

It is obvious that the basic deformations of a bridge, that we are supposed to deal with due to dynamic loading are the following three: **the vertical bending deformation, the horizontal bending deformation, the rotational deformation** and finally, a **combination of all the above deformations.** These deformations may appear in an isolated form only in cases of doubly symmetric cross-sections.

a. Vertical Bending Deformation

This case is met when vertical dynamic loads are acting on bridge and is dealed with the placement of dampers in various positions α_i on the bridge or the bridge axis (Fig. **27b**), or in a symmetric arrangement in the same cross-section (Fig. **27c**).

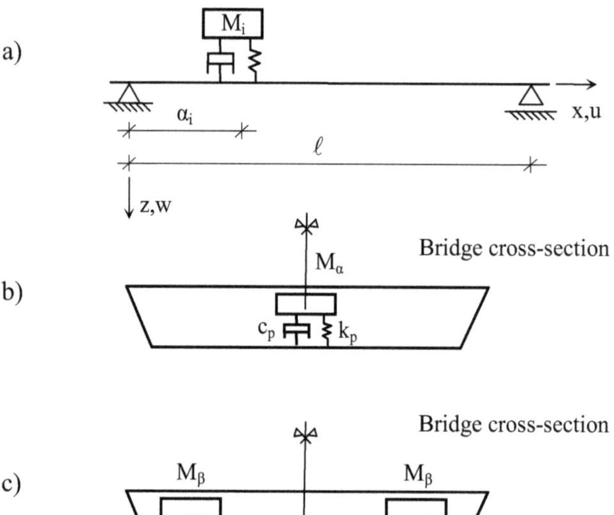

Figure 27: Vertically acting mass dampers

Then, the equation of motion of the bridge takes the form:

$$\frac{\partial^2}{\partial x^2}\left[EJ_y\frac{\partial^2 w}{\partial x^2}\right]+c\frac{\partial w}{\partial t}+m\frac{\partial^2 w}{\partial t^2}+\sum_i M_i\frac{\partial^2 w_p}{\partial t^2}\,\delta(x-\alpha_i)=F_z(x,t) \qquad \textbf{(23)}$$

where w_p is taken from eq (20) by substituting the horizontal deformations υ by the vertical ones w and w_{st}=0 as follows:

$$w_p(x,t) = \frac{1}{\overline{\omega}_p}\int_0^t [\omega_p^2 w(x,\tau) + 2\beta\dot{w}(x,\tau)]\, e^{-\beta(t-\tau)} \sin\overline{\omega}(t-\tau)d\tau \quad \textbf{(24)}$$

In eq (23), $F_z(x,t)$ is the external loading.

Equations (23) and (24) form a system of differential equations of motion with respect to w(x,t) and $w_p(x,t)$, which can be solved numerically.

b. Horizontal Bending Deformation

For a doubly symmetric cross-section, the equation of motion for a beam with dampers at positions α_i will be:

$$\frac{\partial^2}{\partial x^2}\left[EI_z\frac{\partial^2 \upsilon}{\partial x^2}\right] + c\frac{\partial \upsilon}{\partial t} + m\frac{\partial^2 \upsilon}{\partial t^2} + \sum_i M_i \frac{\partial^2 \upsilon_p}{\partial t^2}\delta(x-\alpha_i) = F_y(x,t) \quad \textbf{(25)}$$

where υ_p is taken from eq (17) for the case shown in Fig. **28a** or from eq (20) for the case shown in Fig. **28b**, with υ_{st}=0.

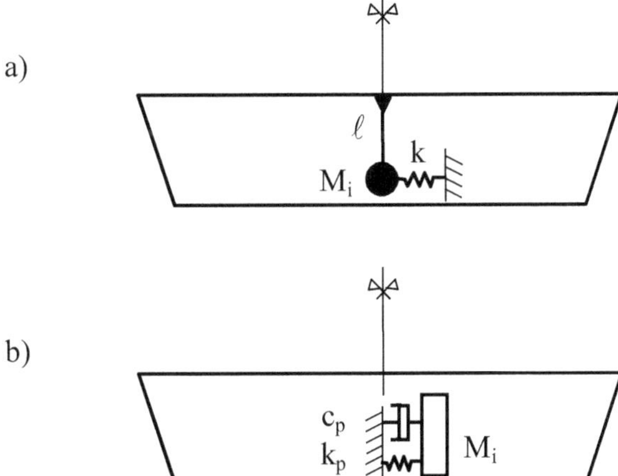

Figure 28: Horizontally active mass dampers

c. Rotational Vibration

The most proper arrangement for dampers is the one shown in Fig. **27c**. We assume that the cross-section is deformed as shown in Fig. **29** and that at points A and B we have placed mass dampers with mass $M_{\beta,}$ as shown in Fig. **27c** each one at distance e from the vertical axis.

Then, we have: $w = w_A = -w_B = e\theta$ and the equation of motion of the bridge is:

$$EC_S\theta'''' + c\dot{\theta} - GJ_d\theta'' + \Theta_x\ddot{\theta} + 2e\sum_i M_{\beta_i}\frac{\partial^2 w_p}{\partial t^2}\cdot\delta(x-\alpha_i) = m_x(x,t) \quad \textbf{(26)}$$

where w_p is taken from eq (20), thus becoming:

$$w_p(x,t) = \frac{1}{\overline{\omega}_p}\int_0^t [\omega_p^2 e\theta(x,\tau) + 2\beta e\dot{\theta}(x,\tau)]\, e^{-\beta(t-\tau)}\cdot \sin\overline{\omega}_p(t-\tau)d\tau \quad \textbf{(27)}$$

with $\overline{\omega}_p$, ω_p , β taken from eq (20b).

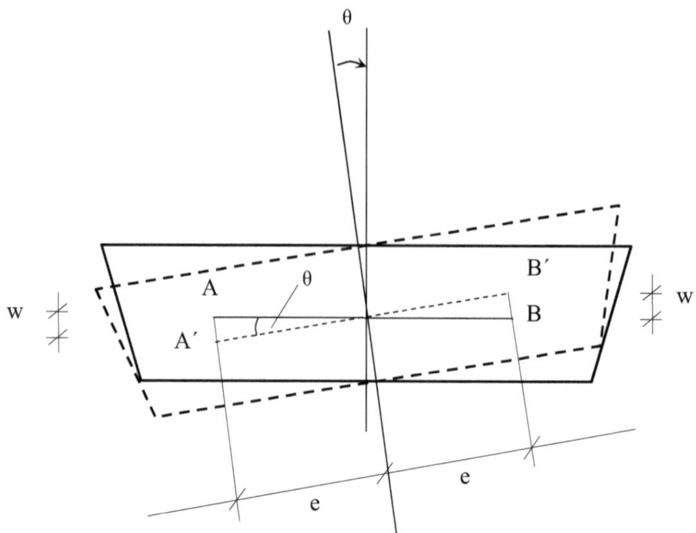

Figure 29: Rotational vibration of a bridge

d. Lateral-Torsional Vibration

In order to deal with the more general case of lateral-torsional vibration of a bridge, we place an arrangement of dampers at positions $x=\alpha_i$ in the cross-section, as shown in Fig. **30**. Damper M_α is supposed to oppose to the horizontal vibration of the beam and the pair of dampers M_β at places A and B in the cross section, for both the vertical and the rotational vibrations. Then, the equations of motion can be written as:

$$\left.\begin{aligned}
&EI_y w'''' + c_y \dot{w} + m\ddot{w} + \sum_i M_{\beta_i} \cdot \left(\frac{\partial^2 w_{p_A}}{\partial t^2} + \frac{\partial^2 w_{p_B}}{\partial t^2} \right) \delta(x-\alpha_i) = F_z(x,t) \\
&EI_z \upsilon'''' - EI_z z_M \theta'''' + c_z \dot{\upsilon} + m\ddot{\upsilon} + \sum_i M_{\alpha_i} \frac{\partial^2 \upsilon_p}{\partial t^2} \delta(x-\alpha_i) = F_y(x,t) \\
&EC_S \theta'''' - EI_z z_M \upsilon'''' + c_\theta \dot{\theta} - GJ_d \theta'' + \Theta_x \ddot{\theta} + \\
&\qquad + e\sum_i M_{\beta_i} \left(\frac{\partial^2 w_{p_A}}{\partial t^2} - \frac{\partial^2 w_{p_B}}{\partial t^2} \right) \delta(x-\alpha_i) = m_x(x,t)
\end{aligned}\right\} \qquad \textbf{(28a,b,c)}$$

where:

$$\left.\begin{aligned}
&w_{p_A}(x,t) = \frac{1}{\bar{\omega}_{p_\beta}} \int_0^t \Big\{ \omega_{p_\beta}^2 e[w(x,\tau) + e\theta(x,\tau)] \\
&\qquad + 2\beta_\beta e[\dot{w}(x,\tau) + e\dot{\theta}(x,\tau)] \Big\} e^{-\beta_\beta (t-\tau)} \sin \bar{\omega}_{p_\beta} (t-\tau) d\tau \\
&w_{p_B}(x,t) = \frac{1}{\bar{\omega}_{p_\beta}} \int_0^t \Big\{ \omega_{p_\beta}^2 e[w(x,\tau) - e\theta(x,\tau)] \\
&\qquad + 2\beta_\beta e[\dot{w}(x,\tau) - e\dot{\theta}(x,\tau)] \Big\} e^{-\beta_\beta (t-\tau)} \sin \bar{\omega}_{p_\beta} (t-\tau) d\tau \\
&\upsilon(x,t) = \frac{1}{\bar{\omega}_{p_\alpha}} \int_0^t [\omega_{p_\alpha}^2 \upsilon(x,\tau) + 2\beta_\alpha \dot{\upsilon}(x,\tau)] e^{-\beta_\alpha (t-\tau)} \sin \bar{\omega}_{p_\alpha} (t-\tau) d\tau
\end{aligned}\right\} \qquad \textbf{(29a,b,c)}$$

and the indexes A and B are referring to points A and B, while the indexes α and β to the dampers related to masses M_α and M_β, respectively.

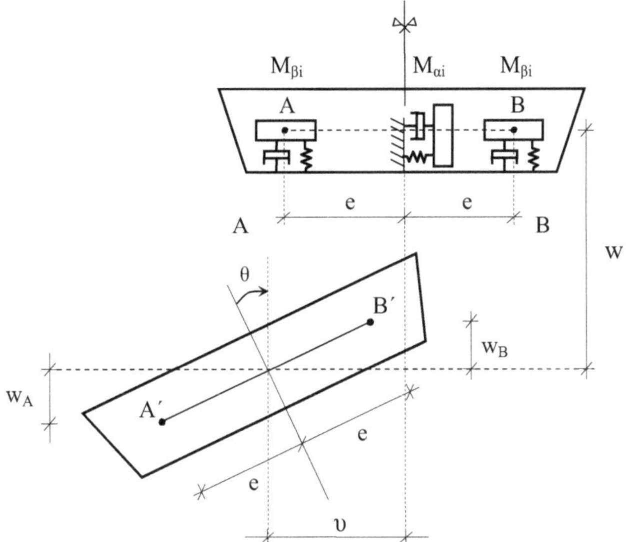

Figure 30: Lateral-torsional vibration of a bridge

e. Seismic Excitation

We will examine the case of seismic excitation of bridge with doubly symmetric cross-section ($y_M = z_M = 0$), where obviously only deformations υ are developed. Then, it is clear that the total deformation is: $\overline{\upsilon} = \upsilon_{st} + \upsilon$, where $\overline{\upsilon}$ is the total deflection of the cross-section, υ_{st} is the seismic displacement (ground displacement) and υ is the elastic deformation of the beam. Introducing the expression for $\overline{\upsilon}$ into eq (25) with $F_y = 0$, we obtain:

$$EI_z\upsilon'''' + c\dot{\upsilon} + m\ddot{\upsilon} + \sum_i \frac{\partial^2 \upsilon_p}{\partial t^2}\delta(x - \alpha_i) = -c\dot{\upsilon}_{st} - m\ddot{\upsilon}_{st} \qquad \textbf{(30)}$$

given that: $\dfrac{\partial^4 \upsilon_{st}(t)}{\partial x^4} = 0$.

Then, deflection $\upsilon_p(x,t)$ is given by eq (20). Equations (30) and (20), constitute a differential system of equations with respect to $\upsilon(x,t)$ and $\upsilon_p(x,t)$.

Control Devices

In Fig. **31**, one can see a typical control device of a semi-active damping system.

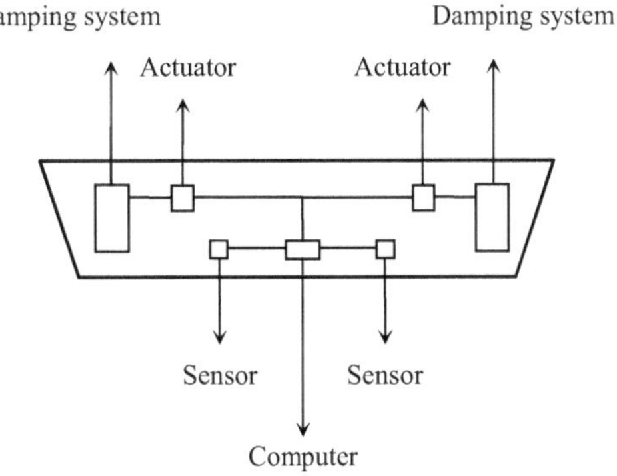

Figure 31: Schematic of a semi-active control device

Finally, it is worthy to present a particularly brilliant system [7] for the control of oscillations against aerodynamic loads shown in Fig. **32**, where a pendulum is used for the excitation of the system as well as the planar or lightly curved surfaces for the creation of reinstitution moments when the system is subjected to wind loads.

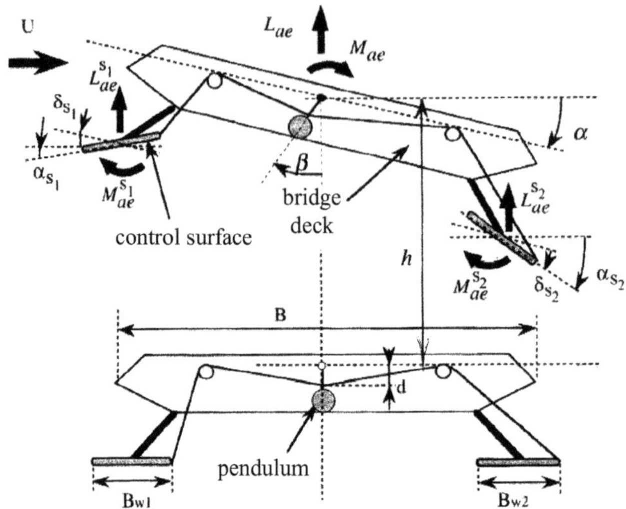

Figure 32: Semi-active control system for aerodynamic loads

REFERENCES

[1] B. Breukelmann *"Good Vibrations"*, 2001, ASCE - Civil Engineering, Vol. 71 (12).

[2] C. Spyrakos *"Structural Strengthening against seismic loads"*, 2004, T.E.E. Publ., Athens.

[3] Min-Yung Chang, L. Librescu *"Vibration Control of Shear deformable laminated plates - Modeling implications"*, 1999, Journal of Sound and Vibration, V. 228, N. 4.

[4] P. Salemi, M.F. Golnaraghi, G.R. Heppler *"Active central of forced and unforced structural Vibrations"*, 1999, Journal of Sound and Vibration, V. 208, N. 1.

[5] S.H. Chen, Z.D. Wang, X.H. Liu *"Active vibration Control and suppression for intelligent Structures"*, 1997, Journal of Sound and Vibration, Vol. 200, N. 2.

[6] I.M. Petalas *"The influence of mass dampers on the behavior of high rise structures under dynamic loads"*, 2004, Diploma Thesis, NTUA., Athens (in Greek).

[7] K. Wilde, Y. Fujino, T. Kawakami *"Analytical and experimental study on passive aerodynamic control of flutter of a bridge deck"*, 1999, Journal of Wind Engineering and Industrial Aerodynamics, No 80.

[8] A. Tondl, R. Nabergoj *"Dynamic absorbers for an externally excited Pendulum"*, 2000, Journal of Sound and Vibration, Vol.234, N. 4.

[9] S.S. Queini, A.H. Nayfeh *"Single-mode control of a cantilever beam under principal parametric excitation"*, 1999, Journal of Sound and Vibration, Vol. 224, N. 1.

[10] Y-G Sung *"Modelling and control with piezoactuators for a simply supported beam under moving mass"*, 2002, Journal of Sound and Vibration, Vol. 250, N. 4.

[11] O.C. Pinto, P.B. Goncalves *"Non-linear control of buckled beams under step loadind"*, 2000, Mechanical Systems and Signal Precessing 14 (6).

[12] Yuh-Yi Lin, C.M. Cheng, C.H. Lee *"Atuned mass dumper for suppressing the coupled flexural and torsional buffeting response of long-span bridges"*, 2000, Engineering Structures 22.

[13] H.C. Kwon, M.C. Kim, I.W. Lee *"Vibration control of bridges under moving Loads"*, 1998, Computers & Structures V. 66, N. 4.

[14] M. Gürgöze *"On the alternative formulations of the frequency equation of a Bernoulli-Euler beam to which several spring-mass systems are attached in-span"*, 1998, Journal of Sound and Vibration, V. 217, N.3.

GENERAL BIBLIOGRAPHY

[1] S.P. Timoshenko, D.H. Young, W. Weaver Jr. *"Vibration problems in Engineering"*, 1974, J. Wiley & Sohns, N. York.

[2] S. Timoshenko, D.H. Young *"Advanced Dynamics"*, 1959, J. Wiley & Sohns, N. York.

[3] G.L. Rogers *"Dynamic of Framed Structures"*, 1959, Wiley & Sohns, N. York.

[4] L. Jacobsen, R. Ayre *"Engineering Vibrations"*, 1958, Mc Graw-Hill, N. York.

[5] C. Norris, R. Hansen, M. Holley, J. Biggs, S. Namyet, J. Minami *"Structural design for dynamic loads"*, 1959, Wiley & Sohns, N. York.

[6] V. Kolousêk *"Dynamics in Engineering Structures"*, 1973, Butterworths, London.

[7] W. Seto *"Theory and problems of mechanical Vibrations"*, 1964, Schaum Publishing Co, N. York.

[8] N.W. McLachlan *"Theory of Vibrations"*, 1970, Dover Publications Inc. N. York.

[9] B.K. Donaldson *"Analysis of Aircraft Structures"*, 1993, McGraw-Hill, N. York.

[10] B.Z. Vlassov *"Piéces longues en voiles minces"*, 1962, Editions Eurolles, Paris.

[11] F. Bleich *"Buckling strength of metal structures"*, 1952, McGraw-Hill, N. York.

[12] S.P. Timoshenko, J.N. Goodier *"Theory of elasticity"*, 1982, McGraw-Hill, London.

[13] A.E.H. Love *"The mathematical theory of elasticity"*, 1944, Dover Publ., N. York.

[14] S.P. Timoshenko, D.H. Young *"Theory of Structures"*, 1965, McGraw-Hill, N. York.

[15] S.P. Timoshenko *"Strength of materials" Part II*, 1956, D. Van Nostrand Comp. Inc., N. York.

[16] M. Denis-Papin, A. Kaufmann *"Cours de calcul Matriciel appliqué"*, 1969, Ed. Albin-Michel, Paris.

[17] M. Denis-Papin, A. Kaufmann *"Cours de calcul Tensoriel appliqué"*, 1961, Ed. Albin-Michel, Paris.

[18] M. Denis-Papin, A. Kaufmann *"Cours de calcul operationnel appliqué"*, 1963, Ed. Albin-Michel, Paris.

[19] I.S. Sokolnikoff *"Tensor analysis"*, 1964, J. Willey & Sohns, N. York.

[20] H. Lass *"Vector and Tensor Analysis"*, 1950, McGraw-Hill, Tokyo.

[21] S.J. McMinn *"Matrices for Structural Analysis"*, 1966, E. & F.N. Spon Limited, London.

[22] E. Festel, F. Leckie *"Matrix methods in elastomechanics"*, 1963, McGraw-Hill, N. York.

[23] M. Krasnov, A. Kiselev, G. Makarenko *"Problems and exercises in integral equations"*, 1971, Mir Publishers, Moscow.

[24] I.G. Petrovsky *"Lectures on the theory of Integral equations"*, 1971, Mir Publishers, Moscow.

[25] F. Ayres Jr. *"Differential equations"*, 1952, Schaum's outline series-McGraw-Hill, N. York.

[26] R. Courant *"Vorlesungen über Differential - und Integralrechnung"*, 1967, Springer Verlag, Düsseldorf.

[27] G. Rakowski, R. Solecki *"Gekrümmte Stäbe"*, 1968, Werner Verlag, Düsseldorf.

[28] G. Bürgermeister, H. Steup, H. Kretzschmar *"Stabilitätstheorie"*, 1966, Akademie Verlag, Berlin.

[29] T. Michaltsos *"Differential Equations"*, 1952, Athens (in Greek).

[30] A. Hawranek, O. Steinhardt *"Theorie und Berechnung der Stahlbrücken"*, 1958, Springer Verlag, Berlin.

[31] M.J. Ryall, G.A. Parke, J.E. Harding *"Manual of bridge engineering"*, 2000, Thomas Telford Ltd, London.

[32] R.C. Wylie *"Advanced Engineering Mathematics"*, 1975, McGraw-Hill, Tokyo.

[33] M. Irvine *"Cable Structures"*, 1981, Dover Publications, N. York.

[34] S. Wigginns *"Introduction to Applied Nonlinear Dynamical Systems and Chaos"*, 1990, Springer Verlag, N. York.

[35] T.D. Burton *"Introduction to dynamic systems Analysis"*, 1994, McGraw-Hill, N. York.

George T. Michaltsos and Ioannis G. Raftoyiannis

Alphabetical Index

A

B

C

D

George T. Michaltsos and Ioannis G. Raftoyiannis

E

F

G

H

I

L

M

N

O

P

R

S

T

U

V

W

www.ingramcontent.com/pod-product-compliance
Lightning Source LLC
LaVergne TN
LVHW070116110826
845147LV00002B/134

9781608054282